Pre-Calculus

Pre-Calculus

George N. Frempong

ISBN: -13: 979-8327035669

Imprint: Independently published

Contents

Preface

1 Sets 1

2 Real Numbers 13

3 Algebraic Expressions 43

4 Linear Equations and Inequalities 73

5 Quadratic Equations 101

6 Systems of Equations 111

7 Radicals 125

8 Functions 137

9 Linear Functions 171

10 Quadratic Functions 193

11 Polynomial Functions 211

12 Rational Functions 221

13 Exponential and Logarithmic Functions 231

14 Variation 249

15 Sequences and Series 257

16 The Binomial Theorem 281

17 Circles and Parabolas 291

18 Matrices 303

19 Trigonometry 331

20 Vectors 371

21 Mathematical Induction 401

Answers to Selected Exercises 405

PREFACE

This book focuses on giving you the tools that you need to succeed in your Pre-Calculus course. Pre-Calculus is ideal for students of a wide range of abilities and especially for those who find mathematics difficult.

Every effort has been made to explain concepts in a clear and simple language which makes the text easy to understand and follow without help. The exercises at the end of each section are varied, providing basic skills in problem solving. Every chapter ends with review exercises. The review exercises help you track your progress in each chapter's topics.

During more than 30 years of teaching mathematics, I have accumulated and absorbed ideas from textbooks, my colleagues and students. It is impossible to look back and recall the origin of most of the material in this book. I am very grateful to all whose work I have benefitted from.

1 Sets

1.1 Sets and Set Notations

The word set is used to describe any collection of objects, such as a collection of books, a class of students and a collection of furniture. Each object in a set is called an element or a member of the set.

The word set is used in mathematics to refer to a well defined collection of objects. A set is well defined if its elements can be clearly determined. For example, the set of mathematics professors in your college is well defined because you can tell whether any professor is or is not in the set.

Describing Sets

You can indicate sets by listing the elements between braces, { }. The braces identify the contents as a set. For example, the set containing the numbers 2, 4, 6 and 8 can be written as, {2, 4, 6, 8} called roster form. A second way to indicate sets is to enclose within braces a rule that determines its elements.

Representing Sets

Capital letters, such as A, B and C, are often used to represent sets and lower case letters, such as a, b and c, to represent elements of sets. For instance, you can use the letter A to name a set that contains the elements a, b, c, d, e and f, and write

$A = \{a, b, c, d, e, f\}$

Listing Elements of a Set

The set of numbers 1, 3, 2, 2, 4, 2, 4, 5 and 6 can be listed as {1, 2, 3, 4, 5, 6}. Notice that identical objects cannot be listed more than ones. The elements can be listed in any order. For example, {1, 2, 3}, {2, 1, 3} and {3, 1, 2} are different listing of the same set.

Example 1

Write set A of prime numbers less than 10 in roster form.

Solution

$A = \{2, 3, 5, 7\}$

Set –Builder Notation

The set of real numbers between 1 and 2 cannot be named by listing all its members because there are infinite number of them. One way to denote this set is to use Set-builder notation, and write the set as

$\{x| \, x \in \text{R and } 1 < x < 2\}$

read " the set of all x such that x is a real number and x is greater than 1 and less than 2." The vertical bar " |," represent such that.

Example 2

Write set A = {1, 2, 3, 4, 5, 6, 7} in set-builder notation.

Solution

Since set A consists of the natural numbers less than 8, we write

$A = \{x| \; x \in N \text{ and } x < 8\}$

Example 3

Write set $B = \{x| \; x \in N \text{ and } 2 \leq x < 10\}$ in roster form

Solution

$B = \{2, 3, 4, 5, 6, 7, 8, 9\}$

Indicating Membership of a Set

We use the symbol $\in$ to indicate that an object is an element of a given set. For example, if A denotes a set and x an element of the set A, we write $x \in A$ read "x belongs to A, or x is a member of A or x is an element of A". However, if x is not an element of A, we write $x \notin A$ read " x does not belong to A or x is not a member of A or x is not an element of A."

Example 4

Classify each of the following as true or false.

(a) $6 \in \{x| \; x \text{ is a multiple of } 2\}$

(b) $2 \notin \{x| \; x \text{ is a prime number}\}$

Solution

(a) Since 6 is a multiple of 2,

$6 \in \{x| \; x \text{ is a multiple of } 2\}$ is true.

(b) Since 2 is a prime number,

$2 \notin \{x| \; x \text{ is a prime number}\}$ is false.

Example 5

Insert the symbol $\in$ or $\notin$ between the following pairs of sets.

(a) $\{9\}$ $\quad$ $\{x| \; x \text{ is a prime number}\}$

(b) -3 $\{x|\ x \text{ is a real number}\}$

Solution

(a) $\{9\} \notin \{x|\ x \text{ is a prime number}\}$

(b) $-3 \in \{x|\ x \text{ is a real number}\}$

The Number of Elements in a Set

The number of elements in set A, called the cardinal number, is denoted by n (A).

Example 6

Let A = {10, 11, 12}

and B = $\{x|\ x \text{ is a prime number less than } 20\}$.

Find:

(a) n (A) (b) n (B)

Solution

(a) Set A has 3 elements, so n (A) = 3

(b) The elements in set B are 2, 3, 5, 7, 11, 13, 17 and 19 so

n (B) = 8

Equal Sets

Two sets A and B are equal, written A = B, if A and B have exactly the same elements.

Example 7

Let A = $\{x|\ x \in \text{N and } x < 4\}$

and B = {1, 2, 3}.

Determine whether the two sets are equal.

Solution

The elements in set A are 1, 2 and 3. Because set A and set B contain exactly the same elements A = B.

Equivalent Sets

Two sets A and B are said to be equivalent if and only if they contain the same number of elements.

Example 8

Let A = {4, 6, 8, 10, 12}

and B = {x| x is a prime number less than 12}.

Determine whether the two sets are equivalent.

Solution

The elements in set B are 2, 3, 5, 7 and 11. Both set A and set B have the same number of elements, so A and B are equivalent sets.

Empty Set

A set that contains no elements is called the empty set or null set, and is denoted by the open braces { } or the Greek letter Ø. It is important to note that the empty set is not written as {Ø}.

Example 9

Given A = {x| $x \in$ N and $x + 3 = 0$} determine whether the set is an empty set or not.

Solution

Because – 3 is not a natural number A is an empty set.

Universal Sets

Often we use a set whose elements are contained in a larger set, called the universal set. For example, the set of integers include whole numbers. We can consider the set of integers as the universal set. In general, the universal set, denoted by U, is a set that contains all the elements for any specific discussion.

Note: When a universal set is given, only the elements in the universal set should be considered.

Example 10

Given, U = {1, 2, 3, 4, 5, 6, 7, 9} list the elements of the set A = {x| x is a multiple of 2}.

Solution

A = {2, 4, 6}

Finite and Infinite Sets

The number of elements of a set can be finite or infinite. A set is said to be finite if it is either empty or if the number of elements in the set can be counted. For example, {– 2, – 1, 0, 1, 2, 3} is a finite set because the set has only 6 elements. A set that is not finite is said to be infinite. For example, the set of real numbers is an infinite set because there is no end in counting its elements.

An infinite set can be described by listing few elements followed by three dots. For instance, the set A = {natural numbers} can be listed as {1, 2, 3, 4, 5 . . .}. The three dots indicate that the elements in the set continue in the same pattern. For instance, the set also contains the numbers 99, 100, 101 and so on.

Subsets

Set A is a subset of set B, written as $A \subset B$ (or $B \supset A$), if and only if each element of set A is also an element of set B. That is, if $x \in A \Rightarrow x \in B$, then $A \subset B$. If a set A is not a subset of a set B, we write $A \not\subset B$.

Proper Subsets

Set A is a proper subset of set B if and only if each element of set A is also an element of set B but there is at least one element in set B that is not in set A.

The definition of a subset allows a set to be a subset of itself. Since the empty set Ø has no elements, every element of Ø is also an element of any given set. Thus, the empty set is a subset of every set. If A and B are two sets then A = B if and only if $A \subset B$ and $B \subset A$.

Example 11

Insert the symbol $\subset$, $\supset$ or = between the following pairs of sets.

(a) {3, 5, 8, 10}   {1, 2, 3, 4, 5, 7, 8, 10, 12}

(b) {2, 4, 6, 8}   {multiples of 2 less than 10}

(c) {prime numbers}   {2, 3, 5, 7, 11}

Solution

(a) {3, 5, 8, 10} $\subset$ {1, 2, 3, 4, 5, 7, 8, 10, 12}

(b) {2, 4, 6, 8} = {multiples of 2 less than 10}

(c) {prime numbers} $\supset$ {2, 3, 5, 7, 11}

Number of Subsets

A subset of a given set can contain any element or all elements of the given set, including the empty set and the given set itself. For example, the set {0, 1} has the following subsets:

{ }, {0}, {1} and {0, 1}.

In general, a set with n elements has 2^n subsets.

Example 12

(a) List all possible subset of {0, 1, 2}.

(b) Determine the number of distinct subsets for a set with 5 elements

Solution

(a) { }, {0}, {1}, {2}, {0, 1}, {0, 2}, {1, 2}, {0, 1, 2}

(b) There are $2^5 = 32$ subsets

Exercise 1.1

In Exercises 1 – 6, name each set using the roster method.

1. The set of natural numbers between 2 and 9.
2. The set of prime numbers between 10 and 20.
3. The set of odd numbers between 30 and 40.
4. The set of multiples of 3 between 12 and 25.
5. $\{x| x$ is an integer and $-3 < x < 3\}$
6. $\{x| x \in \mathrm{N}$ and $7 \leq x < 15\}$

In Exercises 7 – 10, insert the symbol ∈ or ∉ between the following pair of sets.

7. 2 $\{x| x$ is an odd number$\}$
8. 2 $\{x| x$ is a prime number$\}$
9. 0 $\{x| x$ is a natural number$\}$
10. 2 $\{x| x$ is a real number$\}$

In Exercises 11 – 12, use set A = {1, 2, 3, 4, 5, 6, 7, 8, 9, 10} to find:

11. $\mathrm{P} = \{x| x \in \mathrm{A}$ and x is an even number$\}$
12. $\mathrm{Q} = \{x| x \in \mathrm{A}$ and x is an odd number$\}$
13. $\mathrm{R} = \{x| x \in \mathrm{A}$ and x is a prime number$\}$
14. $\mathrm{S} = \{x| x \in \mathrm{A}$ and x is a multiple of 5$\}$

In Exercises 15 – 20, classify each statement as true or false.

15. {5, 6, 7} = {7, 5, 6}

16. $\{5, 3, 4\} = \{8, 9, 10\}$

17. $\{2\} = \{x|\ x$ is an even prime number$\}$

18. $\{\ \} = \{0\}$

19. {multiples of 3} = {odd numbers}

20. $\{x|\ x$ is an integer and $3 \le x < 7\} = \{3, 4, 5, 6\}$

In Exercises 21 – 25, determine whether the following pairs of sets are equal or equivalent.

21. {factors of 6} {1, 2, 3, 6}

22. {1, 3, 5, 7, 8, 12} {factors of 12}

23. $\{x|\ 4x = 12\}$ $\{0\}$

24. $\{2\}$ $\{x|\ x$ is an integer and $1 < x < 3\}$

25. {prime factors of 6} {prime numbers less than 5}

In Exercises 26 – 30, determine whether the following sets are empty set or not empty set.

26. $\{x|\ x$ is an integer and $3x + 2 = 7\}$

27. {a prime number divisible by 2}

28. $\{x|\ x$ is an integer between 4 and 5$\}$

29. $\{x|\ x \in \mathrm{N}$ and $x + 2 = 0\}$

30. $\{x|\ x$ is a real number between 1 and 2$\}$

In Exercises 31 – 34, determine whether each of the following sets are finite or infinite.

31. $\{x|\ x$ is a multiple of 2$\}$

32. $\{x|\ 3x + 2 = 8\}$

33. $\{x|\ x$ is a factor of 20$\}$

34. $\{x|\ x \in \mathrm{Z}$ and $x < 1\}$

In Exercises 35 – 39, insert the correct symbol $\subset$ or $\supset$ between the following pairs of sets.

35. $\{2, 3, 5\}$ $\{5, 2, 3\}$

36. $\{1, 2, 3, 4\}$ $\{\ \}$

37. {2} {even prime numbers}

38. $\{2, 3, 5, 7, 9\}$ $\{x | x \text{ is a prime number and } 1 < x \leq 10\}$

39. $\{x | x \in N \text{ and } x < 6\}$ $\{x | x \in N \text{ and } 1 \leq x \leq 5\}$

40. List all possible subsets of the set $\{3, 4, 5\}$

41. Find the number of subsets of the set $\{-3, -2, -1, 0, 1, 2, 3\}$

42. How many elements has a set that has 512 subsets?

1.2 Operations on Sets

There are three basic set operations; union, intersection and complement.

Union of Sets

The union of set A and set B, written $A \cup B$, is the set containing all elements that are members of A or B or both A and B. That is, x is in $A \cup B$ if it is in A or B, or both. $A \cup B$ is read "the union of A and B." The symbol for union of two or more set is $\cup$.

Example 1

If $A = \{1, 2, 4\}$ and $B = \{2, 3, 5\}$ find $A \cup B$.

Solution

$A \cup B = \{1, 2, 3, 4, 5\}$

Intersection of Sets

The intersection of set A and set B, written $A \cap B$, is the set containing all the elements that are common to both set A and set B. That is, x is in $A \cap B$ if it is in both A and B. $A \cap B$ is read " intersection of A and B." The symbol for set intersection is $\cap$.

Example 2

If $A = \{0, 3, 5, 6, 8, 9\}$ and $B = \{1, 2, 3, 4, 6, 8, 12\}$ find $A \cap B$.

Solution

$A \cap B = \{3, 6, 8\}$

Disjoint Sets

Two sets A and B are said to be disjoint if A and B have no elements in common, that is

$A \cap B = \varnothing$

For example, if $A = \{2, 4, 6, 8\}$ and $B = \{3, 5, 7, 9\}$ then $A \cap B = \varnothing$ and set A and set B are disjoint.

The following properties of union and intersection hold.

If A, B and C are subsets of the universal set U, then

1. $A \cup \emptyset = A$ $\quad\quad$ $A \cap \emptyset = \emptyset$

2. If $A \subset B$ then:

   $A \cup B = B$ $\quad\quad$ $A \cap B = A$

3. $A \cup A = A$ $\quad\quad$ $A \cap A = A$

4. Commutative property

   $A \cup B = B \cup A$ $\quad\quad$ $A \cap B = B \cap A$

5. Associative property

   $(A \cup B) \cup C = A \cup (B \cup C)$

   $(A \cap B) \cap C = A \cap (B \cap C)$

6. Distributive property

   $A \cup (B \cap C) = (A \cup B) \cap (A \cup C)$

   $A \cap (B \cup C) = (A \cap B) \cup (A \cap C)$

The Complement of a Set

The complement of a set A, written A′, is the set of all the elements in the universal set that are not in set A. That is, $A' = \{x \in U \mid x \notin A\}$.

Example 3

If $U = \{1, 2, 3, 4, 5, 6, 7, 8, 9, 10\}$ and

$A = \{1, 3, 5, 7, 9\}$, find A′.

Solution

$A' = \{2, 4, 6, 8, 10\}$

The following properties of complement hold.

1. $A \cup A' = U$

2. $A \cap A' = \emptyset$

3. $(A')' = A$

4. $U' = \emptyset$ and $\emptyset' = U$

5. $(A \cup B)' = A' \cap B'$ De Morgan's law

6. $(A \cap B)' = A' \cup B'$ De Morgan's law

Exercise 1.2

1. Given A = {1, 3, 4, 6} and B = {2, 3, 5} find $A \cup B$.

2. Given A = {1, 2, 4, 5} and B = {1, 3, 4, 6, 8} find $B \cup A$.

3. Let $A = \{x \mid x$ is a multiple of 2 and $x < 12\}$ and $B = \{x \mid x$ is a factor of $20\}$. Find $A \cup B$.

4. Let $A = \{x \mid x \in Z \text{ and } 0 < x < 6\}$ and $B = \{x \mid x \in Z \text{ and } 2 < x \leq 8\}$. Find $A \cup B$.

5. Let $A = \{x \mid x$ is a prime number and $x < 10\}$ and B = {3, 5, 7, 9}. Find $A \cup B$.

6. Given A = {2, 3, 5} and B = {1, 2, 3, 5, 6} find $A \cap B$.

7. Given A = {2, 3, 5, 6} and B = {1, 3, 4, 6, 8} find $A \cap B$.

8. Given A = {3, 6, 9} and B = {1, 2, 3, 4, 6, 12} find $A \cap B$.

9. Given A = {2, 4, 6, 8} and B = {3, 5, 7, 9} find $A \cap B$.

10. Given $A = \{x \mid x \in Z \text{ and } 2 < x < 8\}$ and B = {factors of 6} find $A \cap B$.

In Exercises, 11 – 20, let A = {1, 3, 4, 6}, B = {1, 2, 5, 7} and C = {2, 3, 8} List the elements of the sets:

11. $A \cup B$ 12. $B \cup A$ 13. $A \cup C$

14. $A \cap B$ 15. $B \cap C$ 16. $A \cap C$

17. $(A \cup B) \cup C$ 18. $A \cup (B \cup C)$

19. $(A \cap B) \cap C$ 20. $A \cap (B \cap C)$

In Exercises, 21 – 28, let A = {1, 3, 5, 6}, B = {2, 3, 4, 8} and C = {1, 2, 4, 6}. List the elements of the sets.

21. $A \cup B$ 22. $A \cup C$ 23. $A \cap B$ 24. $A \cap C$

25. $A \cup (B \cap C)$ 26. $A \cap (B \cup C)$

27. $(A \cup B) \cap (A \cup C)$ 28. $(A \cap B) \cup (A \cap C)$

29. Given U = { 1, 2, 3, 4, 5, 6, 7, 8 } and A = { 1, 3, 5, 7}, find A′.

30. Given U = { 1, 2, 3, 4, 5, 6, 7, 8 } and A = { 1, 3, 6, 8}, find (A′)′.

In Exercises, 31 – 36, let U = { 1, 2, 3, 4, 5, 6, 7, 9, 10 }, A = {3, 5, 9, 10} and B = {3, 7, 9, 10}. List the elements of the sets.

31. A′ 32. B′ 33. U′

34. (A ∪ B)′ 35. (A ∩ B)′ 36. (A ∪ B′)′

Review Exercises

In Exercises 1 – 4, list the elements in each set.

1. A = $\{x|\, x$ is a multiple of 3 and $6 < x \leq 18\}$

2. B = $\{x|\, x \in N$ and $1 < x \leq 18\}$

3. C = $\{x|\, x \in Z$ and $-2 < x < 7\}$

4. D = $\{x \mid x \in N$ and x is an even number less than $12\}$

In Exercises 5 – 8, determine whether the pair of sets are equal or equivalent.

5. $\{x|\, x \in N$ and $x < 6\}$ $\{x|\, x \in Z$ and $0 < x < 6\}$

6. $\{x|\, x$ is a prime factor of $20\}$ $\{3, 4\}$

7. $\{x|\, x \in Z$ and $-1 < x < 5\}$ $\{0, 1, 3, 4, 5\}$

8. $\{x \mid x \in N$ and $2 < x \leq 7\}$ $\{x \mid x \in N$ and $3 \leq x < 8\}$

In Exercises 9 – 12, determine whether each set is finite or infinite.

9. $\{x|\, x \in N$ and $0 < x \leq 50\}$ 10. $\{x|\, x \in R$ and $2 < x < 3\}$

11. $\{x|\, x$ is a multiple of $3\}$ 12. $\{x|\, x \in N$ and $3x - 6 = 0\}$

In Exercises 13 – 20, classify each statement as true or false.

13. $\{\ \} \in \{1, 2, 3, 4\}$ 14. $\{\ \} \subset \{1, 2, 3, 4\}$

15. $4 \subset \{3, 4, 5\}$ 16. $5 \notin \{3, 4, 6\}$

17. $\{1, 2, 3\} \in 2$ 18. $\{1, 7, 8\} \not\subset \{1\}$

19. $\{3, 5, 9\} = \{5, 3, 9\}$ 20. $\{6, 7, 10\} \subset \{7\}$

In Exercises 21 – 24, let A = {1, 2, 3, 4, 5}, B = {2, 3, 5} and C = {1, 2, 4}.

Insert the symbol ∈, ∉, ⊂ or ⊄ to make each of the following a true statement.

21. 4 A 22. 6 B 23. B A 24. A C

In Exercises 25 – 35, let A = {1, 2, 4, 5, 7, 8, 9, 10}, B = {1, 4, 7, 8, 9, 10}. List the elements of the sets.

25. $A \cup B$ 26. $B \cup C$ 27. $A \cup C$

28. $A \cap B$ 29. $B \cap C$ 30. $A \cap C$

31. $(A \cup B) \cup C$ 32. $A \cup (B \cup C)$

33. $(A \cap B) \cap C$ 34. $A \cap (B \cap C)$

In Exercises 35 – 42, let A = {2, 3, 5, 6}, B = {1, 3, 4, 6, 8} and C = {1, 2, 5, 8}. List the elements of the sets.

35. $A \cup B$ 36. $A \cup C$ 37. $A \cap B$

38. $A \cap C$ 39. $A \cup (B \cap C)$ 40. $A \cap (B \cup C)$

41. $(A \cup B) \cap (A \cup C)$ 42. $(A \cap B) \cup (A \cap C)$

In Exercises 43 – 51, let U = { 1, 2, 3, 4, 5, 6, 7, 8, 9 }, A = {3, 4, 6, 8} and B = {3, 5, 7, 8}. List the elements of the sets.

43. A' 44. B' 45. $A' \cup B'$

46. $A' \cap B'$ 47. $(A \cup B)'$ 48. $(A \cap B)'$

49. $(A' \cap B)'$ 50. $(A' \cup B)'$ 51. $(A' \cup B')'$

In Exercises, 52 – 57, let U = { 1, 2, 3, 4, 5, 6, 7, 8 }, A = {1, 3, 4, 7} and B = {2, 4, 5, 8}. List the elements of each sets:

52. U' 53. $\emptyset'$ 54. $A' \cap B$

55. $A \cup B'$ 56. $A' \cap B'$ 57. $(A \cap B)'$

Which of the sets in Exercises, 52 – 57 are equal?

In Exercises 58 – 61, let U = { 1, 2, 3, 4, 5, 6, 7 }, A = {1, 3, 5, 6}, B = {2, 3, 6}and C = {4, 6, 7}. List the elements of each sets

58. $A \cap (B \cap C)'$ 59. $(A \cup C)' \cap C$

60. $(A \cup B)' \cap C$ 61. $(A \cap B)' \cap C'$

2 Real Numbers

2.1 The Real Number System

We carry out our mathematical calculations using a number system. The number system had evolved as a result of a process of successive expansion.

The original set of numbers, called the set of natural numbers, consists of the numbers that you use when you count.

$$\{1, 2, 3, 4, 5, \ldots\}$$

The three dots indicate that the pattern continues indefinitely. For example, the set also contains the numbers 12, 13, 14, and so on. The set of natural number is denoted by N.

The set of natural numbers can be expanded to include the additive inverses of the natural numbers and 0. The expanded set is called the set of integers, denoted by Z.

$$\{\underbrace{\ldots, -5, -4, -3, -2, -1}_{Negative\ \ integers}, 0, \underbrace{1, 2, 3, 4, 5, \ldots}_{Positive\ \ integers}\}$$

The subset $\{0, 1, 2, 3, \ldots\}$ of the set of integers is called the set of whole numbers, denoted by W.

The set of rational numbers, denoted by Q, consists of numbers that can be written in the form

$$\frac{p}{q}$$

where p and q are integers and $q \neq 0$.

Examples of rational numbers include

$$\frac{-3}{5}, \quad 2 = \frac{2}{1}, \quad \frac{1}{8} \quad \text{and} \quad \frac{5}{3}$$

The decimal representation of a rational number either repeats or terminates.

Numbers that cannot be written as a ratio of two integers, such as square roots of numbers that are not perfect squares are called irrational numbers. The irrational numbers include:

$$\sqrt{2} = 1.4142135 \ldots \qquad \pi = 3.1415926\ldots \quad \text{and} \quad e = 2.7182818\ldots$$

The number e is called the Euler's number.

The rational numbers together with the irrational numbers make up the real numbers system, denoted by R.

Notice that all numbers are real numbers but not all real numbers are either rational numbers or integers. Figure 2.1 shows the relationships among various kinds of numbers.

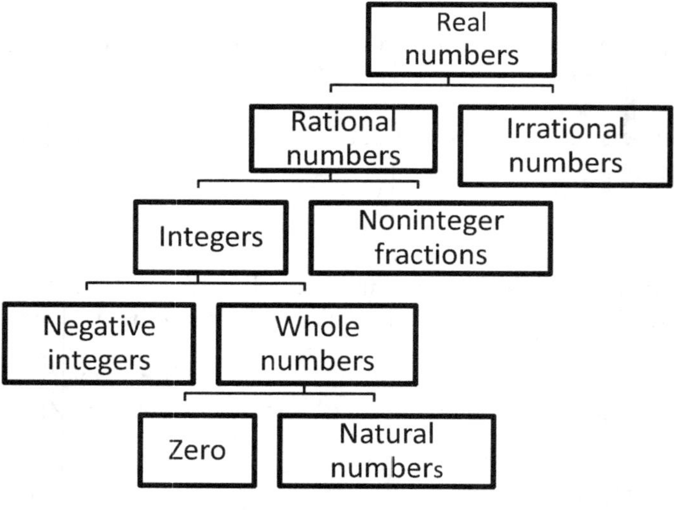

Figure 2.1

Example 1

Determine which numbers in the set $\left\{-9, -5, -\sqrt{2}, -1, -\frac{1}{3}, 0, \frac{3}{4}, \sqrt{3}, \pi, 3, 7, 8\right\}$ are

(a) natural numbers, (b) whole numbers, (c) integers,

(d) rational numbers (e) irrational numbers.

Solution

(a) Natural numbers: {3, 7, 8}

(b) Whole numbers: {0, 3, 7, 8}

(c) Integers: {− 9, − 5, − 1, 0, 3, 7, 8}

(d) Rational numbers: : $\left\{-9, -5, -1, -\frac{1}{3}, 0, \frac{3}{4}, 3, 7, 8\right\}$

(e) Irrational numbers: : $\left\{-\sqrt{2}, \sqrt{3}, \pi\right\}$

The Real Number Line

Real numbers can be represented graphically by a real number line. The number line consists of a horizontal line, which extend in both the positive and negative direction, with centre at 0 and convenient distance chosen as a unit of length. The centre 0 is called the origin.

Numbers lying to the left of 0 are negative, and numbers lying to the right are positive, as shown in Figure 2.2.

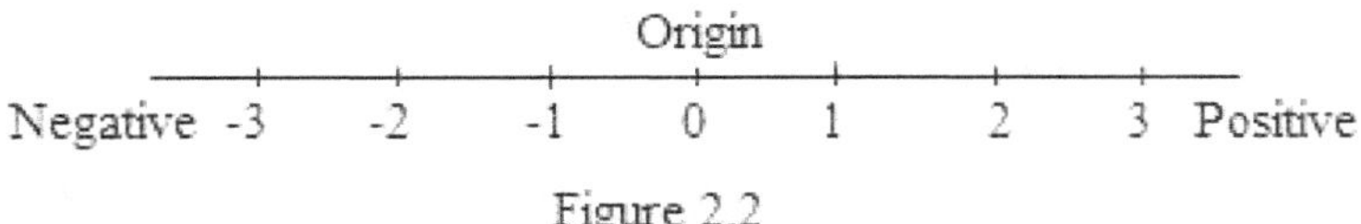

Figure 2.2

Each point on the real number line corresponds to exactly one real number. The point that corresponds to a real number is called the graph of the number. Figure 2.3 shows the graph of $-3/2$ and 2.

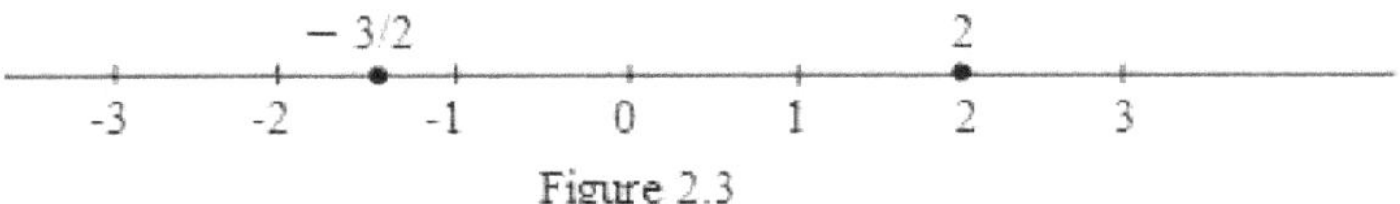

Figure 2.3

Ordering

You may have noticed that a number line has its numbers written in order. Smaller numbers are to the left, and larger numbers are to the right. For any two numbers, the one to the left is less than the one to the right, and the one to the right is greater than the one to the left.

Figure 2.4 shows the graph of -6 and -1.

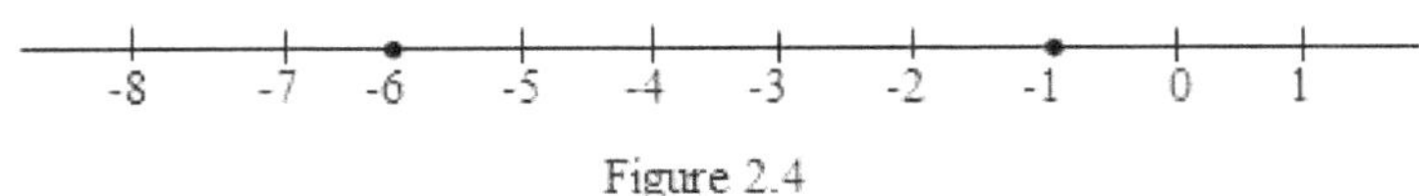

Figure 2.4

Notice that -6 lies on the left of -1, so we say that -6 is less than -1, and write $-6 < -1$, or -1 is greater than -6, and write $-1 > -6$.

In general, if $a < b$, where a and b are real numbers, then a is to the left of b. If $b > a$, then b is to the right of a.

When we write $a \leq b$ we mean $a < b$ or $a = b$ and say a is less than or equal to b. Also, $a \geq b$ means $a > b$ or $a = b$ and we say a is greater than or equal to b.

Notice that for any two real numbers a and b, either $a < b$ or $a > b$, or $a = b$.

The symbols, $<$, $>$, $\leq$ and $\geq$ are called inequality symbols.

Example 2

Place the correct symbol ($<$ or $>$) between each pair of numbers

(a) $-4 \quad -1$ (b) $-2 \quad -7$

(c) $2/3 \quad ½$ (d) $-¼ \quad -½$

Solution

(a) -4 lies to the left of -1 on the number line, so $-4 < -1$.

(b) -2 lies to the right of -7 on the number line, so $-2 > -7$.

(c) 2/3 lies to the right of 1/2 on the number line, so $2/3 > 1/2$

(d) $-1/4$ lies to the right of $-1/2$ on the number line, so $-1/4 > -1/2$

Intervals on the Real Number Line

If $x > a$ and $x < b$, we write $a < x < b$. The set of all real numbers x between a and b is called an interval, and is written in Interval Notation as (a, b). In general, we just write the beginning and ending numbers of the interval, and use the round bracket to indicate that the endpoint is not included. Because the end values are not included (a, b) is called an open interval.

The number line shown in Figure 2.5 is the graph of the interval (a, b). The thick line shows the values included in the interval and the open circle is used to indicate that the end points are not part of the graph.

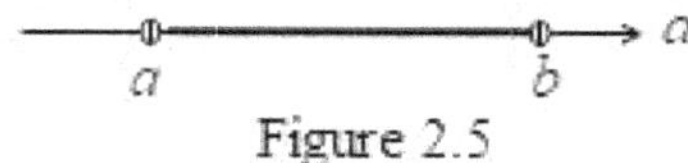

Figure 2.5

If $x \geq a$ and $x \leq b$, we write $a \leq x \leq b$. In interval notation, we write $[a, b]$. The square bracket indicates that the endpoint is included. The interval $[a, b]$ is called a closed interval. The graph of the interval $[a, b]$ is shown in Figure 2.6. The closed circles indicate that the end values are part of the graph.

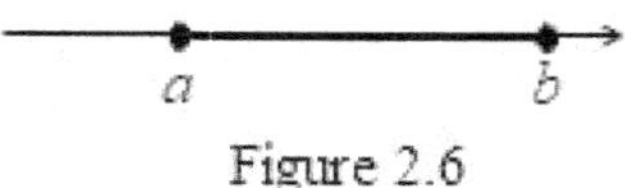

Figure 2.6

Other types of intervals of real numbers are shown in the following table.

Notation	Inequality notation	Interval type
$[a, b]$	$a \leq x \leq b$	Closed
$[a, b)$	$a \leq x < b$	Half - open
$(a, b]$	$a < x \leq b$	Half - open
(a, b)	$a < x < b$	Open
$[a, \infty)$	$x \geq a$	Closed
(a, ∞)	$x > a$	Open
$(-\infty, a]$	$x \leq a$	Closed
$(-\infty, a)$	$x < a$	Open

The symbol ∞, read positive infinity, and $-\infty$ read negative infinity, do not represent real numbers. We write $[a, \infty)$ to represent the interval starting at a and continuing indefinitely to the right.

Note that, we use the round bracket with ∞ because we cannot reach it. The notation $(-\infty, \infty)$ shows that the interval has no limits. Whenever, you write intervals that contains ∞ or $-\infty$ always use a parenthesis because these symbols are never an endpoint of an interval.

Distance on the Real Number Line

If a and b are two real numbers such that $a \leq b$, then the distance between a and b is given by $b - a$. Notice that distances are never negative.

The length of an interval is the distance between its endpoints. An interval that has a finite length is called a bounded interval, and an interval that does not have a finite length is called unbounded interval (or infinite). For instance, the interval $[a, b)$ on the real number line is a bounded interval whiles the interval $[a, \infty)$ is unbounded.

Example 3

Find the distance between each pair of real numbers.

(a) 3 and 7 (b) -5 and 3

Solution

(a) Because $3 < 7$, the distance between 3 and 7 is $7 - 3 = 4$.

(b) Because $-5 < 3$, the distance between -5 and 3 is $3 - (-5) = 3 + 5 = 8$

Absolute Value

The distance a number is from 0 on a number line is called the absolute value of the number. Absolute value is denoted by double vertical bars, $|\ |$. For example, the distance between 3 and 0 is denoted by $|3|$ and the distance between -3 and 0 is denoted by $|-3|$. Because opposite numbers lie the same distance from 0 on the real number line, they have the same absolute value. So $|3| = 3$ and $|-3| = 3$.

Opposite numbers are two real numbers that are the same distance from zero on a number line but on opposite sides of zero.

In general, if a is a real number then

1. $-a$ is the opposite of a.
2. $-(-a) = a$
3. $a + (-a) = 0$

Example 4

Find the absolute value of:

(a) $|-10|$ (b) $|8|$ (c) $-|-3|$

Solution

(a) $|-10| = 10$

(b) $|8| = 8$

(c) $-|-3| = -3$

Exercise 2.1

In Exercises 1 – 4, which of the real numbers in the set are:

(a) natural numbers, (b) integers, (c) rational numbers, and (d) irrational numbers?

1. $\left\{-12, -\sqrt{7}, -\frac{2}{3}, -\frac{1}{4}, 0, \frac{3}{8}, 1, \sqrt{3}, 4\pi, 8\right\}$

2. $\left\{-\frac{9}{2}, -\sqrt{6}, -\frac{\pi}{2}, -\frac{3}{8}, 0, \sqrt{13}, \frac{10}{3}, 8, 12\right\}$

3. $\left\{-3.6, -\sqrt{4}, -1, -\frac{1}{2}, -0.\dot{3} - 0.3, \sqrt{5}, 3\pi, 10, 26.4\right\}$

4. $\left\{-6, -\sqrt{25}, -\frac{5}{2}, -\sqrt{6}, -0.11, 0, 0.75, 3, 20\right\}$

In Exercises 5 – 12, plot the real numbers on the real number line

5. 5 6. 3/2 7. − 7/2 8. − 4.2

9. 7 10. 4/3 11. − 7.65 12. − 9/2

In Exercises 13 – 24, place the correct symbol (< or >) between the pair of numbers

13. 5 9 14. 8 3

15. −7 5 16. 3 − 5

17. − 8 − 2 18. − 1 − 12

19. 0 − 3 20. − 32 − 7

21. ¾ 1/3 22. 0 − ¾

23. − 5/2 − 3/2 24. − 2/3 − 10/3

In Exercises 25 – 32, find the distance between the pair of real numbers.

25. 5 and 11 26. 77 and 22

27. – 13 and 8 28. – 44 and 22

29. 17 and – 33 30. 15 and – 5

31. – 9 and 0 32. – 8 and – 11

In Exercises 33 – 42, evaluate the following expressions.

33. $|12|$ 34. $|85|$

35. $|-12|$ 36. $|-15|$

37. $-|-17|$ 38. $|-18|$

39. $-|16|$ 40. $-|-25|$

41. $-|20|$ 42. $-|-31|$

In Exercises 43 – 16, place the correct symbol <, > or = between the pair of numbers.

43. $|-7| \quad |-2|$ 44. $|4| \quad |-9|$

45. $|-8| \quad |8|$ 46. $|25| \quad |-25|$

47. $|-4| \quad |6|$ 48. $|-28| \quad |-17|$

2.2 Operations with Real Numbers

In arithmetic, we do calculations using four basic operations: addition, subtraction, multiplication and division.

Addition of Real Numbers

The operation of addition assigns a unique real number called the sum to each pair of real numbers. For example, 7 added to 3 produce the sum (3 + 7), i.e. 10.

Adding Integers

Adding two integers with like sign

To add two real numbers with like signs, add their absolute values and attach the common sign to the result.

Example 1

Work out the following additions:

(a) 32 + 48 (b) – 16 + (– 8)

Solution

(a) $32 + 48 = 80$

(b) $-16 + (-8) = -(16 + 8)$

$= -24$

Adding two integers with unlike signs

To add two integers with unlike sign, subtract the smaller absolute value from the greater absolute value, and:

1. if the positive number has the greater absolute value, make the answer positive

2. if the negative number has the greater absolute value, make the answer negative

Example 2

Workout the following additions:

(a) $-20 + 8$ (b) $-7 + 12$

Solution

(a) $-20 + 8 = -(20 - 8)$

$= -12$

(b) $-7 + 12 = +(12 - 7)$

$= 5$

Subtraction of Integers

To subtract two integers, add the opposite of the number being subtracted. For example, $8 - 12 = 8 + (-12)$.

In general, to subtract b from a, where a and b are integers, add the opposite of b to a. That is $a - b = a + (-b)$.

Example 3

Work out the following subtractions.

(a) $7 - 21$ (b) $-17 - 5$ (c) $9 - (-27)$

Solution

(a) $7 - 21 = 7 + (-21)$ Add the opposite of 21

$= -(21 - 7)$ Use rules for adding signed numbers

$= -14$

(b) $-17 - 5 = -17 + (-5)$ Add opposite of 5

$= -(17 + 5)$ Use rules for adding signed numbers

$= -22$

(c) $9 - (-27) = 9 + 27$ Add the opposite of -27

$= 36$

Multiplying Integers

Multiplication of two integers can be regarded as repeated addition. For example, the product of 2×5, can be obtained as follows; $2 + 2 + 2 + 2 + 2 = 10$ or $5 + 5 = 10$.

You can also use dots or parentheses to show multiplication. Thus $2 \times 5 = 2 \cdot 5 = (2)(5)$.

To multiply integers, find the product of their absolute values and then use the following rules to determine the sign of the answer.

1. If the two integers have like signs, the product is positive.
2. If the two integers have unlike signs, the product is negative.

Example 4

Workout the following multiplications:

(a) -8×-3 (b) -7×5

Solution

(a) $-8 \times -3 = 24$

(b) $-7 \times 5 = -35$

Note the following properties:

1. The product of zero and any real numbers is 0.
2. The product of an even number of negative numbers is positive.
3. The product of an odd number of negative numbers is negative.

Dividing Integers

Given two integers a and b, $a \div b$ or a/b, is the number that when multiplied by b gives a. For example, $8 \div 4$ is 2, since $4 \times 2 = 8$.

The sign rules for division are the same as those for multiplication. The quotient of numbers with like signs is positive, the quotient of numbers with unlike signs is negative.

Example 5

Workout the following divisions:

(a) $-24 \div -6$ (b) $36 \div -12$

Solution

(a) $-24 \div -6 = 4$

(b) $36 \div -12 = -3$

Note the following properties:

For any real numbers a and b:

1. $\frac{a}{0}$ is undefined

2. $\frac{-a}{b} = \frac{a}{-b} = -\frac{a}{b}$ and $\frac{-a}{-b} = \frac{a}{b}$

Properties of Real Numbers

Commutative Property

We may add 5 and 9 as follows:

$$5 + 9 = 14 \text{ or } 9 + 5 = 14$$

Notice that changing the order gives the same result. This property of real numbers is called the commutative property of addition. Similarly, we can demonstrate that multiplication is also commutative. Generally, given any two real numbers a and b

$a + b = b + a$ Commutative property of Addition

$a \cdot b = b \cdot a$ Commutative property of Multiplication

Associative Property

We may add 4, 5 and 6 as follows:

$$(4 + 5) + 6 = 9 + 6 = 15 \text{ or } 4 + (5 + 6) = 4 + 11 = 15$$

Although the numbers were ordered and grouped in different ways the result is the same. This property of addition is called associative property of addition. Similarly, we can demonstrate that multiplication is associative. Generally, given any real numbers a, b and c,

$$(a + b) + c = a + (b + c)$$

$$(a \cdot b) \cdot c = a \cdot (b \cdot c)$$

Distributive Property

We can evaluate $8 \cdot (3 + 7)$ in two ways:

We can add 3 and 7 and then multiply the result by 8 to get $8 \cdot 10 = 80$.

You can obtain the same result when you multiply both 3 and 7 by 8 and then add the results;

$$8 \cdot (3 + 7) = 8 \cdot 3 + 8 \cdot 7 = 24 + 56 = 80$$

Generally, for any real numbers a, b and c

$$a\,(b + c) = ab + ac$$

This property of real numbers is called the distributive property.

Additive and Multiplication Identity

For any real number a, $a + 0 = a$. 0 is called the additive identity. For example, $2 + 0 = 2$. Also, for any real number b, $b \times 1 = b$. For example, $-3 \times 1 = -3$. 1 is called the multiplicative identity.

Additive and Multiplication Inverse

For any real number a, there exists a number $-a$ such that $a + (-a) = 0$. $-a$ is called the additive inverse of a. Notice that the additive inverse of $-a$ is a. For example, the additive inverse of -2 is 2, since $2 + (-2) = 0$. -2 and 2 are additive inverses of each other.

Also, for any real number b, where $b \neq 0$, there exists a number $1/b$, such that $b \times (1/b) = 1$. $1/b$ is called the multiplicative inverse (or the reciprocal) of b. For example, the multiplicative inverse of 5 is 1/5, since $5 \times (1/5) = 1$. 1/5 and 5 are multiplicative inverses of each other. The following table summarizes the properties of real numbers.

1. $a + b = b + a$ Commutative property of Addition

   $ab = ba$ Commutative property of Multiplication

2. $(a + b) + c = a + (b + c)$ Associative property of Addition

   $(ab)c = a(bc)$ Associative property of Multiplication

3. $a\,(b + c) = ab + ac$ Distributive property

4. $a + 0 = 0 + a = a$ — Additive Identity property

$a \cdot 1 = 1 \cdot a = a$ — Multiplicative identity property

5. $a + (-a) = 0$ — Additive inverse property

$a \cdot \frac{1}{a} = 1, a \neq 0$ — Multiplicative inverse property

Using the Distributive Property

We may use the distributive property to find the product of two or more real numbers quickly. The method is shown in Example 6.

Example 6

Use the distributive property to evaluate:

(a) 6×79 (b) 17×32

Solution

(a) Begin by rewriting 79 as 80 – 1.

$$6 \times 79 = 6(80 - 1)$$
$$= 480 - 6$$
$$= 474$$

(b) $$17 \times 32 = 17(30 + 2)$$
$$= 510 + 34$$
$$= 544$$

Exercise 2.2

In Exercises 1 – 12, evaluate the following expressions:

1. $7 + (-5)$
2. $-9 + 12$
3. $-15 + 27$
4. $-13 + 20$
5. $-32 + 45$
6. $-17 + 12$
7. $-15 + 7$
8. $-28 + 19$
9. $-7 + (-9)$
10. $-12 + (-16)$
11. $-6 + (-8)$
12. $-6 + (-3)$

In Exercises 13 – 30, evaluate the following expressions:

13. 8 – (– 7)
14. 16 – (– 9)
15. – 6 – 4
16. – 13 – 12
17. – 4 – (– 7)
18. – 6 – (– 4)
19. – 13 – (– 18)
20. – 35 – (– 28)
21. – 11 – (– 7)
22. 8 – 15
23. – 9 + 13
24. 11 – 16
25. 37 – 45
26. – 12 – 18
27. – 23 + 15
28. – 9 – (– 6)
29. – 16 – (– 26)
30. 64 – (– 36)

In Exercises 31 – 42, evaluate the following expressions:

31. – 5 × 6
32. 4 × – 11
33. – 6 × – 7
34. – 8 × – 9
35. 9 × – 4
36. – 7 × 10
37. – 7 × 8
38. 14 × – 5
39. – 5 × – 12
40. 5 × – 9
41. – 7 × – 8
42. – 15 × 11

In Exercises 43 – 54, evaluate the following expressions:

43. – 15 ÷ 3
44. 45 ÷ – 9
45. – 12 ÷ – 6
46. 72 ÷ – 12
47. – 20 ÷ – 5
48. – 21 ÷ 7
49. – 45 ÷ – 15
50. – 20 ÷ 4
51. 48 ÷ – 12
52. – 80 ÷ 10
53. 77 ÷ – 11
54. – 36 ÷ – 9

In Exercises 55 – 66, using the distributive property evaluate:

55. 5 × 7 + 5 × 13
56. 36 × 4 – 11 × 4

57. $64 \times 12 + 36 \times 12$

58. $8 \times 48 - 8 \times 28$

59. $99 \times 25 + 25$

60. $9 \times 148 - 9 \times 28$

61. 7×99

62. 48×7

63. 197×6

64. 999×17

65. 13×57

66. 298×4

2.3 Operations with Fractions

Fractions such as $\frac{3}{5}$ and $\frac{7}{4}$ are called common fractions. The top number in each fraction is called the numerator and the bottom number is called the denominator. A fraction that has its numerator less than its denominator is called a proper fraction, while a fraction that has its numerator greater than its denominator is called an improper fraction. For example, $\frac{4}{5}$ is a proper fraction and $\frac{9}{4}$ is an improper fraction. Fractions such as $2\frac{1}{4}$ are called mixed numbers.

Adding or subtracting fractions with the same denominator

To add or subtract fractions with the same denominators add or subtract their numerators and place the result over the common denominator.

Example 1

Evaluate the following expressions:

$(a)\ \frac{5}{12} + \frac{2}{12}$ $(b)\ \frac{11}{15} - \frac{7}{15}$

Solution

$(a)\ \frac{5}{12} + \frac{2}{12} = \frac{5+2}{12}$ Add numerators

$= \frac{7}{12}$ Simplify

$(b)\ \frac{11}{15} - \frac{7}{15} = \frac{11-7}{15}$ Subtract numerators

$= \frac{4}{15}$ Simplify

Adding or subtracting fractions with different denominators

Fractions with different denominators can only be added or subtracted if the fractions are first changed to equivalent fractions with a common denominator, called the least common denominator (LCD). The fraction is then added or subtracted as described above. It is customary to use the lowest common multiple (LCM) of their denominators as the common denominator.

Example 2

Evaluate the following expressions:

(a) $\frac{3}{8}+\frac{5}{12}$ (b) $\frac{8}{15}-\frac{5}{6}$

Solution

(a) Begin by finding the lowest common multiple.

The lowest common multiple of 8 and 12 is 24.

Then convert $\frac{3}{8}$ and $\frac{5}{12}$ to equivalent fractions with 24 as the common denominator.

$$\frac{3}{8}+\frac{5}{12}=\frac{3(3)}{8(3)}+\frac{5(2)}{12(2)}$$

$$=\frac{9}{24}+\frac{10}{24}$$ Simplify

$$=\frac{9+10}{24}$$ Add numerators

$$=\frac{19}{24}$$ Simplify

(b) $$\frac{8}{15}-\frac{5}{6}=\frac{8(2)}{15(2)}-\frac{5(5)}{6(5)}$$

$$=\frac{16}{30}-\frac{25}{30}$$ Simplify

$$=\frac{16-25}{30}$$ Subtract numerators

$$=-\frac{9}{30}$$ Simplify

$$=-\frac{3}{10}$$ Simplify

Adding or subtracting mixed numbers

To add or subtract mixed numbers change them to equivalent improper fractions and then add or subtract as usual.

Example 3

Evaluate the following expressions:

$(a)\ 4\frac{1}{3}+2\frac{3}{4}$ $\qquad$ $(b)\ 5\frac{4}{5}-2\frac{2}{3}$

Solution

(a) Write each fraction as improper fraction

$$4\frac{1}{3}+2\frac{3}{4}=\frac{13}{3}+\frac{11}{4}$$

$=\frac{52}{12}+\frac{33}{12}$ Least common denominator is 12

$=\frac{52+33}{12}$ Add the numerators

$=\frac{85}{12}$ Simplify

$=7\frac{1}{12}$ Write the fraction as a mixed number

(b) Write each fraction as improper fraction.

$$5\frac{4}{5}-2\frac{2}{3}=\frac{29}{5}-\frac{8}{3}$$

$=\frac{87}{15}-\frac{40}{15}$ Least common denominator is 15

$=\frac{87-40}{15}$ Subtract the numerators

$=\frac{47}{15}$ Simplify

$=3\frac{2}{15}$ Write the fraction as a mixed number

Multiplying Common Fractions

To multiply two fractions multiply the numerator together to give the numerator of the product and then multiply the denominators together to give the denominator of the product. Reduce the answer to the lowest term if possible.

Example 4

Evaluate the following expressions:

$(a)\ \frac{8}{20} \times \frac{6}{9}$ $(b)\ 2\frac{2}{3} \times 5\frac{3}{4}$

Solution

(a) Multiply numerator and denominators.

$$\frac{8}{20} \times \frac{6}{9} = \frac{48}{180}$$

$$= \frac{4}{15} \quad \text{Simplify}$$

(b) Write each fraction as improper fraction, and then multiply numerators and denominators.

$$2\frac{2}{3} \times 5\frac{3}{4} = \frac{8}{3} \times \frac{23}{4}$$

$$= \frac{184}{12}$$

$$= \frac{46}{3} \quad \text{Simplify}$$

$$= 15\frac{1}{3} \quad \text{Write the fraction as a mixed number}$$

Dividing Fractions

To divide fractions multiply the first fraction by the reciprocal of the second fraction. Find the reciprocal by inverting the fraction.

Example 5

Evaluate the following expressions:

$(a)\ \frac{5}{16} \div \frac{3}{8}$ $(b)\ 4\frac{1}{2} \div 1\frac{2}{7}$

Solution

Invert divisor and multiply, and then multiply numerator and denominator.

$$\frac{5}{16} \div \frac{3}{8} = \frac{5}{16} \times \frac{8}{3}$$

$$= \frac{40}{48}$$

$$= \frac{5}{6}$$

(b) Write each fraction as improper fraction, and then invert the divisor and multiply.

$$4\frac{1}{2} \div 1\frac{2}{7} = \frac{9}{2} \div \frac{9}{7}$$

$$= \frac{9}{2} \times \frac{7}{9}$$

$$= \frac{7}{2}$$

$$= 3\frac{1}{2}$$

Exercise 2.3

In Exercise 1 – 10, evaluate the following expressions:

1. $\frac{3}{5} + \frac{1}{5}$
2. $\frac{4}{7} + \frac{2}{7}$
3. $\frac{5}{14} + \frac{3}{14}$
4. $\frac{5}{12} + \frac{3}{12}$
5. $\frac{7}{15} + \frac{11}{15}$
6. $\frac{5}{8} - \frac{3}{8}$
7. $\frac{7}{9} - \frac{5}{9}$
8. $\frac{13}{24} - \frac{5}{24}$
9. $\frac{19}{30} - \frac{7}{30}$
10. $\frac{13}{25} - \frac{8}{25}$

In Exercise 11 – 20, evaluate the following expressions:

11. $\frac{2}{3} + \frac{3}{5}$
12. $\frac{3}{4} + \frac{5}{6}$
13. $\frac{2}{9} + \frac{4}{15}$
14. $\frac{5}{12} + \frac{7}{20}$

15. $\frac{5}{6}+\frac{7}{12}+\frac{1}{4}$

16. $\frac{2}{3}-\frac{3}{5}$

17. $\frac{4}{7}-\frac{3}{8}$

18. $\frac{5}{6}-\frac{3}{8}$

19. $\frac{7}{9}-\frac{5}{12}$

20. $\frac{5}{6}-\frac{3}{8}-\frac{5}{12}$

In Exercise 21 – 32, evaluate the following expressions:

21. $2\frac{1}{2}+3\frac{3}{4}$

22. $3\frac{1}{2}-2\frac{1}{8}$

23. $5\frac{1}{3}+2\frac{1}{2}$

24. $5\frac{5}{8}-3\frac{1}{4}$

25. $3\frac{2}{15}+4\frac{5}{9}$

26. $7\frac{1}{9}-6\frac{1}{2}$

27. $6\frac{7}{8}+5\frac{3}{10}$

28. $8\frac{1}{8}-5\frac{3}{5}$

29. $1\frac{5}{12}+3\frac{7}{20}$

30. $7\frac{5}{8}-3\frac{3}{5}$

31. $2\frac{5}{8}+3\frac{9}{24}$

32. $3\frac{1}{6}-1\frac{2}{3}$

In Exercise 33 – 42, evaluate the following expressions:

33. $2\times\frac{5}{8}$

34. $20\times\frac{3}{16}$

35. $\frac{9}{10}\times\frac{5}{6}$

36. $\frac{15}{24}\times\frac{8}{9}$

37. $\frac{18}{35}\times\frac{7}{12}$

38. $6\times 3\frac{2}{3}$

39. $1\frac{3}{4}\times 8$

40. $2\frac{1}{4}\times 3\frac{1}{3}$

41. $3\frac{2}{9}\times 1\frac{2}{7}$

42. $4\frac{1}{5}\times 7\frac{1}{2}$

In Exercise 43 – 52, evaluate the following expressions:

43. $8\div\frac{4}{9}$

44. $\frac{7}{10}\div 14$

45. $\frac{5}{6} \div \frac{2}{3}$

46. $\frac{7}{8} \div \frac{3}{4}$

47. $\frac{3}{10} \div \frac{9}{25}$

48. $3\frac{1}{8} \div 5$

49. $18 \div 2\frac{1}{4}$

50. $4\frac{1}{2} \div 1\frac{1}{5}$

51. $2\frac{3}{4} \div 1\frac{3}{8}$

52. $6\frac{5}{12} \div 1\frac{5}{6}$

2.4 Exponents and the Order of Operations

In the expression a^n, where $a \neq 0$, a is called a base and n is called an exponent. a^n is read "a raised to the nth power." Expressions like a^2 and a^3 are read "a square" and "a cube" respectively.

For any nonzero real number a and any positive integer n, a^n represents the product of n factors of a. That is

$$a^n = a \cdot a \cdot a \ldots a \qquad n \text{ factors}$$

For example, $2^5 = 2 \cdot 2 \cdot 2 \cdot 2 \cdot 2$.

Multiplying exponential expressions with like base

To evaluate $a^3 \cdot a^5$, you can proceed as follows:

$$\begin{aligned} a^3 \cdot a^5 &= (a \cdot a \cdot a)\,(a \cdot a \cdot a \cdot a \cdot a) \\ &= a \cdot a \cdot a \cdot a \cdot a \cdot a \cdot a \cdot a \\ &= a^8 \end{aligned}$$

Notice that the exponent of product is the same as the sum of the exponent of a^3 and the exponent of a^5. In general, for any nonzero real number a and positive integer m and n

$$a^m \cdot a^n = a^{m+n}$$

Example 1

Evaluate the following exponential expressions:

(a) $2^4 \cdot 2^5$ (b) $3^2 \cdot 3^4$

Solution

(a) $2^4 \cdot 2^5 = 2^{4+5} = 2^9$

(b) $3^2 \cdot 3^4 = 3^{2+4} = 3^6$

Dividing exponential expressions with like base

To evaluate $a^7 \div a^4$, you can proceed as follows:

$$a^7 \div a^4 = \frac{a^7}{a^4}$$

$$= \frac{a \cdot a \cdot a \cdot a \cdot a \cdot a \cdot a}{a \cdot a \cdot a \cdot a}$$

$$= a \cdot a \cdot a$$

$$= a^3$$

Notice that the exponent of the quotient is the same as the difference of the exponents of a^7 and a^4. In general, for any nonzero real number a and positive integer m and n

$$a^m \div a^n = a^{m-n}$$

Example 2

Evaluate the following exponential expressions:

(a) $2^{12} \div 2^7$ (b) $5^8 \div 5$

Solution

(a) $2^{12} \div 2^7 = 2^{12-7} = 2^5$

(b) $5^8 \div 5 = 5^{8-1} = 5^7$

Note that when a number is raised to the first power, we do not normally write the exponent 1.

Raising a power to a power

To evaluate an expression like $(a^2)^3$, we proceed as follows:

$$(a^2)^3 = a^2 \cdot a^2 \cdot a^2$$

$$= a \cdot a \cdot a \cdot a \cdot a \cdot a$$

$$= a^6$$

Notice that the exponent of the result is the product of the two exponents of $(a^2)^3$. In general, for any nonzero real number a and positive integer m and n,

$$(a^m)^n = a^{mn}$$

Example 3

Evaluate the following exponential expressions:

(a) $(7^5)^2$ (b) $(5^4)^3$

Solution

(a) $(7^5)^2 = 7^{5 \times 2} = 7^{10}$

(b) $(5^4)^3 = 5^{4 \times 3} = 5^{12}$

The definition of an exponent can be extended by defining a^n for zero, negative integers and fractional values of n.

The zero Index

For any number a, where $a \neq 0$

$$a^0 \cdot a^2 = a^2$$

Dividing both sides by a^2, we have

$$a^0 = 1$$

Thus, for any nonzero number a, $a^0 = 1$.

For example, $1000^0 = 1$ and $(\frac{1}{2})^0 = 1$.

Negative Indices

$$a^{-2} \cdot a^2 = a^{-2+2}$$

$$= a^0$$

Dividing both sides by a^2, we have

$$a^{-2} = \frac{a^0}{a^2} = \frac{1}{a^2}, \text{ since } a^0 = 1$$

In general, for any nonzero number a and any integer n

$$a^{-n} = \frac{1}{a^n}$$

Example 4

Evaluate the following exponential expressions:

(a) $2^{\div 3}$ (b) $\frac{1}{5^{-3}}$

Solution

(a) $2^{-3} = \frac{1}{2^3} = \frac{1}{8}$

(b) $\frac{1}{5^{-3}} = 5^3 = 125$

Fractional Index

For any nonzero real number a

$$a^{\frac{1}{3}} \cdot a^{\frac{1}{3}} \cdot a^{\frac{1}{3}} = a^{\frac{1+1+1}{3}}$$

$$= a$$

$$\therefore \ a^{\frac{1}{3}} = \sqrt[3]{a}$$

In general, if a is any real number and n is a positive integer then

$$a^{\frac{1}{n}} = \sqrt[n]{a}$$

Example 5

Evaluate the following exponential expressions:

$(a)\ 16^{\frac{1}{2}}$ $(b)\ 27^{\frac{1}{3}}$

Solution

(a) $16^{\frac{1}{2}} = \sqrt{16}$

$= 2$

(b) $27^{\frac{1}{3}} = \sqrt[3]{27}$

$= 3$

You could also obtain the same results as follows:

(a) $16^{\frac{1}{2}} = (2^4)^{\frac{1}{4}}$ Write 16 as a power of 2

$= 2$ Multiply the exponents

(b) $27^{\frac{1}{3}} = (3^3)^{\frac{1}{3}}$ Write 27 as a power of 3

$= 3$ Multiply the exponents

Also, for any nonzero real number a

$$a^{\frac{2}{3}} \cdot a^{\frac{2}{3}} \cdot a^{\frac{2}{3}} = a^{\frac{2+2+2}{3}}$$

$$= a^2$$

$$\therefore \quad a^{\frac{2}{3}} = \sqrt[3]{a^2}$$

In general, if a is any real number and m and n ($n > 1$) then

$$a^{\frac{m}{n}} = \sqrt[n]{a^m} = \left(\sqrt[n]{a}\right)^m$$

Example 6

Evaluate the following exponential expressions:

$(a)\ 27^{\frac{2}{3}}$ $\qquad$ $(b)\ 8^{-\frac{2}{3}}$

Solution

(a) $27^{\frac{2}{3}} = \left(\sqrt[3]{27}\right)^2$

$= 3^2$

$= 9$

(b) $8^{-\frac{2}{3}} = \frac{1}{8^{2/3}}$

$= \frac{1}{\left(\sqrt[3]{8}\right)^2}$

$= \frac{1}{2^2}$

$= \frac{1}{4}$

The laws of exponents are summarized in the list below.

1. $a^m \cdot a^n = a^{m+n}$
2. $a^m \div a^n = a^{m-n}$
3. $(a^m)^n = a^{mn}$
4. $a^{\frac{1}{n}} = \sqrt[n]{a}$
5. $a^{\frac{m}{n}} = \sqrt[n]{a^m} = \left(\sqrt[n]{a}\right)^m$
6. $a^{-n} = \frac{1}{a^n}$

7. $a^0 = 1$

Scientific Notation

You can use scientific notation to write very large and very small numbers in terms of powers of 10.

Consider the number 4,500,000,000.

$$4{,}500{,}000{,}000 = 4.5 \times 1{,}000{,}000{,}000$$
$$= 4.5 \times 10^9$$

This number has been written as a product of two numbers. The first number is a number that is at least 1 but less than 10 and the second number is an integer power of 10. Notice that, multiplying a number by 10^9 has the effect of moving the decimal point of the number nine places to the right. Generally, multiplying a number by 10^n, where n is an integer, has the effect of moving the decimal point of the number n places to right when n is positive, and n places to the left when n is negative.

Scientific notation for a number is an expression of the form $a \times 10^n$, where n is at least 1 but less than 10 and n is an integer.

Example 7

Write each of the following numbers in scientific notation.

(a) 734000 (b) 0.00057

Solution

(a) Move the decimal point 5 places to the left.

$$734000 = 7.34 \times 10^5$$

(b) Move the decimal point 4 places to the right

$$0.00057 = 5.7 \times 10^{-4}$$

Using Scientific Notation

In finding the product of two or more numbers we can sometimes simplify our calculations by rewriting the given numbers in scientific notation before multiplying.

Example 8

Evaluate the expression:

(a) 180000×0.000002

Solution

(a) $180000 \times 0.000002 = 1.8 \times 10^{5} \times 2 \times 10^{-6}$

$= 3.6 \times 10^{-1}$

(b) $\frac{24000 \times 0.003}{0.12}$

Solution

$\frac{24000 \times 0.003}{0.12} = \frac{2.4 \times 10^{4} \times 3 \times 10^{-3}}{1.2 \times 10^{-1}}$

$= 2 \times 3 \times 10^{1} \times 10^{1}$

$= 6 \times 10^{2}$

Order of Operations

How should $8 + 3 \times 10$ be computed? If we multiply 3 by 10 and then add 8, the result is 38. However, if we add 8 and 3 first and then multiply the result by 10 gives 110. The two results differ and to avoid this confusion, the following rules exist for determining how the order of operations should be performed.

1. Perform any operation(s) within grouping symbols.
2. Evaluate all powers
3. Multiply and divide in order from left to right.
4. Finally, add and subtract in order from left to right.

Example 9

Evaluate the expression:

(a) $5 - 28 \div 7 \times 2 + 12$

Solution

$5 - 28 \div 7 \times 2 + 12$

$= 5 - 4 \times 2 + 12$ Evaluate the division

$= 5 - 8 + 12$ Evaluate the multiplication

$= -3 + 12$ Subtract

$= 9$ Add

(b) $12 \div 4 - 2(3 + 2)^{2} + 30$

Solution

$12 \div 4 - 2(3+2)^2 + 30 = 12 \div 4 - 2(5)^2 + 30$	Add the expression in parentheses
$= 12 \div 4 - 2(25) + 30$	Evaluate the power
$= 3 - 2(25) + 30$	Divide
$= 3 - 50 + 30$	Multiply
$= -47 + 30$	Subtract
$= -17$	Add

Exercise 2.4

In Exercises 1 – 20, simplify the expression. Leave the answer in exponent form.

1. $2^3 \cdot 2^4$ 2. $3^3 \cdot 3^2$ 3. $5^7 \cdot 5$ 4. $6^5 \cdot 6^4$

5. $2^{-2} \cdot 2^{-3}$ 6. $7^4 \div 7$ 7. $8^7 \div 8^2$ 8. $2^2 \div 2^4$

9. $3^3 \div 3^{-2}$ 10. $5^{-3} \div 5^{-5}$ 11. $6^{-3} \div 6^5$ 12. $(4^3)^2$

13. $(2^4)^3$ 14. $(3^2)^5$ 15. $(7^5)^4$ 16. $(2^{-3})^2$

17. $(3^2)^{-4}$ 18. $(4^{-3})^{-2}$ 19. $(5^{-1})^3$ 20. $(3^{-4})^{3/2}$

In Exercises 21 – 35, evaluate the expression.

21. $8^{1/3}$ 22. $25^{1/2}$ 23. $256^{1/4}$ 24. $32^{1/5}$ 25. $81^{3/4}$

26. $8^{2/3}$ 27. $625^{3/4}$ 28. $16^{5/4}$ 29. $8^{-1/3}$ 30. $9^{-1/2}$

31. 5^{-2} 32. 2^{-3} 33. $(-1/3)^0$ 34. $(1/4)^{-2}$ 35. $(1/2)^{-1}$

In Exercises 36 – 41, evaluate the expression, writing the answer in exponent form.

36. $(3^{\div 4} \times 3^6)/3^{-2}$ 37. $[(2^2)^3 \times 2^4]/2^7$

38. $(5^3)^4/[5^{-7} \times (5^8)^2]$ 39. $[(4^3)^4 \times 16^3]/[(4^5)^2 \times 64^2]$

40. $(25^{-4} \times 5^6 \times 25^2)/[25^{-3} \times (5^3)^2]$ 41. $[(9^2)^3 \times 81^{-3/4}]/[27^{1/3} \times (3^2)^3]$

In Exercises 42 – 65, write each expression in scientific notation.

42. 4700000 43. 86000 44. 230 45. 76

46. 3640000000 47. 124000 48. 27.5 49. 572

50. 627 51. 9.5 52. 10500 53. 4156

54. 0.0056 55. 0.24 56. 0.0000431 57. 0.000703

58. 0.0643 59. 0.0000085 60. 0.009 61. 0.126

62. 0.03164 63. 0.00002 64. 0.0605 65. 0.00000057

In Exercises 66 – 77, evaluate the expression, writing the answer in scientific notation.

66. 250000×16000 67. 0.00081×0.005

68. 75000×0.003 69. $12100000 \times 0.0000045$

70. $6800 \times 12000 \times 0.00003$ 71. $9700000 \times 0.05 \times 0.00012$

72. $(0.008 \times 90000)/0.12$ 73. $(36000 \times 3.2)/(0.06 \times 0.08)$

74. $(0.125 \times 0.026)/(0.025 \times 0.0013)$ 75. $(35000 \times 5.60)/(0.005 \times 140000)$

76. $(169 \times 7.5 \times 0.036)/(3900 \times 0.18)$ 77. $(45000 \times 1600 \times 81)/(18000 \times 5400)$

In Exercises 78 – 94, evaluate the expression.

78. $18 - 25 + 3$ 79. $8 + 3 \times 4$ 80. $12 + 20 \times - 3$ 81. $15 - 18 \div - 3$

82. $8 \div (10 - 14)$ 83. $9 \times 4 \div 12$ 84. $16 - 2^5 + 8$ 85. $5 \times 6 - 4^3 \div 8$

86. $15 - (3^2 + 32 \div 8)$ 87. $(10 - 4) \div 2 \times 3$ 88. $15 + 6 \div 3 - 18$

89. $8 + 5 \times 3 - 3 + 8 \div 4$ 90. $20 \div (18 - 25 + 3)$

91. $15 \times 2 \div 6 + (12 - 20)$ 92. $16 \div (23 - 5^2) \times 7$

93. $(4 \times 3 - 16) \div (3 - 14 \div 2)$ 94. $5 \times 6 - 2(8 \div 4 \times 3)^2 + 11$

Review Exercises

In Exercises 1 and 2, which of the real numbers in the set are:

(a) natural numbers (b) integers,

(c) rational numbers (d) irrational numbers?

1. $\left\{-3.5, -\sqrt{9},\ -2.3,\ -1,\ -0.75,\ 0,\ \sqrt{3},\ 2\pi,\ \frac{19}{2},\ 12\right\}$

2. $\left\{-\frac{9}{2},\ -3,\ -\sqrt{5},\ -\frac{3}{4},\ 0,\ \sqrt{36},\ \frac{15}{2},\ 9,\ 23\right\}$

In Exercises 3 – 6, plot the real numbers on the real number line.

3. -4 4. ½ 5. -5.2 6. 5/2

In Exercises 7 – 10, place the correct symbol $<$ or $>$ between the pair of numbers

7. $-7 \quad -3$ 8. 5.23 5.0 9. $-12 \quad 0$ 10. ¾ ½

In Exercises 11 – 14, evaluate the expression.

11. $|-8|$ 12. $|7.3|$ 13. $-|-32|$ 14. $-|15|$

In Exercises 15 – 17, place the correct symbol $<$, $>$ or $=$ between the pair of numbers

15. $|-8| \quad |-12|$ 16. $|-13| \quad |7|$ 17. $|-5| \quad |5|$

In Exercises 18 – 32, evaluate the expression.

18. $2-(-9)$ 19. $-5-(-7)$ 20. $-8-(-8)$

21. $12-16$ 22. $15-3-(-12)$ 23. $-42+50-16$

24. $12-(-18)-40$ 25. 8×-3 26. -9×4

27. -6×-13 28. -15×8 29. $24\div -3$

30. $-32\div -4$ 31. $-57\div 19$ 32. $-52\div -13$

In Exercises 33 – 36, use the distributive property to evaluate the expression:

33. $6\times 21+6\times 29$ 34. $83\times 64-83\times 14$

35. 197×7 36. 6×248

In Exercises 37 – 51, evaluate the expression:

37. $\frac{7}{8}+\frac{3}{8}$ 38. $\frac{9}{14}-\frac{3}{14}$ 39. $\frac{2}{5}+\frac{8}{15}$

40. $\frac{7}{9}-\frac{5}{12}$ 41. $2\frac{5}{6}-1\frac{7}{12}$ 42. $4\frac{2}{3}+3\frac{1}{4}$

43. $5\frac{2}{3}-2\frac{3}{4}$ 44. $2\frac{5}{6}+3\frac{7}{10}$ 45. $\frac{7}{15}\times\frac{25}{21}$

46. $2\frac{1}{8}\times 12$ 47. $1\frac{3}{5}\times 4\frac{3}{4}$ 48. $6\div\frac{2}{3}$

49. $\frac{5}{8}\div\frac{3}{4}$ 50. $4\frac{1}{2}\div 3$ 51. $3\frac{5}{8}\div 1\frac{3}{4}$

In Exercises 52 – 63, simplify the expression.

52. $3^7 \cdot 3^{-5}$ 53. $2^{-5} \cdot 2^4$ 54. $5 \cdot 5^{-3}$ 55. $5^8 \div 5^{12}$

56. $3^{-3} \div 3^{-6}$ 57. $3^{-1} \div 3$ 58. $(5^{-2})^{-1}$ 59. $(2^3)^2$

60. $(3^{-10})^{-3/5}$ 61. $(7^{4/3})^{-3/2}$ 62. $(1/3)^0$ 63. $(1/8)^{-2/3}$

In Exercises 64 – 69, evaluate the expression without a calculator:

64. $16^{1/4}$ 65. $216^{1/3}$ 66. $8^{2/3}$

67. $27^{2/3}$ 68. $81^{-3/4}$ 69. $64^{-2/3}$

In Exercises 70 – 77, write expression in scientific notation:

70. 597000 71. 640 72. 67.5 73. 7582

74. 0.65 75. 0.000804 76. 0.02178 77. 0. 00009

In Exercises 78 – 85, evaluate the expression, and write the answer in scientific notation.

78. 65000×0.083 79. 0.0093×0.52

80. $8600 \div 0.0043$ 81. $0.243 \div 0.00027$

82. $(0.625 \times 0.0036)/(0.0025 \times 0.045)$

83. $(0.00035 \times 0.56)/(0.05 \times 0.14)$

84. $(0.169 \times 0.0075 \times 7.2)/(0.00065 \times 0.0018)$

85. $(0.0045 \times 0.16 \times 0.081)/(0.00018 \times 0.54)$

In Exercises 86 – 96 evaluate the expression.

86. $5 + 3 \times 7$ 87. $19 - 5 \times 3 + 3$

88. $12 - 8 \div 4 - 2$ 89. $18 - 6 \div 3 \times 2 + 7$

90. $28 \div 4 + 3 \times 5$ 91. $8 + (15 \div 3) - 7 \times 3$

92. $12 - (14 \div 7 \times 4) + 3$ 93. $16 \div 8 \times 3 - (17 - 23)$

3 Algebraic Expressions

In Chapter 2, you learned how to do calculations with numbers by using the four basic operations of addition, subtraction, multiplication, and division. This chapter introduces some basic concepts in algebra. In algebra, you will use numbers as well as letters. A letter that represents a variety of different numbers is called a variable.

3.1 Polynomials

An expression that consists of a number, a variable with a whole number exponent, the product of a number and variable(s) with whole number exponent(s), and one or more arithmetic operations is called a polynomial. Examples of polynomials include

$$7,\quad 3x,\quad 5x + 2,\quad 2xy - 7xz,\quad x^2 + 6x - 3 \quad \text{and} \quad 3x^3 - 3x^2y + 4xy^2 - 9y^3$$

A term of a Polynomial

A term of a polynomial can be a number, a variable or a product of numbers and variables. For example, the polynomial $3x^2 - 4xy + 2$ has three terms: $3x^2$, $-4xy$ and 2. The first term $3x^2$ and the second term $-4xy$ are called variable terms. The third term 2 is called a constant term. A polynomial with only one term is called a monomial. Polynomials with two terms are called binomials, and those with three terms are called trinomials.

Coefficient of a Variable Term

The numerical factor of a variable term is called the coefficient. For example, the coefficients of the expressions $5x^2$, $-3xy^2$ and xy^2 are 5, -3 and 1 respectively.

Like Terms

Two or more terms having exactly the same variable factors, such as $3x$ and $-7x$, are called like terms. Similarly, $-3x^2y$ and x^2y are like terms because their variable factors are the same. Terms that do not have the same variable(s) or have the same variable(s) with different exponents such as $2y$ and $4y^2$, are unlike terms.

Simplifying Polynomials

You can simplify polynomials by combining like terms. One way of doing this is to use the distributive property in reserve. That is, $ab + ac = a(b + c)$.

Example 1

Simplify: (a) $3x + 5x$ (b) $7y^2 - 12y^2$

Solution

(a) $3x + 5x = (3 + 5)\,x = 8x$

(b) $7y^2 - 12y^2 = (7 - 12)\,y^2 = -5y^2$

With practice, you can leave out the first step. The commutative, associative and distributive properties are useful in simplifying polynomials as illustrated in Example 2.

Example 2

Simplify: (a) $3x + 5y + 2x + 3y$

(b) $6x + 3y - 4x - 8y$

Solution

(a) $3x + 5y + 2x + 3y = 3x + 2x + 5y + 3y$ Use commutative property

$= 5x + 8y$ Combine like terms

(b) $6x + 3y - 4x - 8y = 6x - 4x + 3y - 8y$ Use the commutative property

$= 2x - 5y$ Combine like terms

Multiplying Polynomials

Recall that sometimes we use dot or parentheses to show multiplication of two or more numbers. However, when letters are used to represent numbers, the dots or parentheses are often omitted. For example, $3 \times a = 3 \cdot a = (3)(a) = 3a$.

You can evaluate $2x \cdot 3y$ as follows:

$2x \cdot 3y = 2 \cdot x \cdot 3 \cdot y$ Factor each term

$= 2 \cdot 3 \cdot x \cdot y$ Use commutative property

$= 6xy$ Simplest form

The product of two or more polynomials is the product of their numerical coefficients multiplied by the product of the variables. The product can be written in any order but it is most convenient to write the number first and the letters in alphabetical order.

Example 3

Evaluate the expression: (a) $5x \cdot -2y$ (b) $-4x \cdot -3z$

Solution

(a) $5x \cdot -2y = -10xy$

(b) $-4x \cdot -3z = 12xy$

Recall from the previous chapter that

1. $a^m \cdot a^n = a^{m+n}$

2. $a^m \div a^n = a^{m-n}$

3. $(a^m)^n = a^{mn}$

You can multiply or divide polynomials by using these rules together with the distributive property.

Note that in the expression $2x^5$, the power 5 applies only to the letter x. However, in the expression $(2x)^5$, the power 5 applies to both 2 and x. That is, $(2x)^5 = 2^5 \cdot x^5 = 32x^5$.

Example 4

Multiply: (a) $4x^3y \cdot 2xy^2$ (b) $(-4x^3y)^2$

Solution

(a) $4x^3y \cdot 2xy^2 = 8x^4y^3$

(b) $(-4x^3y)^2 = 16x^6y^2$

Example 5

Divide: (a) $-15x^3y^2 \div -5xy$ (b) $-20a^4b^3 \div 4a^2b^2$

Solution

(a) $-15x^3y^2 \div -5xy = \frac{-15x^3y^2}{-5xy} = 3x^2y$

(b) $-20a^4b^3 \div 4a^2b^2 = \frac{-20a^4b^3}{4a^2b^2} = -5a^2b$

Evaluating Algebraic Expressions

We can find the value of an algebraic expression if numerical values are assigned to the variables. Finding the value of an expression is called evaluating the expression. To evaluate an algebraic expression, replace each variable with a given value and then simplify the numerical expression that results, as illustrated in Example 6.

Example 6

Find the value of $3a + 2b$ when $a = -2$ and $b = 4$.

Solution

Replace a with -2, and b with 4, and use rules for signed numbers.

$$3a + 2b = 3(-2) + 2(4) = -6 + 8 = 2$$

Exercise 3.1

In Exercises 1 – 12, simplify:

1. $4x + 2x$
2. $7y + y$
3. $9x^2 + 3x^2$
4. $8x - 3x$
5. $5y - y$
6. $2a - 3a$
7. $-2x + 7x$
8. $8y + (-5y)$
9. $-x + 3x$
10. $-2x - 7x$
11. $-y - 3y$
12. $4ab - 9ab$

In Exercises 13 – 20, simplify the following:

13. $8x + 2y + 5y$
14. $-6a + 4b + 10a + b$
15. $-a - 3a - 2$
16. $8y^2 - 6y - 5y^2 - 10y$
17. $6xy + 5yz + 3xy + 2yz$
18. $-5x^3 + 2x^2 + 7x^3 + x^2$
19. $3x^2 + 5x - x^2 - 2x$
20. $5ab^2 + 4a^2b - 2ab^2 - 2a^2b$

In Exercises 21 – 29, simplify the following:

21. $3a \cdot 4b$
22. $5x \cdot 2y$
23. $s \cdot 5t$
24. $-2r \cdot 3s$
25. $4p \cdot -2q$
26. $-6x \cdot 4y$
27. $-a \cdot -4b$
28. $-6p \cdot -7q$
29. $-8r \cdot -3s$

In Exercises 30 – 47, simplify the following:

30. $2(-y)^2$
31. $(3x^2y)^2$
32. $(-2xy^2)^3$
33. $3a^2b \cdot (ab)^3$
34. $(-2a)^3 \cdot 3b^2$
35. $-2a^2b^3 \cdot -3ab^2$
36. $5ab \cdot 6a^3b$
37. $4x^3y^2 \cdot 2x^2y$
38. $a^3b \cdot 2bc$
39. $5x^2y \cdot -3y^2$
40. $-6x^2y^3 \cdot -xy^2$
41. $-2a^3b \cdot 3b^3c^2$
42. $5a^2b^3 \cdot 4a^3b^4$
43. $5xy^2 \cdot 3x^2y$
44. $6a^2b \cdot 7ab^3$
45. $-3x^2y^5 \cdot 2x^3y$
46. $-2a^2b \cdot -5a^5b^2$
47. $7b^2c^3 \cdot -b^3c^4$

In Exercises 48 – 59, simplify the following:

48. $\frac{4ab}{a}$
49. $\frac{15ab}{5a}$
50. $\frac{27xyz}{9xz}$
51. $\frac{16xy^2}{8xy}$
52. $\frac{28x^3y^2}{7x^2y}$
53. $\frac{15x^3yz}{18x^2y^2}$

54. $\frac{18xy}{-12xy^2}$ 55. $\frac{-35y^4z^3}{7y^3z^4}$ 56. $\frac{-32a^5b^2z^3}{-8a^3b^3z^2}$

57. $\frac{(-3b)^2 \times 2a^2b}{27ab^2}$ 58. $\frac{(-3a)(-2ab)}{(-ab)^2}$ 59. $\frac{8x(-y)^2}{2(-x)^2y}$

In Exercises 60 – 69, evaluate each of the expressions if $a = -3$, $b = 2$, $c = 6$ and $d = 5$.

60. $7ab$ 61. $4cd$

62. $3a^2c$ 63. $5b^2d$

64. $a + 3c$ 65. $4c - 3d$

66. $4b + d$ 67. $-2a + d$

68. $3d - 2c$ 69. $3ab + 2cd$

70. Find the value of $(-x)^2 - 5x$ when $x = -3$.

71. Find the value of $2x^2 - 6x$ when $x = 2$.

72. Find the value of $a(2b - c)$ when $a = 2$, $b = 3$ and $c = -4$.

73. Evaluate $\frac{3a^2-4a}{a^2+3a}$ for $a = -2$

74. Evaluate $\frac{2a^3+c}{b+3a}$ when $a = -2, b = 3$ and $c = 4$

3.2 Multiplying a Monomial and a Binomial or Two Binomials

The product of a monomial and a binomial or two binomials can be obtained by using the distributive property.

Multiplying a Monomial and a Binomial

Example 1

Multiply:

(a) $3(2a + b)$ (b) $-2(3x - 5y)$

Solution

(a) $3(2a + b) = 3 \cdot 2a + 3 \cdot b$ Use distributive property

$= 6a + 3b$ Simplify

(b) $-2(3x - 5y) = (-2)(3x) - (-2)(5y)$

$= -6x + 10y$ Simplify

Note this, when a set of parentheses preceded by a minus sign is removed the sign of each term inside the parentheses must be changed.

Example 2

Multiply:

(a) $5xy(x^2 - y)$ (b) $-2x^2z(x - 3z^2)$

Solution

(a) $5xy(x^2 - y) = (5xy)(x^2) - (5xy)(y)$

$= 5x^3y - 5xy^2$ Simplify

(b) $-2x^2z(x - 3z^2) = (-2x^2z)(x) - (-2x^2z)(3z^2)$

$= -2x^3z - (-6x^2z^3)$

$= -2x^3z + 6x^2z^3$ Simplify

To multiply a monomial and a polynomial that have three or more terms, you can use the process illustrated in Examples 1 and 2.

Multiplying Binomials

To multiply $(x + 2)(x + 3)$, you can work as follows:

$(x + 2)(x + 3) = (x + 2) \cdot x + (x + 2) \cdot 3$ Use of distributive property

$= x^2 + 2x + 3x + 6$

$= x^2 + 3x + 2x + 6$ Use of commutative property

$= x^2 + 5x + 6$ Simplify

The third step can be obtained if each term of the second binomial is multiplied by the first term and then by the second term of the first binomial.

Example 3

Multiply:

(a) $(x - 3)(x + 2)$ (b) $(x + 5)(x - 3)$ (c) $(y - 3)(y - 4)$

Solution

(a) $(x-3)(x+2) = x \cdot x + x \cdot 2 + (-3) \cdot x + (-3) \cdot 2$

$$= x^2 + 2x - 3x - 6$$

$$= x^2 - x - 6$$

(b) $(x+5)(x-3) = x \cdot x + x \cdot (-3) + 5 \cdot x + 5 \cdot (-3)$

$$= x^2 - 3x + 5x - 15$$

$$= x^2 + 2x - 15$$

(c) $(y-3)(y-4) = y \cdot y + y \cdot (-4) + (-3) \cdot y + (-3) \cdot (-4)$

$$= y^2 - 4y - 3y + 12$$

$$= y^2 - 7y + 12$$

With practice you can write the product directly, and you should try to do so.

Squaring Binomials

Consider the square of a binomial, such as $(x + y)^2$. This can be expressed as $(x + y)(x + y)$.

Since this is the product of two binomials, we have

$$(x + y)^2 = (x + y)(x + y)$$

$$= x^2 + 2xy + y^2$$

Similarly, $(x - y)^2 = x^2 - 2xy + y^2$

The expression obtained when a binomial is squared is called a perfect square trinomial. You may have noticed the following:

1. The first term is the square of the first term of the binomial.

2. The middle term is twice the product of the two terms of the binomial. The middle term can be positive or negative

3. The last term is the square of the last term of the binomial

Example 4

Find: (a) $(x + 5)^2$ (b) $(2x - 3)^2$

Solution

(a) $(x + 5)^2 = (x)^2 + 2(x)(5) + (5)^2$

$$= x^2 + 10x + 25$$

(b) $(2x - 3)^2 = (2x)^2 - 2(2x)(3) + (-3)^2$

$= 4x^2 - 12x + 9$

Again we have shown all the steps. With practice you can write just the square.

Difference of Squares

Consider the product of the sum and difference of the same two terms, such as $(x + y)(x - y)$.

Now $(x + y)(x - y) = x \cdot x + x \cdot (-y) + y \cdot x + y \cdot (-y)$

$= x^2 - xy + yx - y^2$

$= x^2 - xy + xy - y^2$ Use of Commutative property

$= x^2 - y^2$

Notice that the product of two binomials that differ only in the sign between the terms is the difference of their squares. This is called the difference of two squares

Example 5

Find: (a) $(x - 3)(x + 3)$ (b) $(2x - 3)(2x + 3)$

Solution

(a) $(x - 3)(x + 3) = (x)^2 - (3)^2$

$= x^2 - 9$

(b) $(2x - 3)(2x + 3) = (2x)^2 - (3)^2 = 4x^2 - 9$

Exercise 3.2

In Exercise 1 – 40, simplify each expression.

1. $(x + 4)(x + 5)$
2. $(x + 7)(x + 6)$
3. $(x + 1)(x + 9)$
4. $(x + 12)(x + 5)$
5. $(x + 2)(x + 10)$
6. $(x + 3)(x + 15)$
7. $(2x + 3)(x + 4)$
8. $(3x + 2)(2x + 1)$
9. $(5x + 4)(2x + 7)$
10. $(6x + 5)(3x + 4)$
11. $(x - 3)(x + 5)$
12. $(x - 7)(x + 10)$
13. $(x - 4)(x + 8)$
14. $(2x - 5)(x + 3)$

15. $(x-8)(x+9)$
16. $(3x-10)(x+3)$
17. $(7x-15)(x+2)$
18. $(x-9)(x+5)$
19. $(x-10)(x+7)$
20. $(x-12)(x+6)$
21. $(x+9)(x-4)$
22. $(x+8)(x-3)$
23. $(x+7)(x-2)$
24. $(x+13)(x-5)$
25. $(x+5)(x-8)$
26. $(x+4)(x-11)$
27. $(x+6)(x-9)$
28. $(2x+3)(3x-2)$
29. $(4x+3)(2x-7)$
30. $(x+8)(5x-2)$
31. $(x-2)(x-5)$
32. $(x-8)(x-4)$
33. $(x-7)(x-3)$
34. $(x-3)(x-10)$
35. $(x-6)(x-3)$
36. $(x-12)(x-10)$
37. $(3x-4)(5x-3)$
38. $(x-6)(5x-1)$
39. $(3x-5)(2x-7)$
40. $(4x-3)(2x-5)$

In Exercise 41 – 52, simplify each expression.

41. $(x+3)^2$
42. $(x+7)^2$
43. $(x+6)^2$
44. $(x+12)^2$
45. $(2x+3)^2$
46. $(5x+4)^2$
47. $(x-5)^2$
48. $(x-2)^2$
49. $(x-10)^2$
50. $(x-3)^2$
51. $(6x-1)^2$
52. $(7x-2)^2$

In Exercise 53 – 62, simplify each expression.

53. $(x+4)(x-4)$
54. $(x+5)(x-5)$
55. $(x-3)(x+3)$
56. $(x-8)(x+8)$
57. $(x-7)(x+7)$
58. $(2x+3)(2x-3)$
59. $(4x+5)(4x-5)$
60. $(3x-9)(3x+9)$
61. $(7x-4)(7x+4)$
62. $(6x+13)(6x-13)$

3.3 Factorization

In the preceding section, you learned to multiply polynomials using the distributive property. Using the distributive property in reverse, you can write a polynomial as a product of two or more other polynomials. This process is called factoring.

Consider the expression

$$8a^2 + 6a$$

To factor this expression we first identify the largest common factor of $8a^2$ and $6a$. Then remove this factor from each of the terms and place it outside a set of parentheses. To find the largest common factor, look for the largest common numerical coefficient and the largest common power of each common variable. In this case, the largest common factor of $8a^2$ and $6a$ is $2a$. Factoring out $2a$ gives

$$8a^2 + 6a = 2a(4a + 3)$$

Example 1

Factor each expression:

(a) $10x^2 + 15x^3$ (b) $9x^2y - 12xy^2$ (c) $-15x^2 - 6x$

Solution

(a) $10x^2 + 15x^3 = 5x^2(2 + 3x)$

(b) $9x^2y - 12xy^2 = 3xy(3x - 4y)$

(c) $-15x^2 - 6x = -3x(5x + 2)$

Factoring by Grouping

A polynomial having four or more terms can be factored by splitting the terms into groups with a common factor that is not a monomial. This method of factoring is called factoring by grouping.

To factor polynomials with four terms use the following steps.

1. Group the first two terms and the last two terms.

2. Remove a common monomial factor from the first group and the last group.

3. Finally, factor out the common binomial factor.

Example 2

Factor each polynomial by grouping.

(a) $2x^2 + 6x + 3xy + 9y$

(b) $xy - yz + y^2 - xz$

Solution

(a) $2x^2 + 6x + 3xy + 9y$

Group terms and factor each group. Then factor the binomial factor.

$$2x^2 + 6x + 3xy + 9y = (2x^2 + 6x) + (3xy + 9y) \quad \text{Group terms}$$

$$= 2x(x + 3) + 3y(x + 3) \quad \text{Factor each group}$$

$$= (x + 3)(2x + 3y) \quad \text{Factor the binomial}$$

(b) $xy - yz + y^2 - xz$

Observe that grouping this polynomial would not produce a common binomial factor. So, rearrange the terms before you factor.

$$xy - yz + y^2 - xz = xy + y^2 - xz - yz \quad \text{Rearrange terms}$$

$$= (xy + y^2) - (xz + yz) \quad \text{Group terms}$$

$$= y(x + y) - z(x + y) \quad \text{Factor each group}$$

$$= (x + y)(y - z) \quad \text{Factor the binomial}$$

Factoring Trinomials

You may have noticed in the preceding section that when two binomials are multiplied the product is often a trinomial. We will now learn how to factor trinomials of the form $ax^2 + bx + c$, where c is a constant.

Factoring $ax^2 + bx + c$, where $a = 1$

To factor $x^2 + bx + c$ into a product of two binomials we look for two factors of c whose sum is b. If you have trouble finding the two factors, it may be helpful to list all of the distinct pair of factors, and then choose the appropriate pair from the list.

Example 3

Factor each trinomial:

(a) $x^2 + 6x + 8$ (b) $x^2 - 9x + 14$ (c) $x^2 - 2x - 15$

Solution

(a) $x^2 + 6x + 8$

The numbers 2 and 4 have a product of 8 and a sum of 6.

So, $x^2 + 6x + 8 = (x + 2)(x + 4)$

We could also write the answer as $(x + 4)(x + 2)$.

(b) $x^2 - 9x + 14$

The number – 7 and – 2 have a product of 14 and a sum – 9.

So, $x^2 - 9x + 14 = (x - 2)(x - 7)$

(c) $x^2 - 2x - 15$

The numbers – 5 and 3 have a product – 15 and a sum – 2.

So, $x^2 - 2x - 15 = (x - 5)(x + 3)$

You may have observed that:

1. If b and c are both positive the two factors of c are both positive.

2. If b is negative and c is positive the two factors of c are both negative.

3. If b is positive and c is negative the two factors of c are of opposite signs (the value with the larger absolute value is positive).

4. If b is negative and c is negative the two factors of c are of opposite signs (the value with the larger absolute value is negative).

Factoring $ax^2 + bx + c$, where $a > 1$

A trinomial of the form $ax^2 + bx + c$, where $a > 1$, may be factored by using the method of grouping.

To factor $ax^2 + bx + c$ we multiply the coefficient of x^2 and the constant term and then find two factors of the product whose sum is equal to the coefficient of x. Next, we write the middle term as the sum of the factors and then factor the resulting expression by grouping.

Example 4

Factor each trinomial:

(a) $2x^2 + 7x + 6$ (b) $6x^2 - 17x + 7$ (c) $3x^2 + 13x - 10$

Solution

(a) $2x^2 + 7x + 6$

Here $a = 2$ and $c = 6$ so $ac = 12$. The numbers whose product is 12 and whose sum is 7 are 3 and 4. Writing the middle term as $4x + 3x$, we have

$$2x^2 + 7x + 6 = 2x^2 + 4x + 3x + 6$$
$$= 2x(x + 2) + 3(x + 2)$$
$$= (x + 2)(2x + 3)$$

(b) $6x^2 - 17x + 7$

Here the product $ac = 42$. The numbers -14 and -3 have a product 42 and a sum -17.

$$6x^2 - 17x + 7 = 6x^2 - 3x - 14x + 7$$
$$= 3x(2x - 1) - 7(2x - 1)$$
$$= (2x - 1)(3x - 7)$$

(c) $3x^2 + 13x - 10$

Here the product $ac = -30$. The numbers 15 and -2 have a product of -30 and a sum 13. Thus,

$$3x^2 + 13x - 10 = 3x^2 + 15x - 2x - 10$$
$$= 3x(x + 5) - 2(x + 5)$$
$$= (x + 5)(3x - 2)$$

Trial and Error

So far we have use the method of grouping to factor a trinomial of the form $ax^2 + bx + c$, where $a \neq 1$. An alternative method is to try combinations of possible factors until one works. This technique is called the trial and error method. We will not consider this method.

Difference of Two Squares

Recall that the product $(a - b)(a + b) = a^2 - b^2$. To factor the expression on the right hand side we use the expression backwards.

Example 5

Factor: (a) $x^2 - 64$ (b) $9x^2 - 25y^2$ (c) $27x^5y - 12xy^7$

Solution

(a) $x^2 - 64 = x^2 - 8^2$

$$= (x - 8)(x + 8)$$

(b) $9x^2 - 25y^2 = (3x)^2 - (5y)^2$

$$= (3x - 5y)(3x + 5y)$$

(c) $27x^5y - 12xy^7$

Begin by factoring $3xy$ from the expression.

$$27x^5y - 12xy^7 = 3xy(9x^4 - 4y^6)$$

$$= 3xy(3x^2 - 2y^3)(3x^2 + 2y^3)$$

Perfect Square Trinomial

Recall that $x^2 + 2xy + y^2$ and $x^2 - 2xy + y^2$, called the perfect square trinomials, are the products of the square of $(x + y)$ and $(x - y)$ respectively.

Notice that a perfect square trinomial has the following properties:

1. the first term and the last term are the square of the first term and the second term of the binomial respectively.

2. the middle term is twice the product of the two terms of the binomial. (The middle term can be positive or negative)

To factor a perfect square trinomial, find the square roots of the first term and the last term and add if the middle term is positive or subtract if the middle term is negative.

Example 6

Factor:

(a) $x^2 + 6x + 9$ (b) $4x^2 - 20xy + 25y^2$ (c) $2x^3y + 20x^2y^2 + 50xy^3$

Solution

(a) $x^2 + 6x + 9 = (x + 3)^2$

(b) $4x^2 - 20xy + 25y^2 = (2x - 5y)^2$

(c) $2x^3y + 20x^2y^2 + 50xy^3$

Begin by factoring out the largest common factor. This is $2xy$.

$$2x^3y + 20x^2y^2 + 50xy^3 = 2xy(x^2 + 10xy + 25y^2)$$

$$= 2xy(x + 5y)^2$$

Exercise 3.3

In Exercises 1 – 12, factor each of the following polynomials:

1. $3a + 12$ 2. $4y - 8$ 3. $5p + 10q$

4. $7x - 21y$ 5. $8xy - 12yz$ 6. $6p^2 - 15pr$

7. $6p^2q - 18pq^2$ 8. $7xy^2 + 4x^2y$ 9. $-4x^2y - 6xy^2$

10. $-12a^2b + 8ab^2$ 11. $-5xy^2 - 10x^2y$ 12. $-7a^3b^2 + 21a^2b^3$

In Exercises 13 – 24, factor each polynomial by grouping:

13. $ab + ac + bd + cd$ 14. $pr + ps - qr - qs$

15. $5ax + 5ay + 2bx + 2by$ 16. $3ac + 6bc + 4ad + 8bd$

17. $2xy + 4xz - 3y^2 - 6yz$ 18. $6xy - 9xz - 8wy + 12wz$

19. $3ax - ay + 3bx - by$ 20. $10ay - 6b - 2by + 30a$

21. $4pr + 10qs - 5qr - 8ps$ 22. $4au + 6bv + 4bu + 6av$

23. $7x^2 + 6yz - 14xy - 3xz$ 24. $10ax + 8bx + 15ay + 6by$

In Exercises 25 – 40, factor each of the following:

25. $x^2 + 9x + 20$ 26. $x^2 + 10x + 21$

27. $x^2 + 14x + 48$ 28. $x^2 + 12x + 20$

29. $x^2 + 14x + 45$ 30. $x^2 + 15x + 36$

31. $x^2 + 12x + 32$ 32. $x^2 - 9x + 14$

33. $x^2 - 8x + 15$ 34. $x^2 - 13x + 36$

35. $x^2 - 11x + 30$ 36. $x^2 - 17x + 70$

37. $x^2 - 13x + 40$ 38. $x^2 - 12x + 27$

39. $x^2 - 15x + 54$ 40. $x^2 - 15x + 50$

In Exercises 41 – 64, factor each of the following:

41. $x^2 + x - 20$ 42. $x^2 + 4x - 21$

43. $x^2 + 2x - 48$ 44. $x^2 + 5x - 50$

45. $x^2 + 3x - 70$ 46. $x^2 + 3x - 40$

47. $x^2 + 6x - 27$

48. $x^2 + 2xy - 15y^2$

49. $x^2 + xy - 6y^2$

50. $x^2 + 4xy - 32y^2$

51. $x^2 + 7xy - 30y^2$

52. $x^2 + 5xy - 36y^2$

53. $x^2 - 5x - 14$

54. $x^2 - 7x - 30$

55. $x^2 - 4x - 96$

56. $x^2 - x - 56$

57. $x^2 - 2x - 24$

58. $x^2 - 3x - 54$

59. $x^2 - 6x - 27$

60. $x^2 - 5xy - 24y^2$

61. $x^2 - 2xy - 35y^2$

62. $x^2 - xy - 20y^2$

63. $x^2 - 6xy - 16y^2$

64. $x^2 - xy - 42y^2$

In Exercises 65 – 79, factor each of the following:

65. $2x^2 + 9x + 4$

66. $3x^2 + 11x + 6$

67. $5x^2 + 26x - 24$

68. $6x^2 - 5x - 6$

69. $10x^2 - 9x - 9$

70. $21x^2 - 2x - 8$

71. $24x^2 - 10x - 21$

72. $6x^2 + 17x - 45$

73. $2x^2 + 11x + 15$

74. $15x^2 - 19x - 10$

75. $12x^2 - 29x + 15$

76. $8x^2 + 18x - 35$

77. $6x^2 - x - 15$

78. $5x^2 + 2x - 3$

79. $3x^2 + 10x + 8$

In Exercises 80 – 101, factor each of the following:

80. $x^2 + 6x + 9$

81. $x^2 + 10x + 25$

82. $x^2 + 14x + 49$

83. $x^2 + 12x + 36$

84. $x^2 + 18x + 81$

85. $x^2 + 24x + 144$

86. $x^2 + 30x + 225$

87. $4x^2 + 16xy + 16y^2$

88. $9x^2 + 30xy + 25y^2$

89. $16x^2 + 24xy + 9y^2$

90. $25x^2 + 40xy + 16y^2$

91. $4x^2 + 20xy + 25y^2$

92. $x^2 - 4x + 4$

93. $x^2 - 12x + 36$

94. $x^2 - 16x + 64$

95. $x^2 - 20x + 100$

96. $x^2 - 22x + 121$

97. $x^2 - 26x + 169$

98. $x^2 - 12xy + 36y^2$

99. $x^2 - 10xy + 25y^2$

100. $9x^2 - 12xy + 4y^2$ 101. $4x^2 - 20xy + 25y^2$

In Exercises 102 – 116, factor each of the following:

102. $x^2 - 16$ 103. $x^2 - 64$ 104. $x^2 - 144$

105. $9y^2 - 64$ 106. $4x^2 - 25$ 107. $169 - x^2$

108. $25x^2 - 16y^2$ 109. $27x^2 - 12y^2$ 110. $8x^2 - 72y^2$

111. $5x^3 - 45$ 112. $3x^6 - 48$ 113. $1 - 49x^2$

114. $20x^3 - 45xy^2$ 115. $32x^2y - 18y^3$ 116. $50x^3 - 8xy^2$

3.4 Rational Expressions

An algebraic expression of the form P/Q, where P and Q are polynomials is called a rational expression. P is called the numerator and Q is called the denominator. Because division by zero is undefined, Q cannot by zero. When a variable is replaced with a number that makes the denominator 0, the rational expression is undefined. Examples of rational expressions include

$$\frac{3}{x} \qquad \frac{4}{3-x} \qquad \frac{2x+5}{x+4} \qquad \frac{2x^2+3x+5}{x^2-1}$$

Simplifying Rational Expression

A rational expression is said to be in its simplest form if its numerator and denominator have no common factor other than 1. To write a fraction in its simplest form:

1. Factor both the numerator and denominator completely if possible.

2. Divide out the common factors.

Example 1

Simplify each rational expression:

(a) $\frac{9x^2y}{12xy^2}$

Solution

$$\frac{9x^2y}{12xy^2} = \frac{3xy \cdot 3x}{3xy \cdot 4y}$$

$$= \frac{3x}{4y}$$

$(b)\ \frac{6xy}{3x^2-15x}$

Solution

Factor both the numerator and denominator and then divide out the common factor.

$$\frac{6xy}{3x^2-15x}=\frac{3x\cdot 2y}{3x(x-5)}$$

$$=\frac{2y}{x-5}$$

$(c)\ \frac{x^2-2x-15}{x^2-25}$

Solution

Factor both the numerator and denominator and then divide out the common factor.

$$\frac{x^2-2x-15}{x^2-25}=\frac{(x-5)(x+3)}{(x-5)(x+5)}$$

$$=\frac{x+3}{x+5}$$

Multiplying Rational Expressions

To multiply rational expressions, multiply the numerators, and multiply the denominators. Then reduce the result to the lowest term if possible. It will be best to divide by any common factors before you multiply.

Example 2

Evaluate each of the following:

$(a)\ \frac{6x^2z}{5y^2}\cdot\frac{10yz}{9x^3}$

Solution

$$\frac{6x^2z}{5y^2}\cdot\frac{10yz}{9x^3}=\frac{(6x^2z)\cdot(10yz)}{(5y^2)\cdot(9x^3)}$$

$$=\frac{60x^2yz^2}{45x^3y^2}$$

$$=\frac{4z^2}{3xy}$$

The same result is obtained if we divide out common factors before we multiply.

$$\frac{6x^2z}{5y^2} \cdot \frac{10yz}{9x^3} = \frac{2z}{y} \cdot \frac{2z}{3x}$$

$$= \frac{4z^2}{3xy}$$

(b) $\frac{x^2+5x+6}{x^2-x-6} \cdot \frac{x^2+3x-10}{x^2+8x+15}$

Solution

$$\frac{x^2+5x+6}{x^2-x-6} \cdot \frac{x^2+3x-10}{x^2+8x+15} = \frac{(x^2+5x+6)\cdot(x^2+3x-10)}{(x^2-x-6)\cdot(x^2+8x+15)}$$

$$= \frac{(x+2)(x+3)\cdot(x-2)(x+5)}{(x+2)(x-3)\cdot(x+3)(x+5)}$$

$$= \frac{x-2}{x-3}$$

Dividing Rational Expressions

To divide rational expressions multiply by the reciprocal of the divisor. As with fractions, the reciprocal of a rational expression is found by inverting the expression.

Example 3

Evaluate each of the following:

(a) $\frac{8x^2}{10y^3} \div \frac{12x^2}{25y^2}$

Solution

$$\frac{8x^2}{10y^3} \div \frac{12x^2}{25y^2} = \frac{8x^2}{10y^3} \cdot \frac{25y^2}{12x^2}$$

$$= \frac{(8x^2)(25y^2)}{(10y^3)(12x^2)}$$

$$= \frac{200x^2y^2}{120y^3x^2}$$

$$= \frac{5}{3y}$$

(b) $\frac{x^2-x-12}{x^2-5x} \div \frac{x^2-16}{x^2-x-20}$

Solution

$$\frac{x^2-x-12}{x^2-5x} \div \frac{x^2-16}{x^2-x-20} = \frac{x^2-x-12}{x^2-5x} \cdot \frac{x^2-x-20}{x^2-16}$$

$$= \frac{(x-4)(x+3)\cdot(x-5)(x+4)}{x(x-5)\cdot(x-4)(x+4)}$$

$$= \frac{x+3}{x}$$

Adding or Subtracting Rational Expressions

Adding or subtracting rational expressions follow the rules for adding and subtracting fractions.

Adding or Subtracting Rational Expressions with like denominators

To add or subtract rational expressions with like denominators, add or subtract their numerators and then place the result over the common denominator. Simplify the result if possible.

Example 4

Evaluate each of the following:

(a) $\frac{3x}{x+2} + \frac{6}{x+2}$

Solution

$$\frac{3x}{x+2} + \frac{6}{x+2} = \frac{3x+6}{x+2}$$

$$= \frac{3(x+2)}{x+2}$$

$$= 3$$

(b) $\frac{3x+2}{x^2-x-6} - \frac{2x+5}{x^2-x-6}$

Solution

$$\frac{3x+2}{x^2-x-6} - \frac{2x+5}{x^2-x-6} = \frac{3x+2-(2x+5)}{x^2-x-6}$$

$$= \frac{3x+2-2x-5}{x^2-x-6}$$

$$= \frac{x-3}{x^2-x-6}$$

$$= \frac{x-3}{(x-3)(x+2)}$$

$$= \frac{1}{x+2}$$

Adding or Subtracting Rational Expressions with unlike denominators

To add or subtract rational expressions with unlike denominators, you must first write each rational expression as an equivalent rational expression with a common denominator, and then add or subtract as usual. We use the least common denominator (LCD), which is least common multiple of the denominators.

Finding the Least Common Denominator of Rational Expressions

The least common multiple (LCM) of two or more polynomials is the smallest product that contains all the different factors in the polynomials and each of these factors must be the highest power as it occurs in any one of the individual polynomials.

To find the least common denominator, we first factor each denominator and then multiply all the different factors, as shown in Examples 5 and 6.

Example 5

Find the LCM of the following:

(a) $3x$, $2y$ and z

Solution

The three expressions $3x$, $2y$ and z have no common factor, so their least common multiple is simply their product.

The LCM is $3x \cdot 2y \cdot z = 6xyz$.

(b) $18x^2y^2$ and $24xy^3$

Solution

Begin by writing the prime factorization of $18x^2y^2$ and $24xy^3$.

$$18x^2y^3 = 2 \cdot 3 \cdot 3 \cdot x \cdot x \cdot y \cdot y \cdot y$$

$$24xy^3 = 2 \cdot 2 \cdot 2 \cdot 3 \cdot x \cdot y \cdot y \cdot y$$

Thus, the LCM is $2^3 \cdot 3^2 \cdot x^2 \cdot y^3 = 72x^2y^3$.

Example 6

Find the LCM of $3x + 6$, $4x + 12$ and $x^2 + 5x + 6$.

Solution

First factor each expression completely, writing any numerical factor as prime factors

$$3x + 6 = 3(x + 2)$$

$$4x + 12 = 4(x + 3) = 2 \cdot 2 \cdot (x + 3)$$

$$x^2 + 5x + 6 = (x + 2)(x + 3)$$

The LCM is $2^2 \cdot 3 \cdot (x + 2) \cdot (x + 3) = 12(x + 2)(x + 3)$.

Example 7

Evaluate $\frac{3}{x^2+x-6} + \frac{2}{x^2+5x+6}$

Solution

The LCM of the denominators is $(x - 2)(x + 2)(x + 3)$. Rewrite each expression using the LCM as the least common denominator (LCD).

$$\frac{3}{x^2+x-6} + \frac{2}{x^2+5x+6} = \frac{3}{(x-2)(x+3)} + \frac{2}{(x+2)(x+3)}$$

$$= \frac{3(x+2)}{(x-2)(x+2)(x+3)} + \frac{2(x-2)}{(x-2)(x+2)(x+3)}$$

$$= \frac{3(x+2)+2(x-2)}{(x-2)(x+2)(x+3)}$$

$$= \frac{3x+6+2x-4}{(x-2)(x+2)(x+3)}$$

$$= \frac{5x+2}{(x-2)(x+2)(x+3)}$$

There is no need to multiply the expression in the denominator.

Complex Fractions

A fraction that has a fraction in its numerator or denominator or in both is called a complex fraction. Examples include

$$\frac{\frac{3}{4}}{\frac{5}{6}} \qquad \frac{\frac{3}{x^2}}{\frac{4}{x}} \qquad \frac{\frac{a+2}{3}}{\frac{a-3}{2}}$$

Simplifying Complex Algebraic Fractions

We will consider two methods that can be used to simplify complex fractions.

Method 1

Multiply the numerator and the denominator by the LCD of all fractions in the numerator and denominator.

Example 8

Simplify:

(a) $\frac{\frac{x}{y}+2}{\frac{x}{y}-3}$

Solution

$$\frac{\frac{x}{y}+2}{\frac{x}{y}-3} = \frac{\left(\frac{x}{y}+2\right)\cdot y}{\left(\frac{x}{y}-3\right)\cdot y}$$

$$= \frac{\frac{x}{y}\cdot y + 2\cdot y}{\frac{x}{y}\cdot y - 3\cdot y}$$

$$= \frac{x+2y}{x-3y}$$

(b) $\frac{1-\frac{2}{x-1}}{\frac{3}{x-1}-1}$

Solution

$$\frac{1-\frac{2}{x-1}}{\frac{3}{x-1}-1} = \frac{\left(1-\frac{2}{x-1}\right)(x-1)}{\left(\frac{3}{x-1}-1\right)(x-1)}$$

$$= \frac{(x-1)-\left(\frac{2}{x-1}\right)(x-1)}{\left(\frac{3}{x-1}\right)(x-1)-(x-1)}$$

$$= \frac{(x-1)-2}{3-(x-1)}$$

$$= \frac{x-3}{4-x}$$

Method 2

Write the numerator and denominator of the complex fraction as single fractions and then multiply by inverting the denominator as illustrated in Example 9.

Example 9

Simplify:

(a) $\frac{\frac{y}{y+2}}{\frac{5}{3y+6}}$

Solution

The numerator and denominator are already single fractions. Inverting the denominator and multiplying we have

$$\frac{\frac{y}{y+2}}{\frac{5}{3y+6}} = \frac{y}{y+2} \times \frac{3y+6}{5}$$

$$= \frac{y}{y+2} \times \frac{3(y+2)}{5}$$

$$= \frac{3y}{5}$$

(b) $\frac{1-\frac{a^2}{b^2}}{1+\frac{a}{b}}$

Solution

$$\frac{1-\frac{a^2}{b^2}}{1+\frac{a}{b}} = \frac{\frac{b^2-a^2}{b^2}}{\frac{b+a}{b}}$$

$$= \frac{b^2--a^2}{b^2} \times \frac{b}{b+a}$$

$$= \frac{(b-a)(b+a)}{b(b+a)}$$

$$= \frac{b-a}{b}$$

Exercise 3.4

In Exercises 1 – 15, write each rational expression in lowest terms.

1. $\frac{20a^2b}{30ab^2}$
2. $\frac{10x^2y^3}{12x^3y^2}$
3. $\frac{12xy}{9x^2y}$

4. $\frac{3a+12b}{a^2+4ab}$ 5. $\frac{4x^2-1}{2x^2+x}$ 6. $\frac{2xy+8x}{2xy^2-32x}$

7. $\frac{x^3-16x}{x^2-2x-8}$ 8. $\frac{x^2+2x-15}{3x^2-9x}$ 9. $\frac{x^2-x-6}{x^2-9}$

10. $\frac{x^2-3x-28}{x^2-x-42}$ 11. $\frac{2x^2-9x+4}{12+x-x^2}$ 12. $\frac{x^2+4x-21}{x^2+15x+56}$

13. $\frac{x^2-4y^2}{x^2+4xy+4y^2}$ 14. $\frac{x^2-4x+4}{4-x^2}$ 15. $\frac{x+3y}{9y^2-x^2}$

In Exercises 16 – 27, evaluate each of the following expressions.

16. $\frac{6x}{8y} \cdot \frac{20y^2}{18x^2}$ 17. $\frac{5x^3}{3x} \cdot \frac{9}{20x}$

18. $\frac{8xy^5}{5x^3y^2} \cdot \frac{15y^2}{16xy^3}$ 19. $\frac{2x^2y}{3x^3y^2} \cdot \frac{6xy^2}{8x}$

20. $\frac{7xy}{3xz} \cdot \frac{15x^2z^2}{21x^3y^2}$ 21. $\frac{x^2-y^2}{4x} \cdot \frac{4}{x+y}$

22. $\frac{3x^2y^2}{12x^3y} \cdot \frac{8x^2}{xy+2y^2}$ 23. $\frac{4x+12y}{8xy} \cdot \frac{3x^2y}{x^2+3xy}$

24. $\frac{a^2+b^2}{x+2y} \cdot \frac{x^2-4y^2}{a^3+ab^2}$ 25. $\frac{a^2-b^2}{a^2+ab} \cdot \frac{a+b}{(a-b)^2}$

26. $\frac{x^2+3x}{x^2-9} \cdot \frac{x^2-3x}{x^2}$ 27. $\frac{x^2-9}{x^2+4x+3} \cdot \frac{2x+6}{2x-6}$

In Exercises 28 – 39, evaluate each of the following expressions.

28. $\frac{4xy}{5z} \div \frac{8y^2}{15z^2}$ 29. $\frac{4x^2y^2}{9x^3} \div \frac{8y^2}{27xy}$

30. $\frac{8x^3y}{27xy^3} \div \frac{16x^3y}{45y}$ 31. $\frac{6x^2y^2}{14ab^2} \div \frac{12xy^2}{7b^2}$

32. $\frac{3x^3}{2y^4} \div \frac{6x^2}{4y^3}$ 33. $\frac{12xy}{x^2-y^2} \div \frac{3y}{x-y}$

34. $\frac{x}{x-2} \div \frac{x^2}{x^2-4}$ 35. $\frac{3}{x^2-16} \div \frac{6}{2x+8}$

36. $\frac{4x-12}{5x+15} \div \frac{8x^2}{x^2+3x}$ 37. $\frac{a^2+2ab}{a^2} \div \frac{ab+2b^2}{b^2}$

38. $\frac{x^2-xy}{9x^2y} \div \frac{y-x}{6x}$ 39. $\frac{a+b}{a-b} \div \frac{a^2+2ab+b^2}{a-b}$

In Exercises 40 – 49, find the LCM of the expressions.

40. $3x,\ 4x^2,\ 12x$ 41. $2x^2,\ 3xy,\ 6y^2$

42. $a-b, a+b$

43. $3x+3, x^2-1$

44. $2x+y, 2x-y, 4x^2-y^2$

45. $x-2, x^2+x-6$

46. $x+3, x^2+3x, x^2+5x+6$

47. $x^2-5x+6, x^2-7x+12, x^2-6x+8$

48. $x+2, x+1, x^2+3x+2$

49. $x-y, x+y, x^3-xy^2$

ln Exercises 50 – 69, evaluate each of the following expressions.

50. $\frac{x}{2}+\frac{x}{2}$

51. $\frac{3y}{4}-\frac{y}{2}$

52. $\frac{5}{x}+\frac{3}{2x}$

53. $\frac{5}{2y}-\frac{7}{3y}$

54. $\frac{x+y}{4}+\frac{x-y}{3}$

55. $\frac{2x-3}{6}-\frac{3x-2}{4}$

56. $\frac{4x-3y}{4}-\frac{3x-5y}{3}$

57. $\frac{2x+3}{2}+\frac{4-2x}{4}$

58. $\frac{7}{2x-3}+\frac{3}{2x+3}$

59. $\frac{12}{x^2-9}-\frac{2}{x-3}$

60. $\frac{5}{x-y}-\frac{2x}{x^2-xy}$

61. $\frac{3}{x-y}-\frac{2}{x+y}$

62. $\frac{x}{x^2-x-30}-\frac{1}{x+5}$

63. $\frac{3}{x^2+4x+3}-\frac{1}{x^2-9}$

64. $\frac{2}{y^2+y-6}+\frac{3y}{y^2-2y-15}$

65. $\frac{5}{x^2-3x-10}-\frac{4}{x^2+2x-35}$

66. $\frac{x+2}{x^2-x-12}-\frac{x}{x^2+6x+9}$

67. $\frac{x-1}{x^2+4x+4}+\frac{1-x}{x^2-x-6}$

68. $\frac{5}{x^2-4}-\frac{7}{x+2}$

69. $\frac{1}{x-y}-\frac{3}{x+y}+\frac{3x-y}{x^2-y^2}$

ln Exercises 70 – 81, evaluate each of the following expressions.

70. $\frac{1+\frac{1}{y}}{\frac{1}{2}-\frac{1}{2y}}$

71. $\frac{\frac{1}{4}+\frac{3}{4x}}{\frac{1}{2}-\frac{3}{2x}}$

72. $\frac{\frac{1}{x}+1}{1-\frac{1}{x^2}}$

73. $\frac{\frac{x^2}{y^2}-1}{\frac{x}{y}+1}$

74. $\frac{\frac{4}{x^2}-\frac{1}{y^2}}{\frac{2}{x}+\frac{1}{y}}$

75. $\frac{\frac{1}{x^2y}+\frac{2}{x}}{\frac{1}{xy^2}+\frac{2}{y}}$

76. $\frac{3+\frac{2}{x-1}}{\frac{1}{x-1}-2}$

77. $\frac{\frac{5}{x-2}+1}{1-\frac{1}{x-2}}$

78. $\frac{\frac{1}{y^2}-\frac{9}{x^2}}{\frac{1}{y^2}+\frac{5}{xy}+\frac{6}{x^2}}$

79. $\dfrac{\frac{2}{xy}+\frac{1}{y^2}}{\frac{4}{x}-\frac{x}{y^2}}$ 80. $\dfrac{1+\frac{1}{x}-\frac{12}{x^2}}{1+\frac{2}{x}-\frac{8}{x^2}}$ 81. $\dfrac{1-\frac{1}{x}-\frac{6}{x^2}}{1-\frac{3}{x}-\frac{10}{x^2}}$

Review Exercises

In Exercises 1 – 9, simplify each of the following expressions.

1. $5x^3 \cdot 3xy^2$ 2. $-2x^2 \cdot 7x^3$

3. $4x^5 \cdot -3x$ 4. $-3x^4y^2 \cdot -7x^2y^3$

5. $-3x^5y^2 \cdot 2x^4y$ 6. $-4x^3y^2 \cdot -2x^2y^3$

7. $(2x^2y^3)^2$ 8. $(-5ab^2)^3$

9. $(-3x^2y^3)^4$

In Exercises 10 – 15, write each rational expression in lowest terms.

10. $\dfrac{15x^6}{5x^2}$ 11. $\dfrac{-16x^7}{4x^5}$

12. $\dfrac{-45x^5}{-15x^3}$ 13. $\dfrac{40m^3n^2}{-8m^2n}$

14. $\dfrac{-72x^4y^6}{9x^3y^5}$ 15. $\dfrac{35x^5y^2}{7x^2y}$

In Exercises 16 – 21, evaluate each of the expressions if $w = -2$, $x = 5$, $y = -4$ and $z = 6$.

16. $8x + 2y$ 17. $3w^2 + 4y$

18. $4y - 2x$ 19. $4(2w - z)$

20. $y(3w - x)$ 21. $w(x + 3y)$

In Exercises 22 – 27, simplify each of the following expressions.

22. $3x + 4x$ 23. $5a - 7a$

24. $9y - 5y + 4y$ 25. $8y + 7y - y$

26. $3x - 2y + 5x + 3y$ 27. $-2a + 4b - 7a - b$

In Exercises 28 – 33, multiply each of the following expressions.

28. $x(5x + 2)$ 29. $3x(5 - 2x)$

30. $-6x(3y - 2)$ 31. $2x(3y + x)$

32. $-2p(p-3q+4)$ 33. $3x(x-2y+1)$

In Exercises 34 – 41, simplify each of the following expressions.

34. $2x-(5x-3)$
35. $3(3y-1)-2(y-5)$
36. $-3(y^2-2)+y^2(y+3)$
37. $x(x^2+3)-3(x+4)$
38. $5a-(4a-3)$
39. $7y-(2y+5)$
40. $8y^2+y-2(y+3y^2)$
41. $4x-9y-3(2x-y)$

In Exercises 42 – 53, multiply each of the following expressions.

42. $(x+6)(x+3)$
43. $(x+6)(x-2)$
44. $(a-6)(a-7)$
45. $(2x-3)(3x+2)$
46. $(y-4)(y-8)$
47. $(3x+2)(x-3)$
48. $(x-3)(x+3)$
49. $(x+6)(x-6)$
50. $(2x-3)(2x+3)$
51. $(x+3)^2$
52. $(2x-3)^2$
53. $(3x+2)^2$

In Exercises 54 – 59, factor each of the following expressions.

54. $2x^2+6x$
55. $10t^2-5t$
56. $27pq^2+18p^2q$
57. $12x^2+8x$
58. $8x^2y-24xy^2$
59. $3x^2y+6xy^2$

In Exercises 60 – 65, factor the following expressions by grouping.

60. $x^2+2x+xy+2y$
61. $2x^2-4x+xz-2z$
62. $x^2+y+x+xy$
63. $a^2-3y-3a+ay$
64. $x^3+3x^2-4x-12$
65. $2x^3+12x^2-5x-30$

In Exercises 66 – 71, factor each of the following expressions.

66. x^2+6x+5
67. $y^2+11y+28$
68. $y^2+4y-45$
69. $x^2+7x-60$
70. $x^2-2x-15$
71. x^2+x-42

In Exercises 72 – 77, factor each of the following expressions.

72. $2x^2 + 7x - 4$

73. $6x^2 - 23x + 7$

74. $7x^2 + 15x + 2$

75. $9a^2 - 6a - 8$

76. $4a^2 - 4a - 15$

77. $3x^2 - 5x - 2$

In Exercises 78 – 83, factor each of the following expressions.

78. $x^2 - 18x + 81$

79. $x^2 + 14x + 49$

80. $16x^2 - 24x + 9$

81. $4x^2 + 12x + 9$

82. $64x^2 + 16x + 1$

83. $9x^2 - 30x + 25$

In Exercises 84 – 89, factor each of the following expressions.

84. $x^2 - 36$

85. $x^2 - 9$

86. $9y^2 - 4$

87. $16a^2 - 9$

88. $16a^2 - 9b^2$

89. $75xy^6 - 3x^5$

In Exercises 90 – 101, write each rational expression in lowest terms.

90. $\frac{4x^5}{6x^2}$

91. $\frac{10x^2y}{15x^3}$

92. $\frac{12a^2b^5}{16ab^2}$

93. $\frac{18x^5y^4}{24x^3y^6}$

94. $\frac{-15x^3y^4}{-20x^2y^5}$

95. $\frac{-4x^3y}{6xy^2}$

96. $\frac{14x^3y^2}{-21x^2y^5}$

97. $\frac{-8a^2b^4}{-16a^4b^2}$

98. $\frac{x^2+3x+2}{3x+6}$

99. $\frac{5y^2+15y}{y^2-2y-15}$

100. $\frac{x^2-16}{x^2-x-20}$

101. $\frac{x^2-4x+4}{x-2}$

In Exercises 102 – 110, evaluate each of the following expressions.

102. $\frac{3x^4y^2}{10x^2y^4} \cdot \frac{5x^2y^3}{9x^2y^3}$

103. $\frac{8x^2y^6}{5x^4y^3} \cdot \frac{15y^3}{16x^2y^4}$

104. $\frac{-4a^2b^3}{15a^4} \cdot \frac{25a^2b^2}{-16b^4}$

105. $\frac{7x^2y^3}{-12x^3y^2} \cdot \frac{-24x^4y^6}{21x^3y^8}$

106. $\frac{x^2-5x}{3x^2} \cdot \frac{10x}{5x-25}$

107. $\frac{x^2+3x-10}{5x} \cdot \frac{15x^2}{3x+15}$

108. $\frac{x^2+8x}{4x} \cdot \frac{12x}{x^2-64}$

109. $\frac{x^2-81}{x^2+9x} \cdot \frac{5x^2}{x^2-7x-18}$

110. $\frac{2x-6}{x^2+2x} \cdot \frac{x^2+4x+4}{3-x}$

In Exercises 111 – 119, evaluate each of the following expression.

111. $\frac{4x^3y^3}{9x^4} \div \frac{8y^3}{27x^2y^2}$

112. $\frac{8x^3y^2}{27x^2y^4} \div \frac{16x^3y^2}{45y^2}$

113. $\frac{3x-6}{8} \div \frac{5x-10}{6}$

114. $\frac{x^2+2x}{4x} \div \frac{6x+12}{8}$

115. $\frac{4a+12}{5a-15} \div \frac{8a^2}{a^2-3a}$

116. $\frac{6p+18}{9p} \div \frac{3p+9}{p^2-2p}$

117. $\frac{16x}{4x^2-16} \div \frac{4x+12}{x^2+x-6}$

118. $\frac{x^2-9}{2x^2-6x} \div \frac{2x^2+5x-3}{4x^2-1}$

119. $\frac{x^2-2x-8}{9x^2} \div \frac{x^2-16}{3x+12}$

In Exercises 120 – 129, evaluate each of the following expressions.

120. $\frac{3}{7xy} + \frac{4}{7xy}$

121. $\frac{8}{3ab} - \frac{2}{3ab}$

122. $\frac{7x}{x+3} + \frac{21}{x+3}$

123. $\frac{6x-y}{4y} - \frac{2x+3y}{4y}$

124. $\frac{2}{x+1} - \frac{3}{x+3}$

125. $\frac{3}{x^2-9} + \frac{1}{2x+6}$

126. $\frac{2}{y^2-y-6} + \frac{3y}{y^2+2y-15}$

127. $\frac{2a}{a^2+a-12} - \frac{3}{a^2+2a-8}$

128. $\frac{6x}{x^2-9} - \frac{5x}{x^2-x-6}$

129. $\frac{4y}{y^2-6y+5} + \frac{3y}{y^2-1}$

In Exercises 130 – 135, evaluate each of the following expressions.

130. $\frac{y-\frac{y^2}{y-x}}{1+\frac{x^2}{y^2-x^2}}$

131. $\frac{\frac{a^2}{a-b}-a}{\frac{b^2}{a-b}+b}$

132. $\frac{\frac{x}{y}-2+\frac{y}{x}}{\frac{x}{y}-\frac{y}{x}}$

133. $\frac{1+\frac{2}{x}-\frac{15}{x^2}}{1+\frac{4}{x}-\frac{5}{x^2}}$

134. $2-\frac{1}{1-\frac{2}{a+2}}$

135. $1-\frac{1}{1-\frac{1}{1-\frac{1}{x}}}$

4 Linear Equations and Inequalities

An equation is a statement that two algebraic expressions are equal. Examples include

$$3x + 2 = 5, \qquad x^2 + 2x - 8 = 0 \quad \text{and} \quad |x - 2| = 5$$

The solution of an equation in x is all values of x for which the equation is true. An equation that is true for every real number in the domain of the variable is called an identity.

4.1 Linear Equations in One Variable

An equation that can be written in the form $ax + b = 0$, where a and b are real numbers with $a \neq 0$, is called a linear equation. Examples include

$$3x + 5 = -2, \quad 6 = 2 - 3x \quad \text{and} \quad 2x + 7 = 8 - 3x.$$

Equivalent Equations

Two equations that have the same set of solutions are called equivalent equations. For example, $3x - 2 = 4$, $x + 5 = 7$ and $x = 2$ are equivalent equations because each have the same solution $x = 2$.

An equation can be transformed into an equivalent equation by performing one or more of the following operations.

1. The same number can be added to both sides of the equation.

2. The same number can be subtracted from both sides of the equation.

3. Both sides of the equation can be multiplied by the same number except 0.

4. Both sides of the equation can be divided by the same number except 0.

5. Combine like terms on one side or both sides of the equation.

6. Interchange the two sides of the equation.

Solving Linear Equations

To solve the linear equation $ax + b = c$, we transform the equation into the equivalent form $x = k$, where k is a real number. That is, to solve a linear equation in x, isolate x on one side by forming a sequence of equivalent equations, each simpler than the original equation, as shown in Examples 1 and 2.

Example 1

Solve: (a) $2x + 5 = 11$ (b) $3x - 2 = 10$

Solution

(a)

$2x + 5 = 11$	Original equation
$2x + 5 - 5 = 11 - 5$	Subtract 5 from each side
$2x = 6$	Simplify each side
$x = 3$	Divide each side by 2

(b)

$3x - 2 = 10$	Original equation
$3x - 2 + 2 = 10 + 2$	Add 2 to each side
$3x = 12$	Simplify each side
$x = 4$	Divide each side by 3

You may have noticed that when any term is removed from either side of an equation by adding or subtracting a constant number it appears on the other side but with opposite sign. With practice, you can perform some of the steps mentally.

Example 2

Solve: (a) $4x - 3 = 9$ (b) $5x + 2 = 12$

Solution

(a)

$4x - 3 = 9$	Original equation
$4x = 9 + 3$	Add 3 to each side
$4x = 12$	Simplify the right side
$x = 3$	Divide each side by 4

(b)

$5x + 2 = 12$	Original equation
$5x = 12 - 2$	Subtract 2 from each side
$5x = 10$	Simplify the right side
$x = 2$	Divide each side by 5

Solving Linear Equations with variables on each side

Sometimes you will be asked to solve an equation with variable terms on each side. In such cases, isolate all like terms on one side of the equation.

Example 3

Solve: (a) $5x - 3 = 3x + 7$ (b) $4x + 5 = 16 - 7x$

Solution

(a)

(a)	$5x - 3 = 3x + 7$	Original equation
	$5x - 3x = 7 + 3$	Isolate like terms
	$2x = 10$	Combine like terms
	$x = 5$	Divide each side by 2
(b)	$4x + 5 = 16 - 7x$	Original equation
	$4x + 7x = 16 - 5$	Isolate like terms
	$11x = 11$	Combine like terms
	$x = 1$	Divide each side by 11

Solving a Linear Equation Containing Parentheses

To solve equations containing parentheses, use the distributive property.

Example 4

Solve: (a) $3(2x - 1) = 9$ (b) $4 - 3(x - 1) = 1$

Solution

(a)	$3(2x - 1) = 9$	Original equation
	$6x - 3 = 9$	Use distributive property
	$6x = 12$	Add 3 to each side
	$x = 2$	Divide each side by 6

You must be careful if a negative sign precedes parentheses. The sign of each term inside the parentheses must be changed.

(b)	$4 - 3(x - 1) = 1$	Original equation
	$4 - 3x + 3 = 1$	Use distributive property
	$-3x = -6$	Isolate and combine like terms
	$x = 2$	Divide each side by -3

Solving a Linear Equation Containing Fractions

To solve an equation containing fractions, find the least common denominator (LCD) of all the terms and multiply every term by the LCD to clear the original equation of fractions.

Example 5

Solve:

(a) $\frac{1}{3}x + 2 = \frac{1}{2}x$ (b) $\frac{1}{4}(x + 1) = \frac{1}{3}(x + 2)$

Solution

(a)

$\frac{1}{3}x + 2 = \frac{1}{2}x$	Original equation
$2x + 12 = 3x$	Multiply each term by 6
$12 = 3x - 2x$	Isolate like terms
$12 = x$	Combine like terms

It is usually easier to isolate the variable term on the side that will result in a positive coefficient.

(b)

$\frac{1}{4}(x + 1) = \frac{1}{3}(x + 2)$	Original equation
$3(x + 1) = 4(x + 2)$	Multiply each side by 12
$3x + 3 = 4x + 8$	Use distributive property
$-5 = x$	Isolate and combine like terms

Example 6

Solve: $\frac{x+3}{2} - \frac{x-5}{4} = \frac{5}{8}$

Solution

$\frac{x+3}{2} - \frac{x-5}{4} = \frac{5}{8}$	Original equation
$4(x + 3) - 2(x - 5) = 5$	Multiply each term by 8
$4x + 12 - 2x + 10 = 5$	Use distributive property
$2x + 22 = 5$	Combine like terms
$2x = -17$	Subtract 22 from each side
$x = \frac{-17}{2}$	Divide each side by 2
$x = -8\frac{1}{2}$	

Any attempt on the part of the beginner to omit the line with parentheses is likely to cause trouble, particularly where negative signs are involved.

When an equation involves fractions with variable denominators, multiplying the equation by a variable quantity may introduce an extraneous solution. Therefore, it is essential that you check you solutions in the original equation. Because division by zero is undefined reject all numbers that makes one or more of the denominators equal to zero.

Exercise 4.1

In Exercises 1 – 24, solve each of the following equations:

1. $5x + 3 = 18$
2. $3x + 2 = 8$
3. $1 + 2x = 7$
4. $4x - 7 = 13$
5. $6x - 5 = 7$
6. $3x - 2 = 4$
7. $15 = 3 - 2x$
8. $5x + 2 = 17$
9. $1 - 3x = 7$
10. $7x - 2 = 12$
11. $4x + 1 = 2$
12. $2x + 5 = 8$
13. $6x - 15 = 9$
14. $9x - 2 = 7$
15. $4x + 5 = 9$
16. $3x + 4 = 1$
17. $8 + 5x = -2$
18. $9 - 2x = 5$
19. $12 = 3x + 6$
20. $18 = 8 - 5x$
21. $4x + 3 = -9$
22. $7x - 4 = 17$
23. $9x + 2 = 11$
24. $3 - 7x = 24$

In Exercises 25 – 36, solve each of the following equations:

25. $3x = 15 - 2x$
26. $5x - 16 = x$
27. $9x = 2x - 21$
28. $-6y = 12 - 2y$
29. $9x - 7 = 4x + 8$
30. $7x - 4 = 3x + 12$
31. $8 - 5x = 7x - 16$
32. $3x - 7 = 11 - 6x$
33. $8x - 3 - 2x = -3x + 9$
34. $4x + 5 - 3x = 5x - 3$
35. $4x - 8 = 7x + 7$
36. $7x + 1 + 5x = 6x + 25$

In Exercises 37 – 46, solve each of the following equations:

37. $2(x - 4) = 6$
38. $5(x + 3) = 20$
39. $4(x - 2) = 5(x - 3)$
40. $3(2x + 5) = 2(x + 9)$
41. $3(x + 7) = 7x - 3$
42. $8 - (5x - 6) = 2x$
43. $1 - 2(x + 3) = 7$
44. $3 - 4(x - 2) = 2(4 - 3x)$
45. $4(2x - 7) - 3(x - 5) = 2$
46. $5(x - 2) - 6 = 8 + 3(x + 4)$

In Exercises 47 – 58, solve each of the following equations:

47. $\frac{2}{3}x + 5 = 17$

48. $\frac{3}{4}x - 5 = 4$

49. $\frac{1}{2}x + 3 = \frac{5}{6}x$

50. $\frac{1}{4}(x-2) = \frac{1}{3}(x+1)$

51. $\frac{2x-3}{5} = \frac{2+x}{4}$

52. $3 - \frac{1}{2}(x+8) = \frac{3}{4}x$

53. $\frac{1}{5}(2x+3) - \frac{3}{2} = \frac{1}{2}(x+1)$

54. $\frac{x+6}{3} - \frac{x+7}{4} = \frac{x+3}{6}$

55. $\frac{x+6}{5} = \frac{x-4}{10} + \frac{1}{2}$

56. $\frac{2}{3x} + 1 = 3 - \frac{1}{2x}$

57. $5 - \frac{6}{x} = \frac{x+6}{2x}$

58. $\frac{4}{x+5} = \frac{7}{x+8}$

4.2 Literal Equations

An equation which has some or all of the numbers represented by letters is known as literal equations. For example, $ax + b = c$ is a literal equation. It is customary to represent variables by the letters of the later part of the alphabet and constants by the early part of the alphabet. Thus, in the equation $ax + b = c$, x is a variable and a, b and c are constants. A literal equation can be solved using the rules given in Section 4.1.

Example 1

Solve each of the following equations.

(a) $ax - b = c$

Solution

$$ax - b = c$$

$$ax = c + b$$

$$x = \frac{a+b}{a}$$

(b) $ax + d = bx + c$

Solution

$$ax + d = bx + c$$

$$ax - bx = c - d$$

$$(a-b)x = c - d$$

$$x = \frac{c-d}{a-b}$$

Rewriting a Formula

Many real life situations can be modelled by equations with two or more letters known as formulas. For instance, the equation $A = \pi r^2$ gives the area of a circle with radius r. A is called the subject of the formula. A formula is a literal equation, and can be solved in the same way as we solve equations. For example, we can solve $A = \pi r^2$ for r as follows. Dividing each side by π gives

$$\frac{A}{\pi} = r^2$$

And taking positive square root of each side gives

$$r = \sqrt{\frac{A}{\pi}}$$

A formula given in one form can be transformed to an equivalent form with a different subject. The process of rearranging an equation so that a new variable is left alone on one side is called changing the subject of the formula.

Example 2

Make r the subject of the formula $V = \frac{1}{3}\pi r^2 h$.

Solution

$$V = \frac{1}{3}\pi r^2 h$$

$$3V = \pi r^2 h$$

$$\frac{3V}{\pi h} = r^2$$

$$\sqrt{\frac{3V}{\pi h}} = r$$

Example 3

Make u the subject of the formula $\frac{1}{f} = \frac{1}{u} + \frac{1}{v}$

Solution

$$\frac{1}{f} = \frac{1}{u} + \frac{1}{v}$$

$$uv = fv + fu$$

$$uv - fu = fv$$

$$(v - f)u = fv$$

$$u = \frac{fv}{v-f}$$

You can obtain the same result as follows:

$$\frac{1}{f} = \frac{1}{u} + \frac{1}{v}$$

$$\frac{1}{f} - \frac{1}{v} = \frac{1}{u}$$

$$\frac{v-f}{fv} = \frac{1}{u}$$

$$u = \frac{fv}{v-f}$$

Evaluating a Formula

Consider the formula

$$A = h(R^2 - r^2)$$

Given A = 160, h = 2 and R = 12, we can find the value of r as follows:

We substitute the values for A, h and R.

$$160 = 2(12^2 - r^2)$$

$80 = 144 - r^2$ Divide each side by 2

$r^2 = 144 - 80$ Isolate like terms

$r^2 = 64$ Combine like terms

$r = \pm 8$ Take the square root

Since r cannot be negative the value of r is 8.

You may find it easier to first solve for the new variable before making the substitution, as demonstrated in Example 4.

Example 4

If $I = \frac{nE}{R+nr}$, find n when I = 8, E = 32, R = 6 and r = 2.5

Solution

First solve the equation for n.

$$I = \frac{nE}{R+nr}$$

$$I(R + nr) = nE$$

$$IR + Inr = nE$$

$$IR = nE - Inr$$

$$IR = n(E - Ir)$$

$$n = \frac{IR}{E-Ir}$$

Now, we substitute values for I, R, E and r.

$$n = \frac{8\times6}{32-8\times2.5} = 4$$

Exercises 4.2

In Exercises 1 – 15, solve the following equations for x:

1. $cx - d = b$
2. $bx + a = c$
3. $\frac{x}{a} + b = c$
4. $\frac{x}{c} + 1 = \frac{x}{b}$
5. $ex + d = fx$
6. $a(x + 3) = b(2 - x)$
7. $\frac{1}{a} + \frac{x}{c} = \frac{x}{b}$
8. $\frac{x-b}{a} = c$
9. $2a - bx = ax - 3a$
10. $a - c(1 - x) = bx$
11. $x^2 = b + ax^2$
12. $ax^3 + 1 = b$
13. $\frac{3-x^2}{g} = \frac{2+x^2}{f}$
14. $b = \sqrt[3]{\frac{ax}{c}}$
15. $x = \sqrt{\frac{ax^2+c}{b}}$

In Exercises 16 – 30, make the variables indicated the subject of the following formulas

16. $v = u + at, \quad t$
17. $s = \frac{n}{2}(a + l), \quad n$
18. $E = I(R + r), \quad r$
19. $s = \frac{a}{1-r}, \quad r$
20. $I = \frac{PRT}{100}, \quad T$
21. $k = \frac{mv^2}{2g}, \quad v$
22. $A = h(R^2 - r^2), \quad R$
23. $v = \frac{4}{3}\pi r^3, \quad r$
24. $I = \frac{nE}{R+nr}, \quad n$

25. $s = \frac{1}{2}gt^2$, t

26. $T = 2\pi\sqrt{\frac{l}{g}}$, g

27. $\frac{1}{c} = \frac{1}{c_1} + \frac{1}{c_2}$, c_1

28. $E = V + \frac{1}{2}mv^2$, v

29. $k = \sqrt[3]{\frac{mg\cos\propto}{v}}$, v

30. $R = \sqrt{\frac{m}{P-Q}}$, P

31. If A = 2πrh, find h when A = 264, r = 4.

32. If E = ½ mv^2, find v when E = 80 and m = 10.

33. If s = ut + ½ ft^2, find s when u = 10, t = 5 and f = 12.

34. If A = ½ (a + b)h, find a when b = 6, h = 6 and A = 30.

35. The time Ts for the simple pendulum is given by the formula

$T = 2\pi\sqrt{\frac{l}{g}}$ where g ms^{-2} is the acceleration due to gravity.

(a) Make l the subject of the formula

(b) Find l when T = 1.2 s and g = 9.8 ms^{-2}

Give your answer to two decimal places

36. When two electrical resistances R_1 and R_2 are wired in parallel the resulting resistance R is given by the formula

$$\frac{1}{R} = \frac{1}{R_1} + \frac{1}{R_2}$$

Make R_2 the subject of the formula. Hence, find the value of R_2 given that R = 12 ohms and R_1 = 15 ohms

37. The total mechanical energy (E joules) of a particle of mass m kg moving at a speed of v ms^{-1} is given by E = V + ½ mv^2, where V joules is its potential energy. Find the speed of a particle of mass 5 kg when its potential energy is 100 joules and its total mechanical energy is 150 joules

38. The formula gives the amount of heat Q lost through a wall which is D unit thick and insulated with a material whose insulation factor is k, if the temperature on the warmer side of the wall is T_1° and on the cooler side is T_2°.

$$Q = \frac{k(T_1 - T_2)}{D}$$

Given that Q = 1800, k = 108, D = 1.5 and T_2 = 20°, find T_1.

39. Given that $Q = \frac{rk}{P+ms}$

(a) Make P the subject of the formula

(b) Find P when Q = 4/3, m = 15, s = 0.2, k = 4 and r = 10.

40. The formula $\frac{1}{f} = \frac{1}{v} + \frac{1}{u}$ gives the focal length f of a lens

(a) Make v the subject of the formula.

(b). When $f = 3$ and $u = 12$, find the value of v.

4.3 Solving Word Problems

Translating a verbal phrase into a mathematical model is critical in problem solving.

Translating Verbal Statements into Algebraic Statements

Sometimes we find it useful to represent real-life situations with algebraic expressions. To do this, you must be able to translate words and phrases into algebraic expressions. In the following table, we list some verbal expressions that indicate the four basic operations of arithmetic and show their translation into algebraic expression.

Operations	Verbal expression	Algebraic expressions
Addition	The sum of *a* and *b*	$a + b$
	7 plus *x*	$4 + x$
	7 more than *m*	$m + 7$
	y increased by 3	$y + 3$
Subtraction	*a* minus 3	$a - 3$
	The difference of *x* and *y*	$x - y$
	5 less than *b*	$b - 5$
	4 fewer than *y*	$y - 4$
	t decreased by 2	$t - 2$
Multiplication	The product of *a* and *b*	ab
	7 times *x*	$7x$
	Twice *y*	$2x$
Division	The quotient of *m* and *n*	m/n
	x divided by 3	$x/3$
	One-third of *a*	⅓ *a*

Note that each of the algebraic expressions represents one number, just as the sum of 6 and 3 is 9.

Translating Verbal Statements into Algebraic Equations

Most problems in algebra are solved by first translating a condition stated in words into an equation. We start with examples to illustrate the process. Here, we are only interested in writing an equation.

Example 1

The sum of three consecutive even numbers is 12.

Solution

First, assign a variable to the unknown and then translate the problem into an equation. Let x be the smallest even number. Then the next larger even number is $x + 2$ and the largest even number is $x + 4$.

Therefore $x + (x + 2) + (x + 4) = 12$

Example 2

3 less than twice a number is 11

Solution

Let x represent the number. Then twice the number is $2x$.

Therefore $2x - 3 = 11$

Example 3

The sum of the present ages of a man and his son is 50 years. In 5 years time the man will be twice as old as the son

Solution

Let the son's present age be x years

Then the father's age is $(50 - x)$ years.

In five years the son will be $(x + 5)$ years old and the father will be $(50 - x) + 5$ years old.

Therefore $(50 - x) + 5 = 2(x + 5)$

The preceding examples suggest the following steps for translating verbal phrases into algebraic expressions.

1. Read the problem carefully, and make sure you completely understand what the problem is asking you to do.

2. Represent one of the unknown quantities with a variable and try to relate all the other unknown quantities to this variable.

3. Form an equation that will relate known quantities to the unknown quantities.

Forming the correct equation is very important. We hope with practice, you will be able to form the correct equation of any given problem. It is important that you state exactly what the variable represents and in what units it is measured if any.

Using Linear Equations to Solve Problems

We now use our ability to translate problems into equations and our skills in solving equations to solve word problems.

Example 4

5 more than three times a number is 17. Find the number

Solution

Represent the number by x, then three times the number is $3x$.

5 more than three times the number is $3x + 5$.

Thus, $3x + 5 = 17$

Next solve the equation

$$3x + 5 = 17$$

$$3x = 12$$

$$x = 4$$

The number is 4.

Example 5

A boy bought 12 pencils from two shops. He bought a certain number of pencils at 50 ¢ each at one shop and the rest at 20 ¢ each from another shop. How many pencils did he buy from each shop if the total cost of pencils is $ 3.60?

Solution

Represent the number of pencils bought at 50 ¢ each by x.

Then $(12 - x)$ pencils were bought at 20 ¢ each. Hence,

$$50x + 20(12 - x) = 360$$

$$50x + 240 - 20x = 360$$

$$30x = 120$$

$$x = 4$$

He bought 4 pencils at 50 ¢ each and 8 pencils at 20 ¢ each

Example 6

Five hundred tickets were sold for a concert. Adult tickets were \$ 4 while student tickets were \$ 3. If the total sales were \$1650, how many of each type of tickets were sold?

Solution

Let x be the number of adult tickets sold

Then $(500 - x)$ student tickets were sold

The total value of adult tickets sold is $4x$ and the total value of student tickets sold is $3(500 - x)$.

So,
$$4x + 3(500 - x) = 1650$$

$$4x + 1500 - 3x = 1650$$

$$x = 150$$

150 adult tickets and 350 student tickets were sold.

Motion Problems

Motion problems usually involve distance traveled, speed and time. To solve motion problems, we use the relation

$$s = \frac{d}{t}$$

where s = speed, d = distance and t = time.

Example 7

John drove from his house to his father's ranch in 5 hours. In coming back heavy traffic slowed his speed by 10 km h^{-1}, and the trip took 6 hours. Find the average speed in each direction?

Solution

Let x be John's speed to the ranch. Then he returns with a speed of $x - 10$. You may find the use of a table helpful in solving motion problems. Here we have

Speed	Time	Distance
x	5	$5x$
$x - 10$	6	$6(x - 10)$

Since the distance is the same each way, we have

$$5x = 6(x - 10)$$

Solving, we have

$$5x = 6x - 60$$

$$-x = -60$$

$$x = 60$$

So John drove at 60 km h^{-1} to the ranch, and he return with a speed of 50 km h^{-1}.

Some students may find sketching the given information in motion problems helpful.

Example 8

Mary leaves town A for town B at 10 am, driving at 60 km h^{-1}. At 11 am Joseph leaves town B for town A, driving at 65 km h^{-1} along the same route. If the towns are 210 km apart, what time will they meet?

Solution

Let t be the time Mary traveled before they meet. Then $(t - 1)$ is the time Joseph traveled. Again we summarize the given information in a table.

Speed	Time	Distance
60	t	$60t$
65	$t - 1$	$65(t - 1)$

Since the distance traveled by Mary and the distance traveled by Joseph must add to 210 km, we have

$$60t + 65(t - 1) = 210$$

Solving, we have

$$60t + 65t - 65 = 210$$

$$125t = 375$$

$$t = 3$$

Finally, since Ofori left at 10 am, the two will meet at 1 pm.

Exercise 4.3

In Exercises 1 – 20, write an algebraic expression for the following statements.

1. The sum of x and y
2. x increased by 3
3. 2 more than a
4. p increased by 5
5. a minus c
6. 7 less than s
7. z fewer than 7
8. p times q
9. The product of 6 and b
10. Eight less than twice x
11. One more than three times m
12. The product of 2 and the sum of a and b
13. Twice the sum of x and y
14. Three times the difference of x and y
15. 2 more than three times x
16. The difference of a and b all divided by 3
17. 5 less than twice a all divided by 3
18. The sum of y and twice x all divided by 5
19. The product of 3 less than x and 2
20. Three times the difference of x and 4

In Exercises 21 – 34, express the condition stated in words as algebraic expression.

21. 3 more than the product of 2 and x
22. 6 less than the product of 3 and y
23. The product of x and twice y
24. The sum of a and b, all divided by 5
25. A class consists of x students. Five new students joined the class. How many students are now in the class?
26. A boy spent $\$x$ of his pocket money. If his pocket money was \$12, how much did he have left?
27. John is x years old. How old was he 3 years ago?
28. The sum of the ages of Peter and Daniel is 32. If Peter is x years old, how old is Daniel?
29. If the perimeter of a square is x cm, what is the length of a side?
30. Mary is x years old. How old will she be in 5 years time?

31. The smallest of three consecutive odd numbers is x. What is the odd number next to x?

32. John's present age is 5 years more than three times his son's present age. If his son is x years old, how old is John?

33. Joan had $15. She bought 9 pencils at $$x$ per pencil. How much has she left?

34. Johnson has $$x$ and his sister has twice as much in her savings account. If his sister saved a further $ 50, how much has she got in her saving account?

In Exercises 35 – 44, express the condition stated in words as an algebraic equation

35. One number is thrice another and their sum is 12

36. Twice a number exceeds the same number by 5

37. The sum of three consecutive odd numbers is 4 less than 19

38. If 3 is added to 5 times a number the result is 9 less than twice the number

39. The sum of the ages of Eric and Mary is 30 years. Three years ago Mary was twice as old as Eric.

40. Joseph bought two pens and three pencils. The price of a pen is 5 ¢ more than the price of a pencil. Together they cost $ 8.50.

41. A rectangle is twice as long as it is wide and its perimeter is 72 cm

42. One number is four times another number. Their sum is 25

43. Three more than twice a number is nine less than five times the number

44. When 1 is added to a number, and the result is multiplied by 3, you obtained the number you started with.

In Exercises 45 – 50, solve each of the following problems:

45. One number is three times a second number. The sum of the two numbers is 12. Find the number.

46. 3 is added to five times a number and the result is multiplied by 2. The final result is 36. What is the number?

47. One number is 5 more than a second number. If two times the smaller plus three times the larger is 80, find the two numbers.

48. A number is added to 4 and the result divided by 3. The answer is the same as you get by subtracting the number from 6 and dividing by 2. What is the number?

49. A number is three times a second number. The number minus the second number is 18. Find the numbers.

50. A class consists of 40 students. If there are 12 more boys than girls in the class, how many boys are in the class?

51. The sum of three consecutive even numbers is 6 more than 24. Find the numbers.

52. The sum of three consecutive integers is 99. Find the three integers.

53. The sum of three consecutive odd numbers is 4 less than 91. Find the three numbers.

54. The length of a rectangle is 5 cm more than its width. Find the width if the perimeter is 34 cm.

55. The length of a rectangle is 2 cm less than three times the width. If the perimeter is 60 cm, find the dimensions of the rectangle.

56. Joseph builds a fence around a rectangular garden. The perimeter of the garden is 84 cm. The width is 8 cm less than the length. Find the dimensions of the garden.

57. Anita bought 35 ¢ stamps and 15 ¢ stamps at the post office. If she purchased 15 stamps at a cost of $ 3.65, how many of each kind did she buy?

58. Jane has 5 ¢ and 10 ¢ coins. There were 55 coins in all. If the total value of the coins is $ 3.75, how many of 5 ¢ coins does he have?

59. Tickets sales for a music concert total $1,400. Altogether 200 people attended the concert. The prices of the tickets for adults and children are $10 and $5 respectively. Find the number of adult tickets sold.

60. A student paid $18 for 10 exercise books. Some cost $1.50 each and the others cost $ 2 each. How many of each kind did he buy?

61. A man's age is twice his son's age. If the sum of their ages is 60, how old is the father?

62. A father's age now is three times his son's age. Eight years ago he was five times his son's age. Find their ages.

63. A man 40 years old has a son 16 years old. How many years ago was the father five times his son's age?

64. A 36 years old man has a 6 years old son. In how many years will the father be three times his son's age?

65. Sarah drove from home to her farm in 3 hours. On the return trip, heavy traffic slowed her speed by 10 kilometers per hour and the trip took 4 hours. What was her speed each way?

66. A cyclist rode from home to the office in 5 hours. When he returned, his speed was 5 kilometers per hour faster and the trip took 4 hours. What was his speed each way?

67. A bus leaves a terminal at 3 pm and goes north at a rate of 50 kilometers per hour. One hour later, a second bus leaves the terminal and travel south at a rate of 40 kilometers per hour. At what time will the two buses be 320 kilometers apart?

68. John leaves home at 9 am and cycle at a rate of 36 kilometers per hour. Two hours later, Isaac leaves home and drive at the rate of 48 kilometers per hour. If they travel on the same route, at what time will Isaac catch up with John?

69. A bus leaves a station at 1 pm, traveling west at an average rate of 54 kilometers per hour. One hour later a second bus leaves the same station, traveling east at a rate of 68 kilometers per hour. At what time will the two buses be 298 km apart?

70. At 8:00 am Ama leaves on a trip at 63 km h^{-1}. One hour later, Kojo decides to join her and leaves along the same route, traveling at 72 km h^{-1}. When will Kojo catch up with Ama?

71. A train leaves Accra for Kumasi, traveling at 45 kilometers per hours. At the same time, a second train leaves Kumasi for Accra, traveling at 55 kilometers per hour. If the two towns are 300 km apart, how much time must elapse before the two trains meet?

72. Two cars are 500 km apart and move at the same time towards each other. How long will it take the cars to meet, if their average speeds are 120 kilometers per hour and 80 kilometers per hour?

73. Two buses start out 150 km apart and travel east. The speed of the bus on the left was twice as the speed of bus on the right. After six hours the bus on the left catches up with the bus on the right, how fast was each bus moving?

74. Two cyclists John and Isaac start out at the same time from two cities105 kilometers apart and travel toward each other. The average speed of Isaac is 5 kilometers per hour more than the average speed of John. If they meet in 3 hours, what are their respective speeds?

75. A car and a bus left a terminal and together traveled a total distance of 270 kilometers. The time traveled by the bus is 10 minute less than the time traveled by the car. If the average speeds of the car and bus were 80 km h^{-1} and 60 km h^{-1} respectively, how long did each car travel?

76. At 7:00 am, Joan starts walking from Accra to Tema at12 km h^{-1}. At 8:00 am Mary leaves Accra for Tema walking at 15 km h^{-1}. When did Mary catch up with Joan?

4.4 Simple Linear Inequalities

A relation in which one quantity is greater than or is less than another quantity is called an inequality. To write that one number is greater than or less than another number, we use the following symbols.

$<$ means is less than

$\leq$ means is less than or equal to

$>$ means is greater than

$\geq$ means is greater than or equal to

The pass mark m of a test marked out of 10 is 6. If a student passed the test then his mark is 6 or more. In symbol, we write $m \geq 6$. A student who failed the test had less than 6 and we write this as $m < 6$.

Graphs of Inequalities

An inequality has infinitely many solutions. For example, given the inequality $x > 2$, any number to the right of 2 on a number line is a solution of $x > 2$. The set of all solutions of an inequality is the solution set of the inequality. The graph of an inequality is obtained by plotting its solution set on the real number line.

Example 1

Sketch the graph of each inequality.

(a) $x > -2$ (b) $x \leq 1$

Solution

(a) The solution set of $x > -2$ are all real numbers greater than -2. The open dot is used to indicate that -2 is not part of the graph.

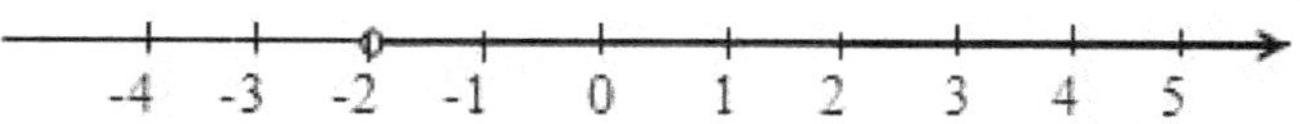

(b) The solution set of $x \leq 1$ are all real numbers less than or equal to 1. The closed dot is used to indicate that 1 is part of the graph.

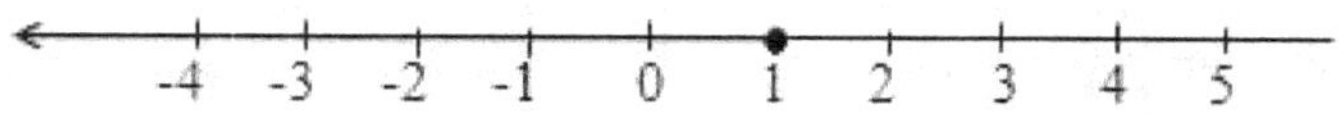

Solving Linear Inequalities

Solving an inequality is much like solving a linear equation. The following properties are similar to the properties you used in solving linear equations, but there are two important exceptions which we will discuss later in this section.

1. Add the same number to each side of the inequality.

2. Subtract the same number from each side of the inequality.

3. Multiply each side by the same positive number.

4. Divide each side by the same positive number.

Example 2

Solve each inequality.

(a) $2x - 3 < 5$ (b) $\frac{2}{3}x + 1 \geq 5$

Solution

(a)

$2x - 3 < 5$	Original inequality
$2x < 8$	Add 3 to each side
$x < 4$	Divide each side by 2

(b)

$\frac{2}{3}x + 1 \geq 5$	Original inequality
$2x + 3 \geq 15$	Multiply every term by 3
$2x \geq 12$	Subtract 3 from both sides
$x \geq 6$	Divide both sides by 2

Example 3

Solve the inequality $5x + 7 \leq 1 + 2x$

Solution

$5x + 7 \leq 1 + 2x$	Original inequality
$5x - 2x \leq 1 - 7$	Isolate like terms
$3x \leq -6$	Combine like terms
$x \leq -2$	Divide each side by 3

Notice that when you multiply or divide each side of an inequality by the same positive number, the inequality symbol stays the same.

Multiplying and Dividing by Negative Numbers

Consider the inequality $3 < 5$. if we multiply both sides by -2, we get $-6 < -10$ which is false. If we reverse the inequality symbol we get $-6 > -10$, which is true. Similarly, if we multiply $-3 > -7$ by -1, we get $3 > 7$ which is false. If we reverse the inequality symbol we get $3 < 7$, which is true. When each side of an inequality is multiplied or divided by a negative number, the direction of the inequality symbol must be reversed. This remain true if the symbol $<$ and $>$ are replaced by $\leq$ and $\geq$.

Example 4

Solve each inequality.

(a) $1 - 2x < 9$ (b) $2 - \frac{1}{3}x \geq 4$

Solution

(a) $1 - 2x < 9$ Original inequality

$-2x < 8$ Subtract 1 from each side

$x > -4$ Divide each side by -2

(b) $2 - \frac{1}{3}x \geq 4$ Original inequality

$-\frac{1}{3}x \geq 2$ Subtract 2 from each side

$x \leq -6$ Multiply each side by -3

Example 5

Solve the inequality $\frac{1}{4}(x + 5) < \frac{1}{3}(x + 2)$

Solution

$\frac{1}{4}(x + 5) < \frac{1}{3}(x + 2)$ Original inequality

$3(x + 5) < 4(x + 2)$ Multiply each side by 12

$3x + 15 < 4x + 8$ Use distributive property

$3x - 4x < 8 - 15$ Isolate like terms

$-x < -7$ Combine like terms

$x > 7$ Multiply each side by -1

Example 6

Solve each inequality.

(a) $-3 < 2x - 5 < 1$ (b) $-3 \leq 1 - 2x < 7$

Solution

(a) $-3 < 2x - 5 < 1$ Original inequality

$-3 + 5 < 2x - 5 + 5 < 1 + 5$ Add 5 to all three parts

$2 < 2x < 6$ Combine like terms

$1 < x < 3$ Divide each part by 2

The inequality can also be solved as follows:

$-3 < 2x - 5$ and $2x - 5 < 1$

$2 < 2x$ $2x < 6$ Add 5 to each side

$1 < x$ $x < 3$ Divide each side by 2

The solution set consists of all real numbers that satisfy both inequalities, that is all numbers greater than 1 and less than 3. This is written as $1 < x < 3$.

(b) $-3 \leq 1 - 2x < 7$ Original inequality

$-3 - 1 \leq 1 - 2x - 1 < 7 - 1$ Subtract 1 from all parts

$-4 \leq -2x < 6$ Combine like terms

$2 \geq x > -3$ Divide each part by -2

We can also write $-3 < x \leq 2$.

Problems in Inequalities

The steps used in solving applications involving inequalities are similar to the steps used in solving word problems discussed in Section 4.3.

The table below lists some phrases that indicate the four inequalities.

Verbal expressions	Examples	Symbols
At least	*b* is at least 2	$b \geq 2$
At most	*s* is at most 5	$s \leq 5$
Cannot exceed	*t* cannot exceed 3	$t \leq 3$
Is less than	*i* is less than 200	$i < 200$
Is more than	*t* is more than 2	$t > 2$
Is between	*n* is between 30 and 60	$30 < n < 60$
Not more than	*d* is not more than 3	$d \leq 3$
Not less than	*b* is not less than 6	$b \geq 6$

Example 7

When two times a certain number is added to 3, the result is greater than 15. What are the possible numbers?

Solution

Let x represent the number, so

$$2x + 3 > 15$$

$$2x > 12$$

$$x > 6$$

The number must be at least 6.

Example 8

A ticket for a concert cost $ 20 for adults and $10 for children. Three times as many adults as children attended the concert. If the gate proceeds were not more than $ 3,500, find the maximum number of children at the concert.

Solution

Let x represent the number of children at the concert. Then $3x$ adult attended the concert.

$$60x + 10x \leq 3500$$

$$70x \leq 3500$$

$$x \leq 50$$

Therefore there were at most 50 children at the concert.

Exercise 4.4

In Exercises 1 – 12, graph each of the following inequalities:

1. $x \geq -1$
2. $x < 3$
3. $x \leq -2$
4. $1 < x$
5. $0 \geq x$
6. $x > -1\frac{1}{2}$
7. $x < 2\frac{1}{4}$
8. $x \leq \frac{3}{4}$
9. $x \leq -\frac{1}{2}$
10. $-1 \leq x < 4$
11. $0 < x \leq 3$
12. $-2 \leq x < 5$

In Exercises 13 – 24, solve the following inequalities:

13. $2x + 1 > 7$
14. $3x - 2 < 4$
15. $5x \leq 3x + 6$
16. $2x + 1 \leq 16 - 3x$
17. $4x + 9 \geq 1 + 2x$
18. $2(x - 1) > 12$
19. $5 \leq 2(x + 1)$
20. $\frac{1}{3}(x + 2) \geq \frac{1}{2}(3 - x)$
21. $5 < 3x + 17$
22. $1 \geq 2x - 3$

23. $7x - 3 > 3(x + 1)$ 24. $\frac{1}{3}(x - 4) + 3 < \frac{1}{4}x$

In Exercises 25 – 36, solve the following inequalities:

25. $3 - 4x < 7$ 26. $5 - 3x > -4$

27. $-3x + 6 \leq 2$ 28. $x > 4 + 3x$

29. $1 - \frac{1}{2}x \leq 5$ 30. $2(3 - x) < 3$

31. $\frac{1}{4}(x + 1) \geq \frac{1}{3}(x + 2)$ 32. $-4x > 27 + 5x$

33. $5(x + 1) < 3(2x + 4)$ 34. $3(x + 5) - 5(x + 4) \geq 0$

35. $1 - 3(x + 2) < 10$ 36. $4x + 3 \leq 7x + 9$

In Exercises 37 – 46, solve the following inequalities:

37. $5 < x - 3 < 7$ 38. $0 \leq 2x + 4 < 2$

39. $-1 < 3 - 2x \leq 9$ 40. $-8 < 1 - 3(x + 2) < 10$

41. $-2 < \frac{1}{2}(x + 1) < 6$ 42. $-2 < x + 2 < 3$

43. $5 < 2 + 3x < 11$ 44. $1 \leq 5 - x \leq 5$

45. $-2 < \frac{1}{3}(x + 2) - \frac{1}{4}(x + 3) < -1$ 46. $4 \leq 2(1 - x) + 3x < 7$

In Exercises 47 – 58, solve the following inequality problems.

47. When three times a certain number is added to 2, the result is less than 8. What are the possible numbers?

48. When 5 is subtracted from a number and the result doubled, the answer is not more than 4. What are the possible numbers?

49. When 7 is added to two times a certain number the result is not less than the number added to 3. What are the possible numbers?

50. The sum of three consecutive even numbers is not less than 36. Find the smallest number.

51. The width of a rectangle is fixed at 8 cm. What lengths will make the perimeter at least 200 cm?

52. A rectangle is twice as long as it is wide. If its perimeter is not greater than 72 cm, find the possible range of values of the width.

53. A rectangle has length x cm, and width 6 cm. The perimeter is less than 48 cm, but greater than 32 cm. What are the possible lengths of the rectangle?

54. A certain number of equal squares sides 4 cm are placed side by side to form a rectangle two squares in width. If the area of the rectangle is at most128 cm^2, what is the maximum number of squares used?

55. A man's age is four times his son's age and his daughter is 5 years younger than her brother. If the sum of their ages is not less than 67 years, find the minimum age of the son.

56. A boy's test grades are 73, 75, 89 and 91. What scores on a fifth test will make his average test grade at least 85?

57. A student took three papers in a physic examination. His marks for two of the papers were 68 and 60 respectively. To obtain a distinction an average of not less than 70 is needed over the three papers. What minimum mark must he obtain in the third paper?

58. A woman bought a certain quantity of oranges at 25 ¢ each and twice as many pine apples at $ 1.00 each. If she did not spend more than $ 45 altogether, what is the maximum quantity of oranges bought?

Review Exercises

In Exercises 1 – 18, solve each equation:

1. $5x + 3 = 18$ 2. $3x + 6 = 21$ 3. $7t - 8 = 6$

4. $6x - 3 = 15$ 5. $-19 = 1 + 4x$ 6. $7 - 3x = 19$

7. $5x - 3 = 7 + 3x$ 8. $3x - 2 = 8 - x$ 9. $5x + 3 = 2x + 15$

10. $4x + 3 = 2x - 5$ 11. $5 - 2x = 25 - 4x$ 12. $10 - 3x = 40 - 6x$

13. $7(2x - 1) = 21$ 14. $35 = 5(3x + 1)$ 15. $13 - 3(2x - 1) = 4$

16. $6x - (3x + 8) = 16$ 17. $5(x + 4) = 7(x - 2)$ 18. $7x - (2x + 8) = 32$

In Exercises 19 – 24, solve each formula for the letter indicated:

19. $A = \frac{1}{2} h(a + b)$, a 20. $A = P + Prt$, P

21. $E = mc^2$, c 22. $A = 2\pi r(r + h)$, h

23. $V = \frac{4}{3}\pi r^3$, r 24. $c = \frac{2ad}{a-d}$, d

25. The formula for the total resistance, R in a parallel circuit is given by the formula

$$\frac{1}{R} = \frac{1}{R_1} + \frac{1}{R_2}$$

Find the total resistance if $R_1 = 6$ ohms and $R_2 = 10$ ohms.

26. The perimeter of a rectangle of length l and width w is given by the formula $P = 2l + 2w$. Find the width when $P = 30$ cm and $l = 10$ cm.

27. The formula that relates Celsius and Fahrenheit temperature is $F = (9/5)C + 32$. If the temperature of the day is 20° C, what is the Fahrenheit temperature?

In Exercises 28 – 31, translate each statement to an algebraic equation. Let x represent the number in each case.

28. 5 more than a number is 9

29. 7 less than a number is 15

30. 4 less than 3 times a number is twice that same number

31. 2 times the sum of a number and 3 is 12 more than that same number

In Exercises 32 – 39, solve each word problem:

32. The sum of twice a number and 7 is 33. What is the number?

33. 4 times a number, decreased by 20, is 44. What is the number?

34. The sum of three consecutive integers is 63. What are the three integers?

35. The sum of three consecutive odd integers is 105. What are the three integers?

36. The sum of three consecutive even integers is 126. What are the three integers?

37. In an election, the winning candidate had 160 more votes than the loser. If the total number of votes cast was 3260, how many votes did each candidate receive?

38. Kofi is 1 year less than twice as old as his sister. If the sum of their ages is 14 years, how old is Kofi?

39. On her vacation in Europe, Ama expenses for food and lodging were \$60 less than twice as much as her airfare. If she spent \$2400 in all, what was the cost of her airfare?

In Exercises 40 – 45, graph each inequality:

40 $x \geq -4$ 41. $x \leq 3$ 42. $x < -2$

43. $x > -4$ 44. $x \leq -3$ 45. $x > -\frac{3}{4}$

In Exercises 46 – 53, solve each inequality:

46. $3x \geq 2x - 4$ 47. $5x < 4x + 7$

48. $6x - 8 < 5x$ 49. $4x - 3 > 3x + 5$

50. $5x - 3 > 3x + 15$ 51. $5x + 7 \leq 8x - 17$

52. $4(x + 7) < 2x + 31$ 53. $2(x - 7) < 5x - 12$

In Exercises 54 – 61, translate each of the following statements into inequalities. Let x represent the number in each case.

54. 5 more than three times a number is less than 7

55. 3 less than a number is greater than 5

56. 7 more than twice a number is greater or equal to 12

57. Between 60 and 80 students attended the concert

58. At least 45 students passed the test

59. At most 1,200 teachers were interviewed.

60. A man's weekly wage is not to exceed $ 100

61. The cost of bread is not less than $ 2.50

In Exercises 62 – 67, solve each of the following problems:

62. A plumber charges $ 25 plus $ 30 per hour for emergency service. A man was billed over $ 100 for an emergency call. How long was the plumber there?

63. A father cannot spend more than $ 2285 on tuition. If a school charges $ 35 registration fee plus $ 375 per course, what is the greatest number of courses for which he can register?

64. A student takes mathematics course in which four tests are given. To get a B, he must average at least 80 on the four tests. He scored 82, 76 and 78 on the first three tests. What scores on the last test will earn him at least a B?

65. The perimeter of a rectangular swimming pool is not to exceed 72 meters. The length is to be twice the width. What widths will meet these conditions?

66. A factory worker earns a daily base pay of $ 15 plus $ 4.50 every hour. How many hours must he work in a day to earn at least $ 42?

67. A man claims that it cost him at least $ 3 to make a call. If a typical call cost 75 ¢ plus 45 ¢ each minute, how long do his calls typically last?

5 Quadratic Equations

A quadratic equation is an equation that can be written in the form $ax^2 + bx + c = 0$, where a, b and c are real numbers with $a \neq 0$. A quadratic equation written in the form $ax^2 + bx + c = 0$ is said to be in standard form. Examples include $3x^2 + 2x - 5 = 0$, $x^2 = 5x + 4$ and $2x^2 - 3x = 0$. You can see that 2 is the highest exponent of the variable in each case.

Three ways of solving quadratic equations algebraically are by factoring, completing the square and using the quadratic formula. The simplest way to solve a quadratic equation is by factoring. This method is based on the Zero-Product Property.

Zero- Product Property

If a and b are real numbers and $a \cdot b = 0$, then either $a = 0$ or $b = 0$ (or both).

5.1 Solving a Quadratic Equation by Factoring

To solve quadratic equations by factoring just follow the following steps:

1. Write the given quadratic equation in standard form.

2. Factor the quadratic expression.

3. Set each factor equal to zero and solve each linear equation.

Example 1

Solve $x^2 + 8x + 15 = 0$.

Solution

$$x^2 + 8x + 15 = 0$$

The equation is already in standard form   Factor the quadratic expression.

$$(x + 3)(x + 5) = 0$$

Finally, equate each factor to 0 and solve each linear equation.

$$x + 3 = 0 \quad \text{or} \quad x + 5 = 0$$

$$x = -3 \qquad\qquad x = -5$$

The solutions are $x = -3$ and $x = -5$.

Observe that the quadratic equation has two solutions.

Example 2

Solve $x^2 + 7x = 30$

Solution

$x^2 + 7x = 30$ Original equation

First, write the equation in standard form

$x^2 + 7x - 30 = 0$ Subtract 30 from each side

$(x - 3)(x + 10) = 0$ Factor the quadratic expression

$x - 3 = 0$ or $x + 10 = 0$ Set each factor equal to 0

$x = 3$ $x = -10$ Solve each equation

The solutions are $x = 3$ and $x = -10$.

Example 3

Solve $(x + 5)(x - 2) = -6$.

Solution

It will be incorrect to set each factor equal to – 6. First, multiply the expression on the left side and correct all terms to left side to make the right side of the equation zero.

$(x + 5)(x - 2) = -6$ Original equation

$x^2 + 3x - 10 = -6$ Multiply the expression on the left

$x^2 + 3x - 4 = 0$ Write the equation in the standard form.

$(x - 1)(x + 4) = 0$ Factor the quadratic expression.

$x - 1 = 0$ or $x + 4 = 0$ Set each factor equal to 0

$x = 1$ $x = -4$ Solve each equation

The solutions are $x = 1$ and $x = -4$.

Example 4

Solve $4x^2 - 9 = 0$.

Solution

$4x^2 - 9 = 0$ Original equation

$(2x - 3)(2x + 3) = 0$ Factor the quadratic expression

$2x - 3 = 0$ or $2x + 3 = 0$ Set each factor equal to 0

$x = \frac{3}{2} = 1\frac{1}{2}$ $\quad$ $x = -\frac{3}{2} = -1\frac{1}{2}$ $\quad$ Solve each equation

The solutions are $x = 1\frac{1}{2}$ and $x = -1\frac{1}{2}$.

You can obtain the same result by extracting square roots.

$4x^2 - 9 = 0$ $\quad$ Original equation

$4x^2 = 9$ $\quad$ Add 9 to each side

$x^2 = \frac{9}{4}$ $\quad$ Divide each side by 4

$\sqrt{x^2} = \pm\sqrt{\frac{9}{4}}$ $\quad$ Take the square root of each side

$x = \pm\frac{3}{2}$ $\quad$ Simplify

The $\pm$ sign represents the two solutions of the equation.

The two solutions are $x = \frac{3}{2} = 1\frac{1}{2}$ and $x = -\frac{3}{2} = -1\frac{1}{2}$.

Example 5

Solve $x^2 + 5x = 0$.

Solution

$x^2 + 5x = 0$ $\quad$ Original equation

$x(x + 5) = 0$ $\quad$ Factor the quadratic expression

$x = 0$ or $x + 5 = 0$ $\quad$ Set each factor equal to 0

$x = -5$

The solutions are $x = 0$ and $x = -5$.

Example 6

Solve $3x^2 - 5x - 12 = 0$.

Solution

You can factor the quadratic expression by the method of grouping. We will not give details of the factorization.

$3x^2 - 5x - 12 = 0$ $\quad$ Original equation

$(3x + 4)(x - 3) = 0$ $\quad$ Factor the quadratic expression

$3x + 4 = 0$ or $x - 3 = 0$ Equate each factor to 0

$x = -\frac{4}{3}$ $x = 3$

The solutions are $x = -4/3$ and $x = 3$

Exercise 5.1

Solve each equation:

1. $x^2 + 4x - 12 = 0$	2. $x^2 - 8x + 15 = 0$	3. $x^2 + 7x - 18 = 0$
4. $x^2 - x - 30 = 0$	5. $x^2 - x - 20 = 0$	6. $x^2 + 12x + 32 = 0$
7. $x^2 + 15x + 56 = 0$	8. $x^2 - 14x + 45 = 0$	9. $x^2 + 3x - 28 = 0$
10. $42 + x - x^2 = 0$	11. $15 - 2x - x^2 = 0$	12. $x^2 + 7x = 0$
13. $3x^2 - 4x = 0$	14. $4x^2 - 25 = 0$	15. $9x^2 - 16 = 0$
16. $x^2 - 2x = 18 + 5x$	17. $x^2 + 20 = 9x$	18. $x(x - 7) = -12$
19. $(x + 2)(x + 3) = 20$	20. $(x - 3)(x - 4) = 42$	21. $(x - 4)(x + 4) = -6x$
22. $3x^2 - 20x - 7 = 0$	23. $8x^2 + 14x + 3 = 0$	24. $6x^2 - 7x - 5 = 0$
25. $3x^2 - 13x + 12 = 0$	26. $7x^2 - 37x + 10 = 0$	27. $2x^2 + 13x - 24 = 0$
28. $3x^2 + 20 = 4x + 35$	29. $x(2x - 17) = -35$	30. $3x^2 - 2x = 5$
31. $(x - 1)(5x + 4) = 2$	32. $(6x + 1)(x + 1) = 21$	33. $30 = 8x(x + 1)$

A quadratic equation cannot always be solved by factoring.

5.2 Solving Quadratic Equations by Completing the Square

Consider the two expressions below called the perfect square trinomials:

$$x^2 + 2ax + a^2 \quad \text{and} \quad x^2 - 2ax + a^2$$

You may noticed that the last term in each case is the square of half the coefficient of x. In solving quadratic equation by completing the square we use this relationship between the coefficient of x and the constant term to write an equivalent equation that has a perfect square trinomial on one side.

Example 1

Solve $x^2 + 8x - 20 = 0$ by completing the square

Solution

$$x^2 + 8x - 20 = 0$$

First isolate the constant term

$$x^2 + 8x = 20$$

Next add the square of half the coefficient of x to each side in order to maintain equality. In this case, dividing 8 by 2 and squaring the result gives 16.

$$x^2 + 8x + 16 = 36$$

Factor the quadratic expression.

$$(x + 4)^2 = 36$$

Taking the square root of each side gives

$$x + 4 = \pm 6$$

So $$x = -4 \pm 6$$

Simplifying this expression we get $x = -4 + 6 = 2$ and $x = -4 - 6 = -10$.

So, the solutions of the equation are $x = 2$ and $x = -10$.

Example 2

Solve $2x^2 + x - 6 = 0$ by completing the square

Solution

First, divide each side of the equation by the leading coefficient before completing the square.

$$2x^2 + x - 6 = 0$$ Original equation

$$2x^2 + x = 6$$ Add 6 to each side

$$x^2 + \frac{1}{2}x = 3$$ Divide each term by 2

$$x^2 + \frac{1}{2}x + \frac{1}{16} = \frac{49}{16}$$ Add1/16 to each side

$$\left(x + \frac{1}{4}\right)^2 = \frac{49}{16}$$ Factor the quadratic expression

$$x + \frac{1}{4} = \pm\frac{7}{4}$$ Take square root of each side

So $$x = -\frac{1}{4} \pm \frac{7}{4}$$

The solutions of the equation are $x = -\frac{1}{4} + \frac{7}{4} = \frac{3}{2} = 1\frac{1}{2}$ and $x = -\frac{1}{4} - \frac{7}{4} = -2$

Exercise 5.2

Solve the following quadratic equations by completing the square. Give your answer to two decimal places if possible

1. $x^2 + 2x - 3 = 0$
2. $x^2 - x - 12 = 0$
3. $x^2 - 4x - 6 = 0$
4. $x^2 + 10x + 22 = 0$
5. $x^2 - 3x - 11 = 0$
6. $3x^2 - 6x - 31 = 0$
7. $2x^2 - 3x + 1 = 0$
8. $3x^2 - 9x + 2 = 0$
9. $2x^2 + 10x + 5 = 0$
10. $5x^2 - 7x - 6 = 0$
11. $2x^2 = 7x + 15$
12. $4x(x - 1) = 5$

5.3 Solving Quadratic Equations by Use of the Formula

The general quadratic equation is $ax^2 + bx + c = 0$, where $a \neq 0$. The solution to this equation is given by the formula below, called the Quadratic Formula.

$$x = \frac{-b \pm \sqrt{b^2 - 4ac}}{2a}$$

We will derive the Quadratic Formula in a later chapter.

Example 1

Solve $x^2 - 6x - 16 = 0$.

Solution

Identify the values of a, b and c in the quadratic equation. Here $a = 1$, $b = -6$ and $c = -16$.

Substituting these values into the Quadratic Formula we get

$$x = \frac{6 \pm \sqrt{(-6)^2 - 4(1)(-16)}}{2(1)}$$

$$= \frac{6 \pm \sqrt{100}}{2}$$

$$= \frac{6 \pm 10}{2}$$

So $x = \frac{6+10}{2} = 8$ or $x = \frac{6-10}{2} = -2$

The solutions of equation are $x = 8$ and $x = -2$.

Example 2

Solve $3x^2 = -7x + 6$.

Solution

$$3x^2 = -7x + 6$$

First, write the equation in standard form before you apply the Quadratic Formula.

$$3x^2 + 7x - 6 = 0$$

Here $a = 3$, $b = 7$ and $c = -6$.

Substituting these values into the quadratic formula we get

$$x = \frac{-7 \pm \sqrt{7^2 - 4(3)(-6)}}{2(3)}$$

$$= \frac{-7 \pm \sqrt{121}}{6}$$

$$= \frac{-7 \pm 11}{6}$$

So $x = \frac{-7+11}{6} = \frac{2}{3}$ or $x = \frac{-7-11}{6} = -3$

The solutions of the equation are ⅔ and – 3.

Any quadratic equation can be solved by completing the square or by using the Quadratic Formula. However, if the quadratic expression can be factored, solve the equation by factoring.

Exercise 5.3

Solve the following quadratic equation by use of the formula. Give your answer to two decimal places if possible

1. $x^2 - 4x + 3 = 0$
2. $x^2 - 3x - 10 = 0$
3. $x^2 + 6x + 7 = 0$
4. $x^2 - 5x + 2 = 0$
5. $x^2 + 7x - 30 = 0$
6. $3x^2 - 5x - 2 = 0$
7. $7x^2 - 5x - 2 = 0$
8. $2x^2 + 8x + 3 = 0$
9. $5x^2 - 9x + 3 = 0$
10. $4x^2 = 6 - 5x$

11. $8x^2 + 6x = 3$

12. $3x(x - 2) = 1$

13. $(3x - 2)^2 = 25$

14. $3x^2 - x = 6$

15. $(2x + 3)^2 = 16x^2$

16. $7x^2 = 2x + 3$

5.4 Application of Quadratic Equations

Some application problems will require us to solve quadratic equations. The two examples below illustrate this.

Example 1

The length of a rectangular field is 3 kilometers longer than its width. If the area of the field is 108 square kilometers, find the dimensions of the field.

Solution

Let x be the width of the field, so $(x + 3)$ will be the length of the field. Hence, the area of the field is $x(x + 3)$. So

$$x(x + 3) = 108$$

$$x^2 + 3x = 108$$ Multiply the expression on the left side

$$x^2 + 3x - 108 = 0$$ Subtract 108 from both sides

$$(x - 9)(x + 12) = 0$$ Factor the quadratic expression

$$x - 9 = 0 \quad \text{or} \quad x + 12 = 0$$ Set each factor equal to 0

$$x = 9 \qquad\qquad x = -12$$

The solutions of the equation are 9 and – 12. Since the width must be positive, we will reject the negative solution. Therefore, the width of the field is 9 kilometers. The length is 3 kilometers longer than this and so the length of the field is 12 kilometers. Thus, the field is 12 kilometers long and 9 kilometers wide.

Example 2

The length of a rectangular picture frame is 1 meter longer than the width. If the diagonal of the frame is 5 meters, what are the dimensions of the rectangle?

Solution

Let x centimeters be the width of the frame. So, the length of the frame is $(x + 1)$ centimeters.

Using the Pythagorean Theorem, we have

$$x^2 + (x + 1)^2 = 5^2$$

$$2x^2 + 2x - 24 = 0 \qquad \text{Simplify the equation}$$

$$x^2 + x - 12 = 0 \qquad \text{Divide each term by 2}$$

$$(x + 4)(x - 3) = 0 \qquad \text{Factor the quadratic expression}$$

$$x + 4 = 0 \quad \text{or} \quad x - 3 = 0 \qquad \text{Equate each factor to 0}$$

$$x = -4 \qquad x = 3$$

So, the frame is 3 m wide and 4 m long

Exercise 5.4

1. The product of the page number on two consecutive pages of a book is 240. Find the page numbers

2. The product of two consecutive odd integers is 195. Find the integers.

3. The product of two consecutive even integers is 168. Find the integers.

4. Twice a number is 8 less than its square. Find all such numbers

5 A rectangular picture frame is twice as long as it is wide. If the area of the frame is 338 square centimeters, find its dimensions.

6. One number is 3 more than another. The sum of their squares is 89. What are the numbers?

7. The length of a rectangle is 2 centimeters longer than its width. If the diagonal of the rectangle is 10 centimeters, what are the dimensions of the rectangle?

8. The length of the base of a triangle is 3 centimeters more than the height of the triangle. If the area of the triangle is 35 square centimeters, find the height and the length of the base.

9. A triangular traffic island has a base half as long as its height. Find the base and the height if the island has an area of 81square meters.

10. A 15 meters ladder leaning against a building touches the bottom of a window. The foot of the ladder from the building is 3 meters shorter than the height of the window above the ground. Find the height of the window above the ground.

11. A 10 meters rope is fastened from the top of a vertical pole to a peg on the ground. The distance of the peg from the pole is 2 meters longer than the height of the pole. Find the height of the pole.

12. Ama starts at a point and walks north at 2 km h^{-1}. One hour later, Esi starts at the same point and walks east at 3 km h^{-1}. How long after Ama starts walking will the two girls be 5 kilometers apart?

Review Exercises

In Exercises 1 – 12, solve each equation by factoring:

1. $x^2 - 7x + 6 = 0$
2. $x^2 + 6x + 5 = 0$
3. $x^2 + 4x - 21 = 0$
4. $x^2 + 7x - 18 = 0$
5. $x^2 + 9x + 14 = 0$
6. $x^2 + 8x + 15 = 0$
7. $4x - x^2 = 0$
8. $x^2 = 2x - 1$
9. $x^2 + 16 = 8x$
10. $(x - 7)(x + 1) = -16$
11. $(x + 2)(x - 7) = -18$
12. $x^2 - 3x - 4 = 10 - x^2$

In Exercises 13 – 22, use any applicable method to solve the following equations:

13. $3x^2 - 7x = 20$
14. $3x^2 - 2x = 5$
15. $12x^2 - 5x = 2$
16. $x(3x + 1) = 2$
17. $(x - 1)(5x + 4) = 2$
18. $2x^2 + 3x - 20 = 0$
19. $2x^2 + 9x - 35 = 0$
20. $5x^2 - 26x + 5 = 0$
21. $2x^2 + 5x - 12 = 0$
22. $3x^2 + 8x - 16 = 0$
23. The product of two consecutive positive even integers is 224. Find the integers
24. The product of twice a number and 3 less than the number is 80. Find the numbers
25. A picture frame is 4 centimeters taller than it is wide and has an area of 192 square centimeters. What are the dimensions of the picture frame?
26. The length of one leg of a right triangle is 3 centimeters more than the other. If the length of the hypotenuse is 15 centimeters, what are the lengths of the two legs?
27. The length of a rectangle is 7 centimeters longer than its width. If the diagonal of the rectangle is 13 centimeters, what are the dimensions of the rectangle?
28. A10-meter long ladder leans against a wall. The top of the ladder touches the top of the wall, and the distance of the foot of the ladder from the wall is 2 centimeters shorter than the height of the wall. Find the height of the wall.
29. John starts out from a point and walk north at 2 kilometers per hour. Eric leave the same point one hour later and walk east at 6 kilometers per hour. How long after John start walking will the two boys be 26 kilometers apart?
30. Two buses leave a terminal at approximately the same time and drive in opposite directions. How far apart are the buses after $3\frac{1}{2}$ hours if their average speeds are 48 kilometers per hour and 60 kilometers per hour?

6 Systems of Equations

An equation that can be written in the form $ax + by = c$, where a, b and c are real numbers and $a, b \neq 0$, is called a linear equation in two variables. Examples include

$$-2x + 3y = 5, \quad y = 3x - 2 \quad \text{and} \quad 2y - 3x + 5 = 0$$

Systems of Linear Equations in Two Variables

A system of equations in two variables is a set of equations in the same two variables. To solve a linear equation in two variables, you need to solve two equations. A solution to a system of equations in two variables x and y is any ordered pair (x, y) that makes each equation in the system true.

Consider the two equations below;

$$2x + y = 5$$

$$2x - y = -1$$

If we substitute 1 for x and 3 for y in both equations we get

$$2(1) + 3 = 5$$

$$2(1) - 3 = -1$$

Since (1, 3) satisfies both equations, it is the solution of this system of equations.

You can solve systems of equations by:

1. Graphing

2. Method of elimination

3. Method of substitution

Sections 6.1 and 6.2 discuss the methods of elimination and substitution.

6.1 Solving System of Linear Equations by Method of Elimination

To solve a system of equations by method of elimination, we eliminate one of the variables by either adding or subtracting the two equations.

Example 1

Solve the system of equations

$$3x + 2y = 12$$

$$x + 2y = 8$$

Solution

$$3x + 2y = 12$$

$$x + 2y = 8$$

The y terms in both equations have the same coefficient. Thus, subtracting the second equation from first equation eliminate y giving

$$2x = 4$$

The equation $2x = 4$ has only one variable, for which you can solve.

$$x = 2 \qquad \text{Divide each side by 2}$$

To find the y - value, we substitute 2 for x in either of the original equations. We choose the second equation.

$$x + 2y = 8 \qquad \text{Second equation}$$

$$2 + 2y = 8 \qquad \text{Substitute 2 for } x$$

$$2y = 6 \qquad \text{Subtract 2 from each side}$$

$$y = 3 \qquad \text{Divide each side by 2}$$

The solution is (2, 3).

Example 2

Solve the system of equations

$$2x + 3y = 12$$

$$-2x + y = -4$$

Solution

Here the coefficients of each of the x-terms are opposite. So, you can eliminate the x- term by adding the terms of the two equations. The working can be presented briefly as shown below.

First write the two equations and number them for easy reference.

$$2x + 3y = 12 \qquad (1)$$

$$-2x + y = -4 \qquad (2)$$

$$(1) + (2) \qquad 4y = 8$$

$$y = 2$$

By substituting $y = 2$ into equation (1) we get

$$2x + 3(2) = 12$$

$$2x = 6$$

$$x = 3$$

The solution of the equation is (3, 2).

With practice you will noticed that a term with equal coefficient in both equations can be eliminated by:

1. Subtracting the equations if they have like signs

2. Adding the equations if they have unlike signs

Example 3

Solve the system of equations

$$3x + 2y = 13$$

$$2x - 5y = -4$$

Solution

Adding or subtracting the equations in this form will not eliminate the x-term or the y-term. We need to multiply each term of an equation by a value that will make the coefficients of either the x-terms or the y-terms the same or opposite of each other.

$$3x + 2y = 13 \quad (1)$$

$$2x - 5y = -4 \quad (2)$$

$$(1) \times 5 \qquad 15x + 10y = 65 \quad (3)$$

$$(2) \times 2 \qquad 4x - 10y = -8 \quad (4)$$

$$(3) + (4) \qquad 19x = 57$$

$$x = 3$$

By substituting $x = 3$ into equation (1) we get

$$3(3) + 2y = 13$$

$$2y = 4$$

$$y = 2$$

The solution is (3, 2).

Exercise 6.1

Solve each of the following system of equations:

1. $x + 2y = 8$	2. $5a + 2b = 3$	3. $2x + 3y = 12$
$x + y = 5$	$4a + 2b = 2$	$2x - y = 4$
4. $3n + 4m = 6$	5. $2x - 5y = 17$	6. $4n - 3m = 10$
$-2n + 4m = 16$	$x - 5y = 16$	$2n - 3m = 8$
7. $3a + 2b = 5$	8. $6x - 4y = 2$	9. $3x + 2y = 4$
$3a - 2b = 7$	$3x + 4y = 7$	$2x - y = 5$
10. $5x + 2y = 6$	11. $3a - 2b = 12$	12. $-6a + 7b = -9$
$7x - 3y = 20$	$4a - 3b = 18$	$2a - 5b = 11$
13. $3x - 4y = 1$	14. $5x - 2y = 7$	15. $-5x + 4y = 5$
$4x + 2y = 19$	$2x - 3y = -6$	$-3x + 2y = 2$

6.2 Solving System of Linear Equations by Method of Substitution

Another method for solving systems of equations is known as the method of substitution. This method is useful if one variable in an equation in the system is expressed in terms of the other.

Example 1

Solve the system of equations

$$3x + 2y = 12$$

$$y = 2x - 1$$

Solution

In this case, the second equation has a variable alone on one side. We can substitute $2x - 1$ for y in the first equation.

$$3x + 2(2x - 1) = 12$$

This is an equation in x for which you can solve.

$3x + 4x - 2 = 12$ Use distributive property

$7x = 14$ Isolate and combine like terms

$$x = 2 \qquad \text{Divide each side by 7}$$

To find y we substitute $x = 2$ into either of the original equations. Because the second equation is already solved for x, it is easier to substitute into this equation.

$$y = 2(2) - 1$$

$$y = 3$$

The solution is (2, 3).

Sometimes you can solve one of the equation for a specific value before you use the method of substitution, as shown in Example 2. Consider carefully the equations of a system to determine which equation to rewrite. Try to avoid fractions if it is possible to do so.

Example 2

Solve the system of equations

$$5x + 3y = 4$$

$$x + 2y = 5$$

Solution

Since the coefficient of x is 1 in the second equation, it is easier to solve that equation for x.

$$x = 5 - 2y$$

Now, substitute $5 - 2y$ for x in the first equation and then solve.

$$5(5 - 2y) + 3y = 4$$

$$25 - 10y + 3y = 4 \qquad \text{Use distributive property}$$

$$-7y = -21 \qquad \text{Isolate and combine like terms}$$

$$y = 3 \qquad \text{Divide each side by } -7$$

Substitute 3 for y in the second equation of the original equations.

$$x + 2(3) = 5$$

$$x = -1$$

The solution is $(-1, 3)$.

Exercise 6.2

Solve the simultaneous equations by substitution

1. $2x + 3y = 4$

$y = 3 + x$

2. $5a - 2b = 7$

$b = 13 - 3a$

3. $2s - 5t = 12$

$s = 7 + 3t$

4. $2m - 3n = 5$

$n = 5 + 4m$

5. $3a + 2b = 5$

$a + b = 2$

6. $a - 4b = -5$

$2a + 5b = 16$

7. $3x + 2y = 6$

$5x + y = 17$

8. $2p - 3q = 9$

$4p + q = 11$

9. $5x + 3y = 1$

$3x = 13 + 3y$

10. $7x - 5y = -1$

$y = 7 - 2x$

11. $3x + 2y = 7$

$x + 2y = 5$

12. $-3x + 2y = 5$

$x + 2y = 9$

6.3 Using System of Equations to Solve Problems

Often it is easier to solve a real-life problem using a system of equations. We can apply the technique discussed in Section 4.3 to translate a problem to a system of equations.

Example 1

Mary bought 3 pens and 5 pencils and paid a total of \$ 3. John bought 5 pens and 8 pencils and paid \$ 4.90. Find the cost of a single pen and a single pencil.

Solution

Let x = the cost of a pen in dollars and y = the cost of a pencil in dollars.

The total cost of 3 pens is \$$3x$ and the total cost of 5 pencil is \$$5y$. Since, the total cost of 3 pens and 5 pencils is \$ 3, we have

$$3x + 5y = 3 \qquad (1)$$

The second equation is obtained in a similar manner.

$$5x + 8y = 4.90 \qquad (2)$$

Now, we solve the system of equations using the method of elimination.

$$(1) \times 5 \qquad 15x + 25y = 15 \qquad (3)$$

$$(2) \times 3 \qquad 15x + 24y = 14.70 \qquad (4)$$

Subtracting equation (4) from equation (3), we get

$$y = 0.30$$

Substituting 0.30 for y in equation (1) gives

$$3x + 5(0.30) = 3$$

$$3x + 1.5 = 3$$

$$x = 0.50$$

So a pen cost 50 ¢ and a pencil cost 30 ¢.

Example 2

400 tickets for a musical concert were sold out. Adult tickets were \$ 25 each, while children tickets were \$ 12 each. If the total ticket sales were \$ 6750, how many tickets of each type were sold?

Solution

Let x = the number of adult ticket sold and y = the number of children ticket sold. Then

$$x + y = 400 \qquad (1)$$

$$25x + 12y = 6750 \qquad (2)$$

Solve the simultaneous equations by using the method of elimination.

$$(1) \times 25 \qquad 25x + 25y = 10{,}000 \qquad (3)$$

Subtracting equation (3) from equation (2), we get

$$-13y = -3250$$

$$y = 250$$

Substituting 250 for y in equation (1), gives

$$x + 250 = 400$$

$$x = 150$$

So, 150 adult tickets were sold and 250 children tickets were sold.

Example 3

Two hundred milliliters of 44% acid solution is obtained by mixing a 20% solution with a 60% solution. How many milliliters of each solution must be used to obtain the desired mixture?

Solution

Let x = the volume of 20% acid solution and y = the volume of 60% acid solution

Since, the total volume of the solution is 200 milliliters, we have

$$x + y = 200 \qquad (1)$$

The amount of acid in the solution is the percentage of acid × the volume of the solution.

So, the amount of acid in the 20% acid solution is $0.20x$ and that of the 60% acid solution is $0.60y$. The amount of acid in the 200- milliliter solution is 0.44(200) i.e. 88 milliliters. Since the amount of acid in the 200-milliliter solution comes from the amount of acid in the 20% solution and the 60% solution, we have

$$0.20x + 0.60y = 88 \qquad (2)$$

Solve the simultaneous equations.

You can multiply the terms of equation (1) by 20 and equation (2) by 100

$$20x + 20y = 4000 \qquad (3)$$

$$20x + 60y = 8800 \qquad (4)$$

Subtracting the terms of equation (4) from equation (3), we have

$$-40y = -4800$$

$$y = 120$$

By substituting $y = 120$ into equation (1), we get

$$x + 120 = 200$$

$$x = 80$$

The amounts to be mixed are 80 milliliters of the 20% acid solution and 120 milliliters of the 60% acid solution.

Exercise 6.3

1. The sum of two numbers is 25.Their difference is 5. Find the two numbers

2. The sum of two numbers is 17. Their difference is 1. Find the two numbers.

3. The difference of two numbers is – 1. When twice the second number is subtracted from three times the first number the result is 1. Find the two numbers.

4. Twice a number is five more than thrice another number. The sum of the numbers is 10. Find the numbers.

5. Two pencils and three erasers cost \$ 3.50. Five pencils and four erasers cost \$ 7.00. How much does each pencil and each eraser cost?

6. Eight compact disks and two pen drives cost \$ 2.80. Three compact disks and four pen drives cost \$ 2.35. Find the cost for one of each?

7. A 30 -metre rope is cut into two pieces so that one piece is 6 meters longer than the other. How long is each piece?

8. The perimeter of a certain rectangle is 26 meters. Also the difference between the length of the rectangle and its width is 5 meters. Find the length and width of the rectangle.

9. The length of a rectangle is 3 centimeters more than twice its width. If the perimeter of the rectangle is 36 centimeters, find the dimensions of the rectangle.

10. The sum of the masses of two parcels is 30 kilograms. Thrice the mass of the heavier parcel is 6 kilograms more than four times the mass of the lighter parcel. Find the masses of the parcels.

11. The sum of the ages of a father and his son is 48 years. The father's age is three times that of the son. Find their ages.

12. Three years ago Eric was four times as old as his son, but in one year time he will be only three times as old as his son. What are their ages now?

13. Johnson has 22 coins with a total value of $ 2.45. If the coins are all 5 ¢ and 20 ¢, how many of each type of coin does he have?

14. Tickets sales for a school play total $3100. The prices of the tickets for student and adult are $7 and $9 respectively. If 400 people bought tickets, how many of each ticket were sold?

15. A concert ticket is $ 10 for adults and $ 5 for children. If the total receipts from 700 tickets were $ 5000, how many adults and how many children attended the concert?

16. A labourer charges a certain fixed amount plus a certain amount per hour for doing a job. If he was paid $ 160 for 3 hours job and $ 200 for 5 hours job, find the fixed charge and the charge per hour.

17. A chemist has a 25% and a 50% acid solution. How much of each solution should be used to form 200 milliliters of a 35% acid solution?

18. You have two alcohol solutions, one a 15% solution and one a 45% solution. How much of each solution should be used to obtain 300 milliliters of a 25% solution?

19. Kojo has a total of $12,000 invested in two accounts. One account pays 8 percent simple interest and the other 9 percent. If the interest for 1 year is $ 1010, how much is invested in each account?

20. Esi invested a part of $ 8000 in bonds paying 12 percent simple interest. She invested the remainder in a saving account paying 8 percent interest. If the interest for 1 year is $ 840, how much did she invest in bonds and in the saving account?

6.4 Systems of Equations with One Linear Equation

The method of substitution already explained in Section 6. 2 can be used to solve systems of equations in two variables with one linear equation.

Example

Solve the simultaneous equations:

(a) $3x + 2y = 12$ (b) $x^2 + 3xy + y^2 = -5$

$xy = 6$ $2x + y = 1$

Solution

(a) $$3x + 2y = 12 \quad (1)$$

$$xy = 6 \quad (2)$$

Begin by solving for y in Equation (1) to obtain $y = \frac{12-3x}{2}$

Next substitute $y = \frac{12-3x}{2}$ into (2)

$$x\left(\frac{12-3x}{2}\right) = 6$$

$$12x - 3x^2 = 12$$

Simplify the equation and write it in the standard form.

$$x^2 - 4x + 4 = 0$$

Factoring the quadratic expression we get

$$(x - 2)^2 = 0$$

$$x = 2$$

Now substitute $x = 2$ into Equation (2) to solve for y.

$$2y = 6$$

$$y = 3$$

The solution of the system of equation is (2, 3).

(b) $$x^2 + 3xy + y^2 = -5 \quad (1)$$

$$2x + y = 1 \quad (2)$$

Begin by solving for y in Equation (2) to obtain $y = 1 - 2x$.

Next substitute $y = 1 - 2x$ into (1).

$$x^2 + 3x(1 - 2x) + (1 - 2x)^2 = -5$$

Simplify the equation and write it in the standard form.

$$x^2 + x - 6 = 0$$

Factoring the quadratic expression we get

$$(x - 2)(x + 3) = 0$$

$$x - 2 = 0 \quad \text{or} \quad x + 3 = 0$$

$$x = 2 \qquad\qquad x = -3$$

Substitute these values of x into (2).

For $x = 2$

$$2(2) + y = 1$$

$$4 + y = 1$$

$$y = -3$$

For $x = -3$

$$2(-3) + y = 1$$

$$-6 + y = 1$$

$$y = 7$$

Check each pair of values in (1)

For $x = 2$, $y = -3$

$$2^2 + 3(2)(-3) + (-3)^2 = 4 - 18 + 9 = -5$$

For $x = -3$, $y = 7$

$$(-3)^2 + 3(-3)(7) + 7^2 = 9 - 63 + 49 = -5$$

So, the system of equation has two solutions; $(2, -3)$ and $(-3, 7)$.

Exercise 6.4

Solve the following systems of equations.

1. $3x + 2y = 18$
   $xy = 12$

2. $2x - 3y = 4$
   $xy = 10$

3. $5x - 3y = -16$
   $xy = -4$

4. $4x - y = 7$
   $xy = 15$

5. $xy = 30$
   $3x + y = 21$

6. $xy = -12$
   $3x + 2y = 6$

7. $x^2 + 2xy + y^2 = 9$
   $2x + y = 5$

8. $2x^2 + y^2 = 19$
   $3x + y = 10$

9. $2x - y = 2$
   $y^2 - 2x^2 - 14 = 0$

10. $3y - x = -1$
    $x^2 - 2xy - y^2 = 7$

11. $2x - y = 1$
    $3x^2 - xy + y^2 = 15$

12. $x^2 + 2xy - 4y^2 = -4$
    $3x - 2y = 8$

13. $2x + y = 3$
    $2x^2 + 3xy + 2y^2 = 9$

14. $3x - 2y = 1$
    $9x^2 - 4y^2 = 17$

15. $x^2 + 2xy = 8$
    $x + 2y = 2$

16. $x^2 + y^2 + 2xy + 6x = -2$
    $2x - y = -5$

17. $4x + 3y = -3$
    $2x^2 + 2xy + y^2 = 13$

18. $x + 2y = 4$
    $x^2 + xy + y^2 = 7$

6.5 Systems of Linear Equations in Three Variables

The method of elimination can be applied to solve system of linear equations in more than two variables.

Example

Solve the system of linear equation.

$$2x + y - z = 1 \quad (1)$$

$$3x - y + 2z = 10 \quad (2)$$

$$4x - 2y - 3z = 9 \quad (3)$$

Solution

First, take two equations and eliminate one of the variables. Adding equations (1) and (2) we get

$$5x + z = 11 \quad (4)$$

Next, take another pair of equations and eliminate y. Multiplying equation (1) by 2 and then adding the result to equation (3) yields

$$8x - 5z = 11 \quad (5)$$

Now we solve equations (4) and (5) for values of x and z in the usual way.

$$(4) \times 5 \qquad 25x + 5z = 55 \quad (6)$$

$$(5) + (6) \qquad 33x = 66$$

$$x = 2$$

Substitute $x = 2$ in equation (4) to obtain

$$5(2) + z = 11$$

$$z = 1$$

Finally, substitute $x = 2$ and $z = 1$ in equation (1) to obtain

$$2(2) + y - 1 = 1$$

$$y = -2$$

So, the solution is $(2, -2, 1)$.

Exercise 6.5

Solve each of the following simultaneous equations:

1. $x + 2y - z = 2$
   $2x - 3y + z = -1$
   $4x + y + 2z = 12$

2. $x + 2y + 3z = -4$
   $x - y - 3z = 8$
   $2x + y + 6z = -14$

3. $4x + 3y - z = 12$
   $3x - y = 5$
   $x + 2y + 2z = 2$

4. $x + y + z = 6$
   $2x + 3y + z = 11$
   $3x + 2y + 2z = 13$

5. $2x + 3y - 5z = 0$
   $3x + 2y - 4z = -2$
   $4x + y - z = 2$

6. $x + 2y + 3z = -1$
   $4x - 3y + 2z = 2$
   $3x - 8y - 5z = 11$

Review Exercises

In Exercises 1 – 6, solve the simultaneous equations using the elimination method

1. $2x + y = 3$
   $2x + 3y = 5$

2. $5x + 3y = -4$
   $-5x + 2y = 14$

3. $4x - 3y = 5$
   $2x - 3y = 1$

4. $3x + 5y = 6$
   $-2x + 3y = 15$

5. $2x + 3y = 8$
   $3x + 4y = 13$

6. $2x - y = 9$
   $3x + 4y = -14$

In Exercises 7 - 12, solve the simultaneous equations using the substitution method

7. $x + y = 5$
   $y = x - 1$

8. $x + y = 9$
   $x = y + 3$

9. $3x + 4y = 6$
   $y = 3x + 9$

10. $x = 3y + 11$ | 11. $5x - 4y = 5$ | 12. $5x - 6y = 21$

$2x - 5y = 19$ | $y = 4x + 7$ | $x = 5 + 2y$

13. The sum of two numbers is 50. The second number is 2 more than three times the first. What are the two numbers?

14. The larger of two supplementary angles is 75° more than twice the size of the smaller angle. Find size of each angle.

15. Kojo has 22 coins with a total value of $1.70. If the coins are all 10 ¢ and 5 ¢, how many of each type of coin does he have?

16. Tickets for a concert sell for $5 and $8. One night, 500 people bought tickets. The receipts from tickets sales were $3100. How many of each tickets were sold?

17. The length of a rectangle is 2 centimeters less than three times its width. If the perimeter of the rectangle is 36 centimeters, find the dimensions of the rectangle.

18. The perimeter of an isosceles triangle is 19 centimeters. The lengths of the two equal legs are 3 centimeters less than twice the length of the base. Find the lengths of the three sides.

19. Ama invested a part of $800 in bonds paying 12 percent simple interest. She invested the remainder in a saving account paying 8 percent interest. If the interest for 1 year is $84, how much did she invest in bonds and in the saving account?

20. A plane flies 540 kilometers with the wind in 3 hours. Flying back against the wind, the plane takes 9 hours to make the trip. What was the speed of the plane in still air? What was the speed of the plane in the wind?

In Exercises 21 – 26, solve each of the following system of equations:

21. $x^2 + y^2 = 10$ | 22. $x^2 + y^2 - 4 = 0$ | 23. $2x^2 + y^2 = 18$

$y - x = 4$ | $x + y = 2$ | $y = x - 3$

24. $2x^2 + xy + y^2 = 2$ | 25. $x^2 - 2xy + y^2 = 36$ | 26. $2y = x + 3$

$x + y = 1$ | $x + y = 2$ | $x^2 + y^2 - 2x + 6y = 15$

In Exercises 27 – 30, solve each of the following system of equations:

27. $x - y + 2z = 7$ | 28. $x - 2y + 3z = 7$ | 29. $x + y - z = 6$ | 30. $x + 2y - z = -1$

$2x + y - z = 3$ | $2x + y + z = 4$ | $3x - 2y + z = -5$ | $2x - 4y + z = -9$

$x + y + z = 9$ | $-3x + 2y - 2z = -10$ | $x + 3y - 2z = 14$ | $-2x + 2y - 3z = 10$

7 Radicals

The positive or principal square root of a number a, denoted by $\sqrt{a}$ is a positive number whose square is a. For example, 3 is the square root of 9, because $3^2 = 9$. The symbol $\sqrt{\ }$ is called the radical sign, and is used to indicate the principal root. The expression under the radical is called radicand. An expression that involves a radical sign is called a radical expression.

Some square roots are rational numbers. For example, $\sqrt{4}$, $\sqrt{\frac{16}{25}}$ and $\sqrt{144}$ represent the rational numbers 2, $\frac{4}{5}$ and 12 respectively. A real number, like 4 or 16, whose square root is a rational number is called a perfect square. Most real numbers, such as 2 and 3, are not perfect squares and do not have rational square roots. The square root of any whole number which is not a perfect square is irrational. For example, $\sqrt{2}$, $\sqrt{3}$ and $\sqrt{5}$ are irrational numbers.

Decimal representation of an irrational number would be non-terminating and non-repeating. The decimal approximation of an irrational number found from a calculator can be given to as many decimal places as one required.

7.1 Simplifying Radicals

The radical expression $\sqrt{a}$, where a is a real number, is said to be in the simplest form if a has no perfect square factor other than 1. For example, $\sqrt{10}$ is in its simplified form, since 10 has no perfect square factor. However, $\sqrt{8}$ is not simplified because one of its factors, 4, is a perfect square.

To simplify radical expressions you may need to apply one or more of the following properties:

1. $\sqrt{a^2} = a$

2. $\sqrt{a} \times \sqrt{a} = \left(\sqrt{a}\right)^2 = a$

Product Rule

3. $\sqrt{ab} = \sqrt{a} \times \sqrt{b}$

Quotient Rule

4. $\sqrt{\frac{a}{b}} = \frac{\sqrt{a}}{\sqrt{b}}$

Example

Simplify:

$(a)\ \sqrt{18}$ $(b)\ \sqrt{108}$

Solution

(a) First factor the perfect square. Then use the product rule in reverse.

$$\begin{aligned}\sqrt{18} &= \sqrt{9 \times 2} \\ &= \sqrt{9} \times \sqrt{2} \\ &= 3\sqrt{2}\end{aligned}$$

Alternatively, we could write a prime factorization of 18, i.e. $3 \cdot 3 \cdot 2$, and pair each group of like factors. Note that each pair of like factors is a perfect square.

$$\begin{aligned}\sqrt{18} &= \sqrt{3^2 \times 2} \\ &= \sqrt{3^2} \times \sqrt{2} \\ &= 3 \times \sqrt{2} \\ &= 3\sqrt{2}\end{aligned}$$

(b)

$$\begin{aligned}\sqrt{108} &= \sqrt{36 \times 3} \\ &= 6\sqrt{3}\end{aligned}$$

Exercise 7.1

Simplify each of these radicals:

1. $\sqrt{20}$ 2. $\sqrt{28}$ 3. $\sqrt{27}$ 4. $\sqrt{40}$

5. $\sqrt{45}$ 6. $\sqrt{48}$ 7. $\sqrt{50}$ 8. $\sqrt{54}$

9. $\sqrt{60}$ 10. $\sqrt{72}$ 11. $\sqrt{75}$ 12. $\sqrt{80}$

13. $\sqrt{96}$ 14. $\sqrt{98}$ 15. $\sqrt{147}$ 16. $\sqrt{150}$

17. $\sqrt{162}$ 18. $\sqrt{180}$ 19. $\sqrt{192}$ 20. $\sqrt{245}$

21. $\sqrt{288}$ 22. $\sqrt{320}$ 23. $\sqrt{405}$ 24. $\sqrt{588}$

25. $\sqrt{605}$ 26. $\sqrt{675}$ 27. $\sqrt{726}$ 28. $\sqrt{540}$

29. $\sqrt{847}$ 30. $\sqrt{720}$ 31. $\sqrt{338}$ 32. $\sqrt{567}$

7.2 Adding and Subtracting Radicals

You can add and subtract radicals in much the same way that you will add and subtract polynomials.

Example 1

Evaluate:

$(a)\ 5\sqrt{3}+2\sqrt{3}$ $(b)\ 7\sqrt{2}-4\sqrt{2}$

Solution

(a)
$$5\sqrt{3}+2\sqrt{3}=(5+2)\sqrt{3}$$
$$=7\sqrt{3}$$

(b)
$$7\sqrt{2}-4\sqrt{2}=(7-4)\sqrt{2}$$
$$=3\sqrt{2}$$

The example shows that we can add or subtract radicals by multiplying the sum or difference of their coefficients by the radical expression. In practice, you can perform the first step mentally.

Example 2

Evaluate

$(a)\ \sqrt{18}+\sqrt{50}$ $(b)\ \sqrt{27}-\sqrt{12}$

Solution

(a) Begin by writing $\sqrt{18}$ and $\sqrt{50}$ in the simplest form.

$$\sqrt{18}+\sqrt{50}=3\sqrt{2}+5\sqrt{2}$$
$$=8\sqrt{2}$$

(b)
$$\sqrt{27}-\sqrt{12}=3\sqrt{3}-2\sqrt{3}$$
$$=\sqrt{3}$$

Exercise 7.2

Evaluate each of the following:

1. $5\sqrt{3} + 2\sqrt{3}$
2. $3\sqrt{2} + 5\sqrt{2}$
3. $10\sqrt{5} - 3\sqrt{5}$
4. $6\sqrt{7} - 4\sqrt{7}$
5. $2\sqrt{3} - 5\sqrt{3}$
6. $2\sqrt{5} - 4\sqrt{5}$
7. $2\sqrt{2} + \sqrt{2} + 3\sqrt{2}$
8. $3\sqrt{3} + 2\sqrt{3} + \sqrt{3}$
9. $5\sqrt{5} - 2\sqrt{5} + \sqrt{5}$
10. $2\sqrt{7} + 3\sqrt{7} - 8\sqrt{7}$
11. $2\sqrt{3} + \sqrt{12}$
12. $5\sqrt{2} + \sqrt{18}$
13. $3\sqrt{12} - \sqrt{48}$
14. $2\sqrt{45} - 2\sqrt{20}$
15. $5\sqrt{8} + 2\sqrt{18}$
16. $\sqrt{50} + \sqrt{32} - \sqrt{8}$
17. $\sqrt{20} + 2\sqrt{5} - \sqrt{45}$
18. $5\sqrt{8} + 3\sqrt{18} - 4\sqrt{32}$
19. $9\sqrt{2} - 20\sqrt{2} + 11\sqrt{2}$
20. $\sqrt{27} - 6\sqrt{5} + \sqrt{48}$
21. $\sqrt{63} - 2\sqrt{28} + 5\sqrt{7}$
22. $2\sqrt{50} + 3\sqrt{18} - \sqrt{32}$
23. $\sqrt{96} - \sqrt{150} + \sqrt{6}$
24. $4\sqrt{8} + 2\sqrt{50} - 3\sqrt{32}$
25. $\frac{3}{5}\sqrt{125} - \sqrt{20}$
26. $\sqrt{243} - \frac{2}{3}\sqrt{27}$
27. $\frac{2}{3}\sqrt{18} - \frac{1}{2}\sqrt{72}$
28. $\frac{3}{4}\sqrt{128} - \frac{3}{5}\sqrt{50}$
29. $3\sqrt{125} - 5\sqrt{20} - 3\sqrt{45}$
30. $\sqrt{54} - 2\sqrt{24} + \sqrt{216}$
31. $3\sqrt{12} - 5\sqrt{147} + 4\sqrt{75}$
32. $\frac{2}{3}\sqrt{18} - \frac{3}{4}\sqrt{32} - \sqrt{50}$
33. $9\sqrt{8} + \sqrt{72} - 8\sqrt{8}$
34. $7\sqrt{12} - 2\sqrt{27} - \sqrt{75}$
35. $4\sqrt{20} + \sqrt{125} - 2\sqrt{75}$
36. $5\sqrt{18} - 2\sqrt{32} - \sqrt{50}$

7.3 Multiplying Radicals

To multiply two or more radicals use the product rule in reverse. That is

$\sqrt{a} \cdot \sqrt{b} = \sqrt{a \cdot b}.$

Example 1

Evaluate:

(a) $\sqrt{5} \times \sqrt{10}$ (b) $3\sqrt{5} \times 4\sqrt{3}$ (c) $\sqrt{8} \times \sqrt{24}$

Solution

(a) Multiply the radicals and then simplify

$$\sqrt{5} \times \sqrt{10} = \sqrt{50}$$

$$= 5\sqrt{2}$$

(b)

$$3\sqrt{5} \times 4\sqrt{2} = 3 \times \sqrt{5} \times 4 \times \sqrt{2}$$

$$= 3 \times 4 \times \sqrt{5} \times \sqrt{2}$$

$$= 12 \times \sqrt{5 \times 2}$$

$$= 12\sqrt{10}$$

Note that, to multiply radical expressions multiply the numbers outside the radical sign together, and multiply the radicand. Simplify the result if possible.

(c) Begin by writing the radical in the simplest form

$$\sqrt{8} \times \sqrt{24} = 2\sqrt{2} \times 2\sqrt{6}$$

$$= 4\sqrt{12}$$

$$= 4 \times 2\sqrt{3}$$

$$= 8\sqrt{3}$$

Multiplying radical expressions containing parentheses

You can multiply radical expressions inside parentheses by using the Distributive Property and the product rule for radicals.

Example 2

Evaluate:

(a) $3\sqrt{2}(2\sqrt{2} + \sqrt{3})$ (b) $(3\sqrt{2} - \sqrt{3})(2\sqrt{3} - 3\sqrt{2})$

Solution

(a)

$$3\sqrt{2}(2\sqrt{2} + \sqrt{3}) = 3\sqrt{2} \cdot 2\sqrt{2} + 3\sqrt{2} \cdot \sqrt{3}$$

$$= 6 \cdot 2 + 3 \cdot \sqrt{6}$$

$$= 12 + 3\sqrt{6}$$

(b) $(3\sqrt{2} - \sqrt{3})(2\sqrt{3} - 3\sqrt{2}) = 3\sqrt{2} \cdot 2\sqrt{3} - 3\sqrt{2} \cdot 3\sqrt{2} - \sqrt{3} \cdot 2\sqrt{3} + \sqrt{3} \cdot 3\sqrt{2}$

$$= 6\sqrt{6} - 18 - 6 + 3\sqrt{6}$$

$$= 9\sqrt{6} - 24$$

Example 3

Evaluate: (a) $(2\sqrt{3} + 3\sqrt{2})(2\sqrt{3} - 3\sqrt{2})$ (b) $(3 + 2\sqrt{3})^2$

Solution

(a) $(2\sqrt{3} + 3\sqrt{2})(2\sqrt{3} - 3\sqrt{2}) = (2\sqrt{3})^2 - (3\sqrt{2})^2$

$$= 4 \cdot 3 - 9 \cdot 2$$

$$= 12 - 18$$

$$= -6$$

(b) $(3 + 2\sqrt{3})^2 = 3^2 + 2 \cdot 3 \cdot 2\sqrt{3} + (2\sqrt{3})^2$

$$= 9 + 12\sqrt{3} + 12$$

$$= 21 + 12\sqrt{3}$$

Exercise 7.3

Evaluate each of the following radical expressions:

1. $\sqrt{2} \times \sqrt{6}$
2. $\sqrt{5} \times \sqrt{15}$
3. $3\sqrt{2} \times \sqrt{3}$
4. $5\sqrt{3} \times 2\sqrt{5}$
5. $3\sqrt{2} \times 2\sqrt{5}$
6. $5\sqrt{3} \times 3\sqrt{2}$
7. $2\sqrt{5} \times 3\sqrt{3}$
8. $\sqrt{3} \times \sqrt{12}$
9. $\sqrt{8} \times \sqrt{10}$
10. $\sqrt{5} \times \sqrt{32}$
11. $\sqrt{15} \times \sqrt{27}$
12. $\sqrt{12} \times \sqrt{18}$
13. $\sqrt{50} \times \sqrt{30}$
14. $\sqrt{98} \times \sqrt{75}$
15. $\sqrt{60} \times \sqrt{8}$
16. $\sqrt{3} \times \sqrt{2} \times \sqrt{10}$

17. $\sqrt{6} \times 2\sqrt{2} \times \sqrt{15}$

18. $\sqrt{20} \times \sqrt{15} \times \sqrt{6}$

19. $\sqrt{30} \times \sqrt{2} \times \sqrt{27}$

20. $\sqrt{72} \times \sqrt{75} \times \sqrt{18}$

21. $\sqrt{3}(\sqrt{5} - \sqrt{3})$

22. $\sqrt{3}(2\sqrt{5} - 3\sqrt{3})$

23. $\sqrt{3}(\sqrt{3} + \sqrt{5})$

24. $\sqrt{7}(2\sqrt{3} + 3\sqrt{7})$

25. $\sqrt{6}(\sqrt{6} + \sqrt{5})$

26. $\sqrt{2}(\sqrt{6} - \sqrt{2})$

27. $5\sqrt{2}(3\sqrt{2} - 2\sqrt{3})$

28. $2\sqrt{12}(\sqrt{3} + \sqrt{2})$

29. $(\sqrt{3} + 5)(\sqrt{3} + 2)$

30. $(\sqrt{5} - 2)(\sqrt{5} - 1)$

31. $(3\sqrt{2} - \sqrt{6})(2\sqrt{3} + \sqrt{2})$

32. $(2\sqrt{2} + \sqrt{3})(\sqrt{12} - \sqrt{8})$

33. $(1 - \sqrt{3})(2 + 3\sqrt{2})$

34. $(\sqrt{6} - \sqrt{5})(\sqrt{8} + \sqrt{15})$

35. $(\sqrt{5} + \sqrt{15})(2\sqrt{3} - 1)$

36. $(2\sqrt{2} + \sqrt{5})(\sqrt{10} - 2\sqrt{2})$

37. $(\sqrt{5} - 2)(\sqrt{5} + 2)$

38. $(\sqrt{7} + 5)(\sqrt{7} - 5)$

39. $(\sqrt{10} + 5)(\sqrt{10} - 5)$

40. $(\sqrt{8} - 2)(\sqrt{8} + 2)$

41. $(\sqrt{12} - 3)(\sqrt{12} + 3)$

42. $(2\sqrt{3} + 4\sqrt{2})(2\sqrt{3} - 4\sqrt{2})$

43. $(\sqrt{3} + 2)^2$

44. $(\sqrt{5} - 3)^2$

45. $(\sqrt{3} + \sqrt{2})^2$

46. $(\sqrt{3} - 2\sqrt{2})^2$

47. $(2 + 3\sqrt{2})^2$

48. $(3\sqrt{2} - 2\sqrt{3})^2$

49. $(\sqrt{6} - 2\sqrt{3})^2$

50. $(\sqrt{3} + 2\sqrt{2})^2$

7.4 Rationalizing the Denominator

A radical expression whose denominator contains a radical can be simplified by rationalizing the denominator. Note that a radical expression is in the simplest form if no denominator contains a radical.

To rationalize the denominator, we multiply both the numerator and denominator of the fraction by a number that will make the denominator a rational number.

Example 1

Rationalize the denominator of the fraction:

(a) $\frac{4}{\sqrt{2}}$ (b) $\frac{3}{2\sqrt{3}}$

Solution

(a) Multiply both numerator and denominator by $\sqrt{2}$.

$$\frac{4}{\sqrt{2}} = \frac{4}{\sqrt{2}} \times \frac{\sqrt{2}}{\sqrt{2}}$$

$$\frac{4}{\sqrt{2}} = \frac{4\sqrt{2}}{2}$$

$$= 2\sqrt{2}$$

Note that the fraction does not change because we have multiplied by 1

(b) $$\frac{3}{2\sqrt{3}} = \frac{3}{2\sqrt{3}} \times \frac{\sqrt{3}}{\sqrt{3}}$$

$$= \frac{3\sqrt{3}}{2\times 3}$$

$$= \frac{1}{2}\sqrt{3}$$

Rationalizing fractions whose denominators are sums or difference of radicals

Using the identity $(a - b)(a + b) = a^2 - b^2$, we can rationalize the denominators of fractions with denominators like $\sqrt{a} + b$ or $\sqrt{a} - \sqrt{b}$. For example, you can rationalize a fraction whose denominator is $\sqrt{3} + 2$ by multiplying both the numerator and denominator by $\sqrt{3} - 2$. The number $\sqrt{3} - 2$ and $\sqrt{3} + 2$ are said to be conjugates of each other. To rationalize the denominator involving such radicals, multiply the numerator and the denominator by the conjugate of the expression in the denominator.

Example 2

Rationalize the denominator of $\frac{1}{2-\sqrt{3}}$

Solution

Multiply both the denominator and numerator by $2 + \sqrt{3}$

$$\frac{1}{2-\sqrt{3}} = \frac{1}{2-\sqrt{3}} \times \frac{2+\sqrt{3}}{2+\sqrt{3}}$$

$$= \frac{2+\sqrt{3}}{(2-\sqrt{3})(2+\sqrt{3})}$$

$$= \frac{2+\sqrt{3}}{4-3}$$

$$= 2 + \sqrt{3}$$

Exercise 7.4

Rationalize the denominator of the following radical expressions.

1. $\frac{2}{\sqrt{3}}$
2. $\frac{10}{\sqrt{5}}$
3. $\frac{4\sqrt{3}}{\sqrt{6}}$
4. $\frac{9}{\sqrt{3}}$
5. $\frac{3\sqrt{5}}{\sqrt{6}}$
6. $\frac{12\sqrt{2}}{\sqrt{6}}$
7. $\frac{15}{\sqrt{10}}$
8. $\frac{18}{\sqrt{12}}$
9. $\frac{14}{\sqrt{7}}$
10. $\frac{3}{4\sqrt{2}}$
11. $\frac{8}{\sqrt{32}}$
12. $\frac{5\sqrt{6}}{\sqrt{15}}$
13. $\frac{3\sqrt{6}}{\sqrt{12}}$
14. $\frac{\sqrt{5}}{\sqrt{3}}$
15. $\sqrt{\frac{3}{2}}$
16. $\frac{\sqrt{12}}{\sqrt{27}}$
17. $\frac{\sqrt{18}}{\sqrt{12}}$
18. $\sqrt{\frac{12}{5}}$
19. $\frac{\sqrt{15}}{2\sqrt{3}}$
20. $\frac{6}{\sqrt{27}}$
21. $\frac{1}{\sqrt{2}+1}$
22. $\frac{2}{\sqrt{3}-1}$
23. $\frac{1}{2-\sqrt{3}}$
24. $\frac{4}{3+\sqrt{7}}$
25. $\frac{\sqrt{5}}{\sqrt{15}-\sqrt{10}}$
26. $\frac{5}{4+\sqrt{6}}$
27. $\frac{3\sqrt{7}}{5-\sqrt{7}}$
28. $\frac{2\sqrt{6}}{2\sqrt{3}-1}$
29. $\frac{2-\sqrt{3}}{2+\sqrt{3}}$
30. $\frac{3+\sqrt{2}}{3-\sqrt{2}}$
31. $\frac{\sqrt{6}+\sqrt{2}}{\sqrt{6}-\sqrt{2}}$
32. $\frac{2\sqrt{3}-\sqrt{5}}{2\sqrt{3}+\sqrt{5}}$
33. $\frac{3-2\sqrt{2}}{3+2\sqrt{2}}$
34. $\frac{\sqrt{3}+\sqrt{5}}{\sqrt{3}-\sqrt{5}}$
35. $\frac{\sqrt{7}-3\sqrt{2}}{\sqrt{7}-\sqrt{2}}$
36. $\frac{\sqrt{6}+2\sqrt{2}}{\sqrt{6}-2\sqrt{2}}$
37. $\frac{\sqrt{5}+3}{4-\sqrt{20}}$
38. $\frac{\sqrt{3}+\sqrt{6}}{3\sqrt{2}-3}$
39. $\frac{2\sqrt{5}+\sqrt{3}}{5\sqrt{3}-3\sqrt{5}}$
40. $\frac{\sqrt{7}+3\sqrt{3}}{2\sqrt{7}+\sqrt{3}}$

7.5 Radical Equations

An equation which contains a radical expression is called an irrational equation. Examples include $\sqrt{3x-2} = 4$ and $\sqrt{4x+3} = \sqrt{2x-3} + 5x$.

Solving Radical Equations

To solve a radical equation with one radical expression, we first isolate the radical expression, and square each side of the equation. Then solve the new equation. If the equation contains more than one radical expression, you may have to repeat the process more than once. Squaring each side of an equation can introduce extraneous solutions, so always check your solutions in the original equation.

Example

Solve:

(a) $\sqrt{2x-3}+x=3$ (b) $\sqrt{x+5}+\sqrt{5-x}=4$

Solution

(a) $$\sqrt{2x-3}+x=3$$

$$\sqrt{2x-3}=3-x$$

$$2x-3=9-6x+x^2$$

$$x^2-8x+12=0$$

$$(x-2)(x-6)=0$$

$$x-2=0 \quad \text{or} \quad x-6=0$$

$$x=2 \qquad\qquad x=6$$

Check these values in the origin equation $\sqrt{2x-3}+x=3$.

If $x=2$, $\sqrt{4-3}+2=1+2=3$

and if $x=6$, $\sqrt{12-3}+6=3+6=9$

Only $x = 2$ satisfies the original equation, so the solution is $x = 2$.

(b) $$\sqrt{x+5}+\sqrt{5-x}=4$$

Rewrite the equation with a radical term on each side.

$$\sqrt{x+5}=4-\sqrt{5-x}$$

$$x+5=16-8\sqrt{5-x}+5-x$$

$$x-8=-4\sqrt{5-x}$$

$$x^2-16x+64=16(5-x)$$

$$x^2=16$$

$$x=\pm 4$$

Check these values in the original equation

$\sqrt{x+5} + \sqrt{5-x} = 4.$

If $x = 4$, $\sqrt{4+5} + \sqrt{5-4} = 3 + 1 = 4$

and $x = -4$, $\sqrt{-4+5} + \sqrt{5+4} = 1 + 3 = 4.$

Both $x = -4$ and $x = 4$ satisfy the original equation.

The solutions are $x = -4$ and $x = 4$.

Exercise 7.5

Solve each of these irrational equations:

1. $\sqrt{4x-3} = 3$
2. $\sqrt{5x+1} = 4$
3. $\sqrt{5-4x} = 5$
4. $x = \sqrt{3x^2 - 18}$
5. $\sqrt{4x^2+3} = 2$
6. $\sqrt{3x-5} = x - 3$
7. $\sqrt{10-3x} + 2 = x$
8. $\sqrt{7+x^2} - x = 1$
9. $\sqrt{20-x^2} - x = 6$
10. $x - \sqrt{5x-9} = 1$
11. $\sqrt{2x+5} + 5 = x$
12. $\sqrt{x+5} - \sqrt{x} = 1$
13. $\sqrt{x-5} + 1 = \sqrt{x}$
14. $\sqrt{x+1} = 2 - \sqrt{x}$
15. $1 + 2\sqrt{x-3} = \sqrt{4x-3}$
16. $\sqrt{x-6} + 3 = \sqrt{x+9}$
17. $\sqrt{3x-2} + 1 = \sqrt{2x+5}$
18. $\sqrt{2x-4} = \sqrt{3x-1} - \sqrt{x-1}$
19. $\sqrt{2x+2} - \sqrt{x-6} = \sqrt{x+2}$
20. $\sqrt{2x+1} - \sqrt{4-2x} = \sqrt{4x-5}$

Review Exercises

In Exercises 1 – 10, simplify each radical:

1. $\sqrt{108}$
2. $\sqrt{90}$
3. $\sqrt{18}$
4. $\sqrt{135}$
5. $\sqrt{507}$
6. $\sqrt{578}$
7. $\sqrt{216}$
8. $\sqrt{243}$
9. $\sqrt{384}$
10. $\sqrt{845}$

In Exercises 11 – 16, evaluate each of the following:

11. $3\sqrt{6} - 8\sqrt{6}$
12. $2\sqrt{2} + 3\sqrt{2} - \sqrt{2}$

13. $\sqrt{5} - 3\sqrt{5} - 4\sqrt{5}$

14. $2\sqrt{2} + 3\sqrt{2} - 7\sqrt{2}$

15. $\sqrt{18} - \sqrt{72} + \sqrt{32}$

16. $2\sqrt{8} - 3\sqrt{50} + 5\sqrt{18}$

In Exercises 17 – 25, evaluate each of the following:

17. $\sqrt{18} \times \sqrt{6}$

18. $\sqrt{8} \times \sqrt{10}$

19. $\sqrt{14} \times \sqrt{8}$

20. $\sqrt{20} \times \sqrt{27}$

21. $\sqrt{75} \times \sqrt{50}$

22. $\sqrt{12} \times \sqrt{45}$

23. $\sqrt{72} \times \sqrt{27}$

24. $\sqrt{12} \times \sqrt{3} \times \sqrt{6}$

25. $2\sqrt{5} \times \sqrt{20} \times \sqrt{45}$

In Exercises 26 – 31, rationalize the denominator:

26. $\frac{1}{5\sqrt{2}}$

27. $\frac{8}{3\sqrt{2}}$

28. $\frac{8}{3\sqrt{18}}$

29. $\frac{9}{4\sqrt{6}}$

30. $\frac{\sqrt{7}}{\sqrt{21}}$

31. $\frac{2+\sqrt{3}}{\sqrt{3}}$

In Exercises 32 – 37, evaluate:

32. $(\sqrt{3} + 2\sqrt{2})(\sqrt{3} + 5\sqrt{2})$

33. $(3\sqrt{2} - \sqrt{7})(3\sqrt{2} + 2\sqrt{7})$

34. $(\sqrt{3} + \sqrt{2})^2$

35. $(5\sqrt{3} - 2)^2$

36. $(2\sqrt{3} + \sqrt{2})(\sqrt{3} - 2\sqrt{2})$

37. $(2\sqrt{5} + 3\sqrt{2})(2\sqrt{5} - 3\sqrt{2})$

In Exercise 38 – 43, rationalize the denominator:

38. $\frac{1}{3\sqrt{2}-\sqrt{5}}$

39. $\frac{3\sqrt{2}}{2\sqrt{2}-3}$

40. $\frac{\sqrt{3}+2}{2\sqrt{3}-1}$

41. $\frac{\sqrt{3}+\sqrt{2}}{\sqrt{3}-\sqrt{2}}$

42. $\frac{2-\sqrt{3}}{\sqrt{6}+\sqrt{2}}$

43. $\frac{3\sqrt{2}-2\sqrt{3}}{3\sqrt{3}-2\sqrt{2}}$

In Exercises 44 – 48, solve each of the following irrational equation:

44. $\sqrt{x+1} + 3 = 5$

45. $x + 2 = \sqrt{8x+1}$

46. $7 - \sqrt{x-1} = 2$

47. $\sqrt{3x+4} - \sqrt{x+1} = 3$

48. $\sqrt{2x+7} + \sqrt{7-3x} = 5$

8 Functions

Functions are important in modern mathematics. Functions have numerous applications, and an understanding of concepts would help you succeed in your calculus course.

8.1 Functions

A function from a set A to set B, written f: A → B, is a rule that assigns to each element of set A exactly one element of set B. The set A is called the domain of the function f and the set B is called the co-domain, and it contains the range of the function.

One way a function can be represented is to a draw a diagram, called an arrow diagram. Examples are shown in Figure 8.1(a) and Figure 8.1(b).

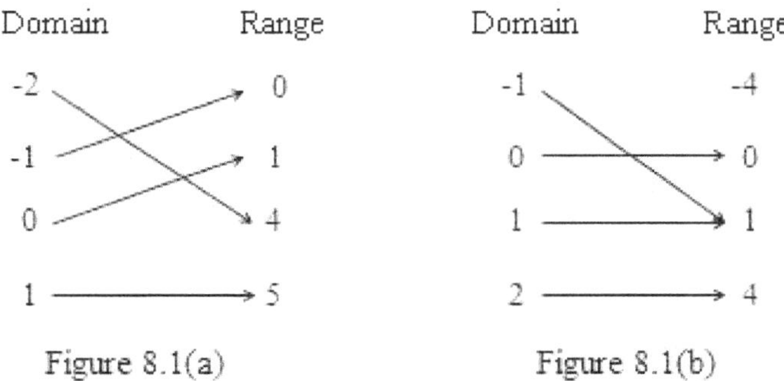

Figure 8.1(a) Figure 8.1(b)

For any element x in the domain the corresponding element y in the co-domain is called the image of x. The set of all images of the elements of the domain is called the range of the function. The range is usually a proper subset of the co-domain.

The arrow diagrams shown in Figure 8.1(a) and Figure 8.1(b) illustrate the following properties of functions.

1. Each element in the domain must be matched with an element in the co-domain.
2. Some element in the co-domain may not be matched with any element in the domain.
3. Two or more elements in the domain may be matched with the same element in the co-domain.
4. An element in the domain cannot be matched with two different elements in the co-domain.

A function can be represented by the following.

1. A table or a list of ordered pairs.
2. A rule or an equation in two variables.
3. Points on a graph in a coordinate plans.

Example 1

Determine whether each of the following relations is a function:

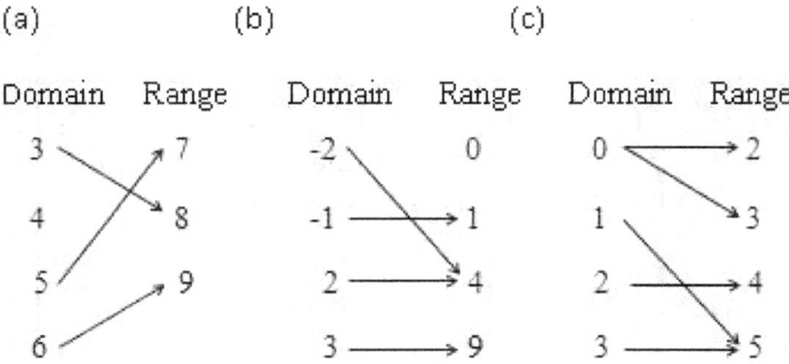

Solution

(a) The element 4 is not matched with an element in the range. So the relation does not represent a function.

(b) Each element in the domain is matched with exactly one element in the co-domain. So the relation represents a function.

(c) 0 is paired with two elements, so the relation does not represent a function.

Example 2

Find the range of each function represented by the following relations.

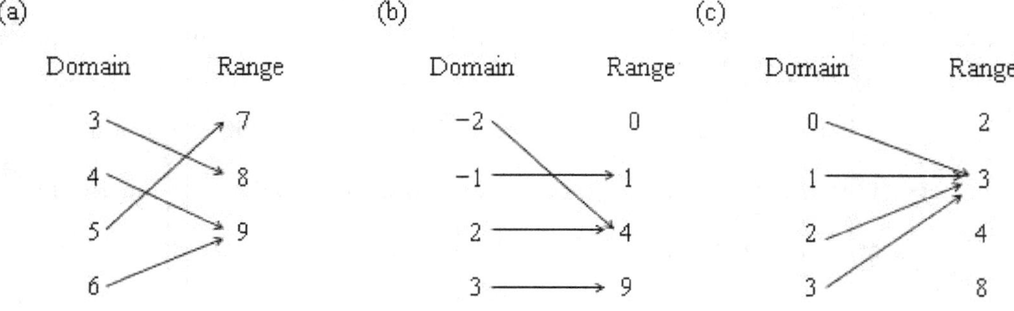

(a) The range is {7, 8, 9}

(b) The range is {1, 4, 9}

(c) The range is {3}

By definition of a function, any y-value corresponds to exactly one x-value. This means that, the graph of a function cannot have two or more different points with the same x-

coordinate. It follows, that a vertical line can intersect the graph of a function at most once. This observation suggests the following vertical line test.

The Vertical Line Test

A graph represents a function if and only if a vertical line intersects the graph in no more than one point.

Example 3

Which of the following graphs represent a function?

(a)

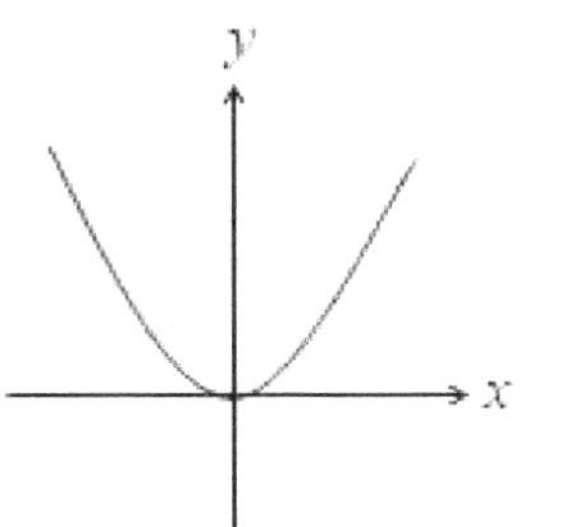

(b)

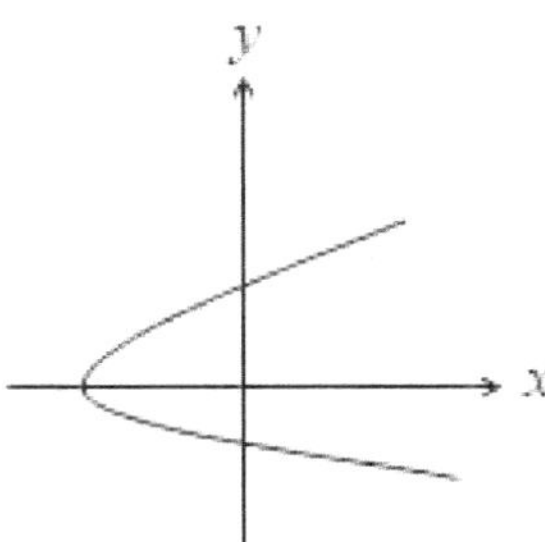

Solution

(a)

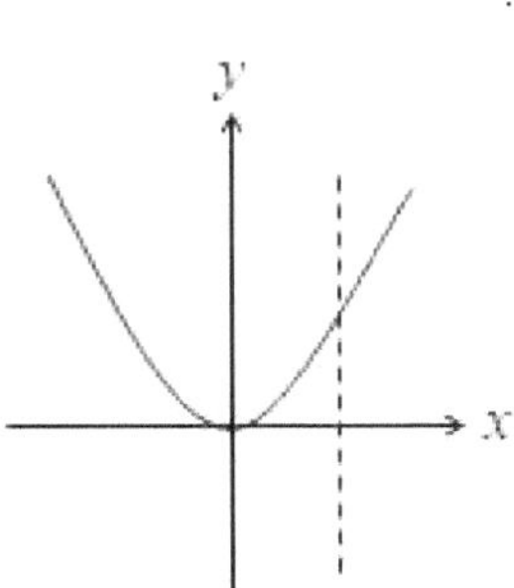

(b)

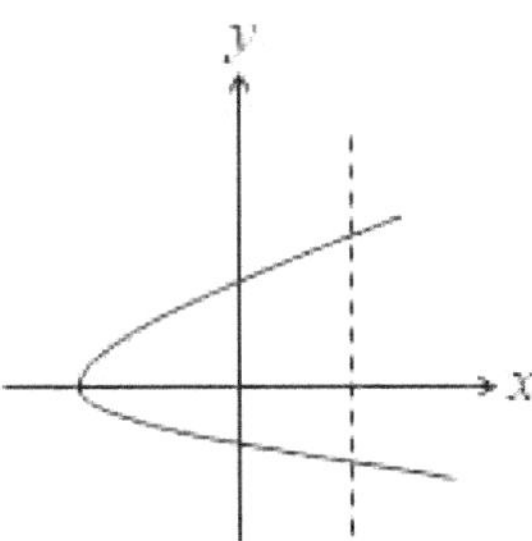

(a) A vertical line drawn anywhere in the graph intersects the graph in at most one point, so this is a graph of a function.

(b) A vertical line drawn in the graph intersects the graph twice, so this is not a graph of a function.

Function Notation

Functions are often named by letters, such as f, g and h. For instance the function represented by the equation $y = 3x + 2$ can be named f and written as $f(x) = 3x + 2$. The symbol $f(x)$ is read as " the value of f at x or simply f of x".

Evaluating a Function

The process of finding the value of a function $f(x)$ at x is called evaluating a function. This involves substituting a given x-value into the equation to obtain the value of $f(x)$, as illustrated in Example 4.

Example 4

Let $f(x) = 3x^2 - 5$. Find each value of the function:

(a) $f(-2)$ (b) $f(a + 1)$

Solution

(a) Replace x with -2

$$f(-2) = 3(-2)^2 - 5$$
$$= 3(4) - 5$$
$$= 12 - 5$$
$$= 7$$

(b) Replace x with $a + 1$.

$$f(a + 1) = 3(a + 1)^2 - 5$$
$$= 3(a^2 + 2a + 1) - 5$$
$$= 3a^2 + 6a - 2$$

Example 5

Let $h(x) = \dfrac{5x+8}{x+7}, x \neq -7$. If $h(x) = x$, find the values of x.

Solution

$$\frac{5x+8}{x+7} = x$$
$$5x + 8 = x(x + 7)$$
$$x^2 + 2x - 8 = 0$$
$$(x - 2)(x + 4) = 0$$
$$x - 2 = 0 \quad \text{or} \quad x + 4 = 0$$
$$x = 2 \qquad\qquad x = -4$$

The values of x are 2 and -4.

Finding the Domain of a Function

Often the domain of a function defined by an equation is the set of all real numbers for which the equation is defined. For example, the function $f(x) = 2x - 3$ is defined for the set of all real numbers. The domain is $\{x | x \in R\}$, written in interval notation as $(-\infty, \infty)$.

In some cases, x cannot take on all real numbers. For example, the domain of the function $f(x) = 2x/(x - 3)$ excludes x-value(s) that result in division by zero. Thus, the value $x = 3$ is excluded from the domain because division by zero is undefined. So the domain of f is the set of all real numbers x except $x = 3$, written in set notation as $\{x | x \in R, x \neq 3\}$. The function $f(x) = \sqrt{x}$ is defined for all nonnegative real numbers, i.e. $x \geq 0$. So, in set notation we write the domain of this function as $\{x | x \in R, x \geq 0\}$ and in interval notation we write $[0, \infty)$.

Example 6

Find the largest possible domain for the function $f(x) = \frac{5x+4}{x^2-x-6}$

Solution

Set the denominator equal to zero and solve for x.

$$x^2 - x - 6 = 0$$

$$(x - 3)(x + 2) = 0$$

$$x = 3 \text{ or } x = -2$$

The domain is the set of all real numbers except $x = -2$ and $x = 3$, written in set notation as $\{x | x \in R, x \neq -2, x \neq 3\}$.

Example 7

Find the largest possible domain for the function $g(x) = \sqrt{5 - x}$

Solution

Set the expression under the radical sign to greater than or equal to zero

$$5 - x \geq 0$$

$$x \leq 5$$

The domain of the function is the interval $(-\infty, 5]$.

Finding the Range of a Function from Graphs

One way of finding the range of a function would be to examine its graph to determine all y values within the domain. From the graph identify the minimum and maximum values.

Example 8

Find the range of the function represented by each of the following graphs.

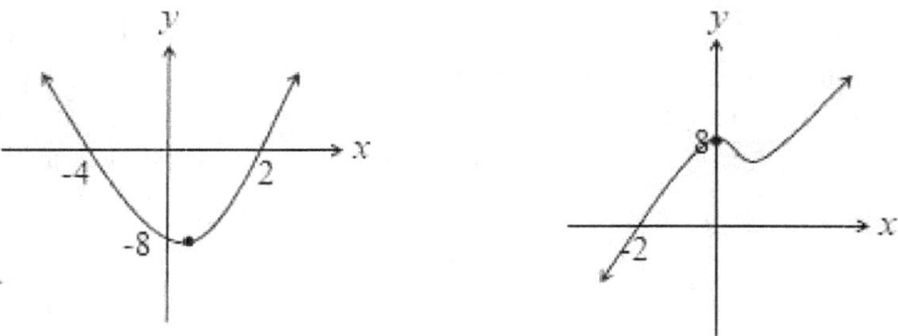

(a) The minimum y-value is at the vertex. The graph opens upwards and extends infinitely from -8. The range of the function is the interval $[-8, \infty)$.

(b) The graph extends infinitely in both the positive and negative y-direction. Thus, the range of the function is the interval $(-\infty, \infty)$.

Example 9

Find the range of the function represented by the following graph.

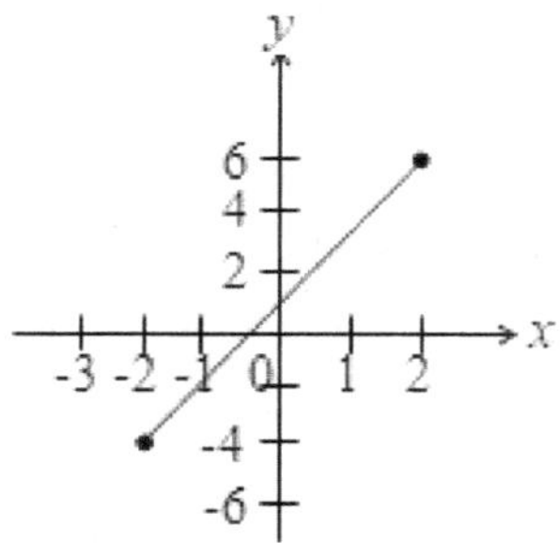

Solution

You can see from the graph that the x values go from -2 to 2. By looking at the values of y, we see that the minimum value of the function is -4 and the maximum value is 6. So, the range of the function is the interval $[-4, 6]$.

Exercise 8.1

In Exercises 1 – 4, determine whether or not each of the following relations is a function.

1. Domain Range

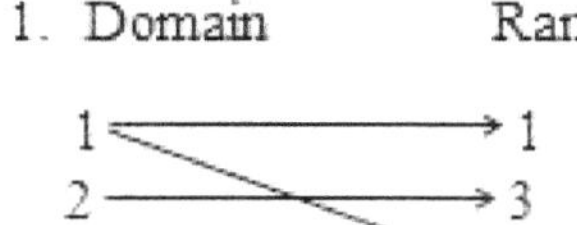

2. Domain Range

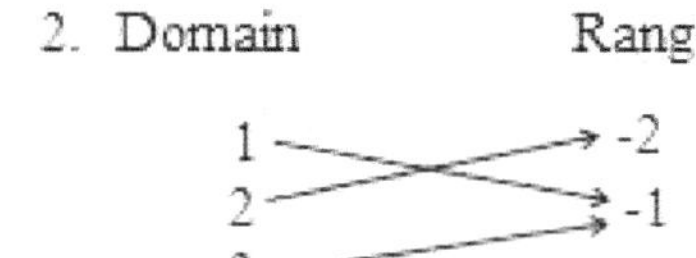

3. Domain Range

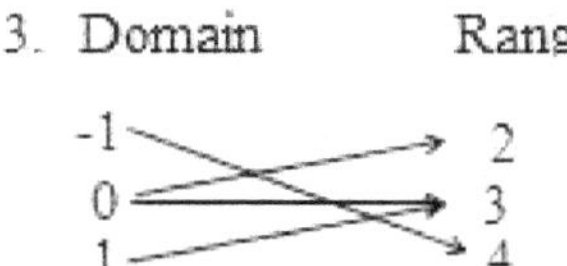

4. Domain Range

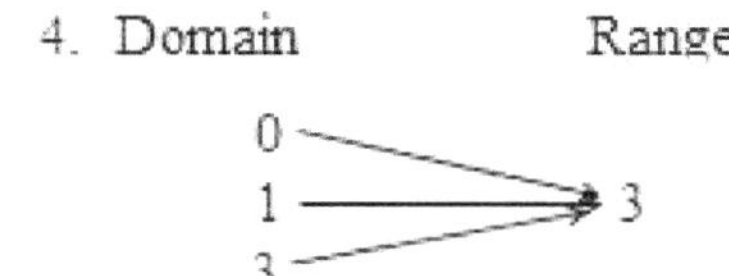

In Exercises 5 – 8, determine whether each graph represents a function.

5.

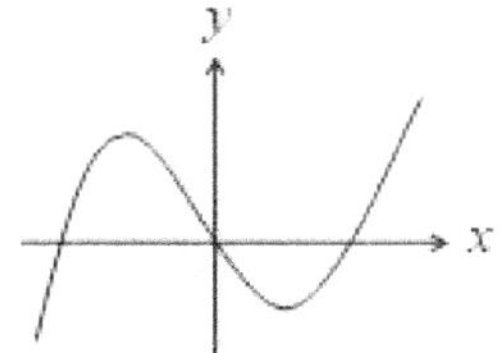

6

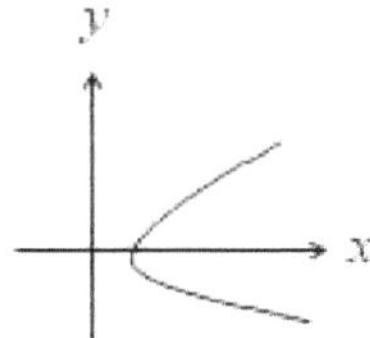

7.

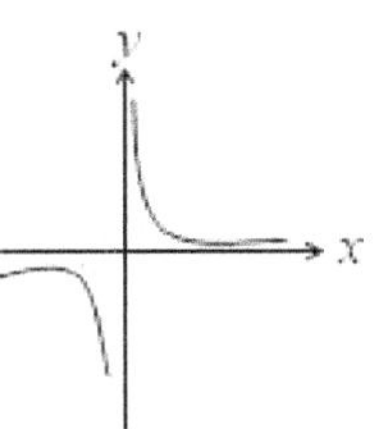

8.

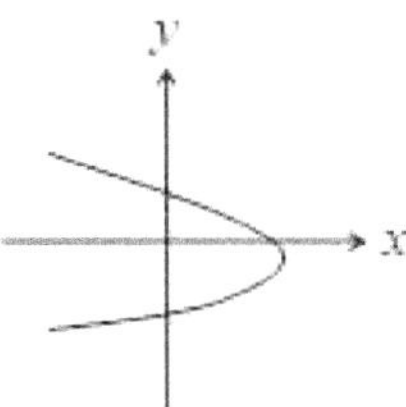

In Exercises 9 – 12, find the domain and range of each function.

9. Domain Range

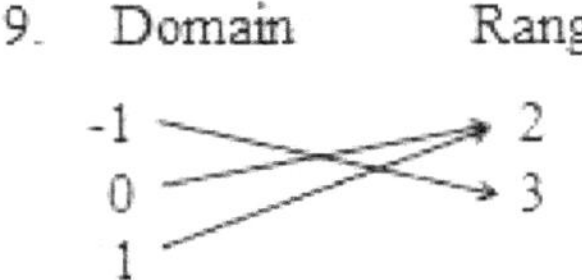

10. Domain Range

11. Domain Range

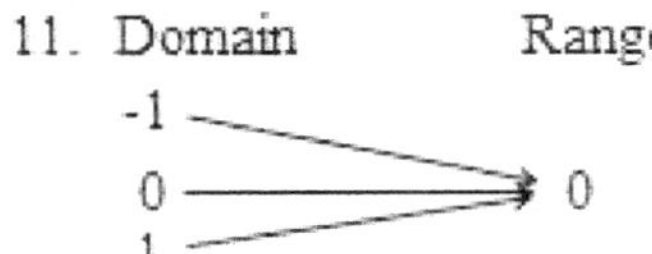

12. Domain Range

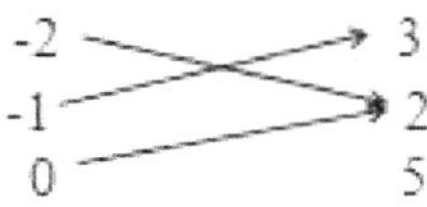

In Exercises 13 – 18, find the domain and range of each function.

13.

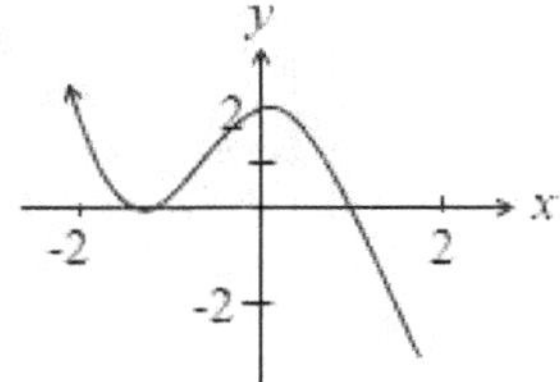

14.

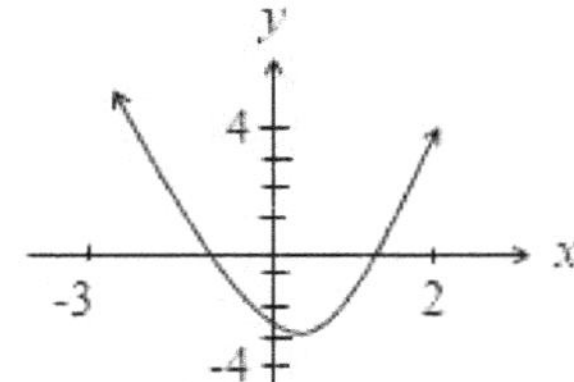

15.

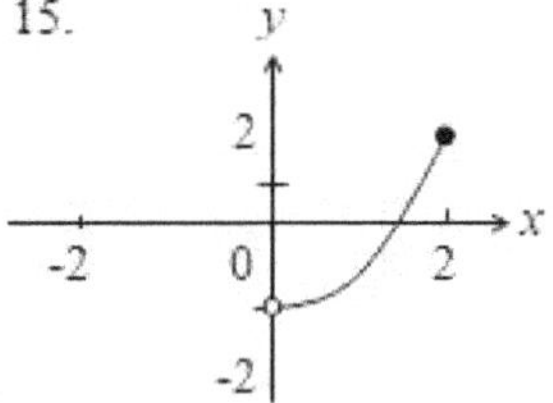

16.

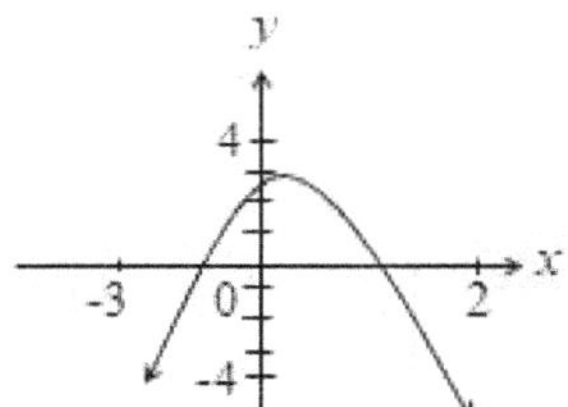

17.

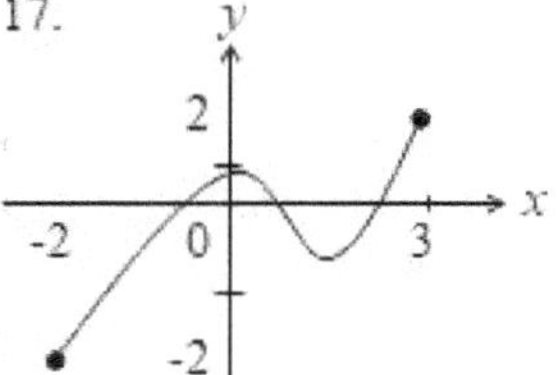

18.

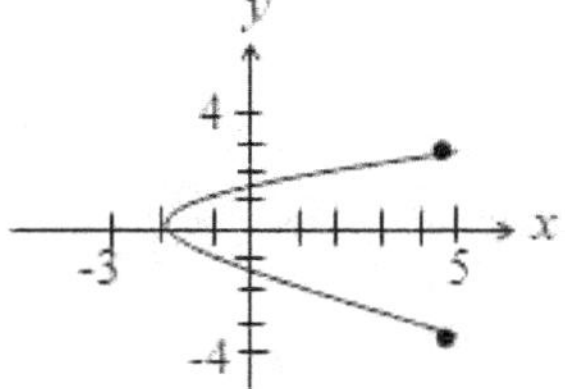

8.2 One-to-One and Inverse Functions

One-to-One Functions

A function f from set A to set B is said to be one- to- one if each element in the set A corresponds to exactly one element in the set B. In other words, f is one-to-one if $f(a) = f(b)$ implies $a = b$ for all a and b in A.

The diagram shown in Figure 8.2 is an illustration of a one-to-one function.

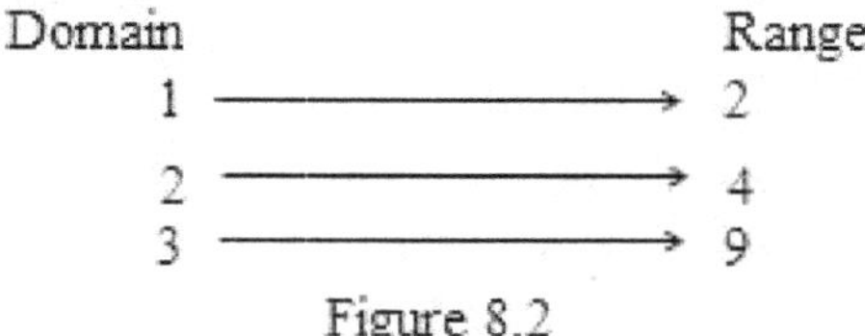

Figure 8.2

Horizontal Line Test

A graph represents a one-to-one function f, if and only if no horizontal line intersects the graph of f at more than one point.

Example 1

Determine whether each graph represents a one-to-one function.

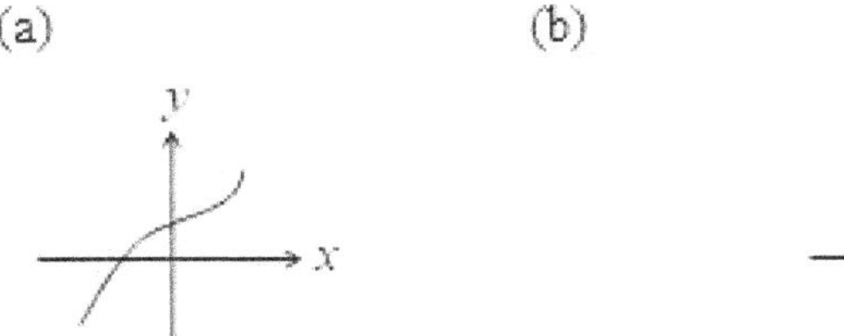

Solution

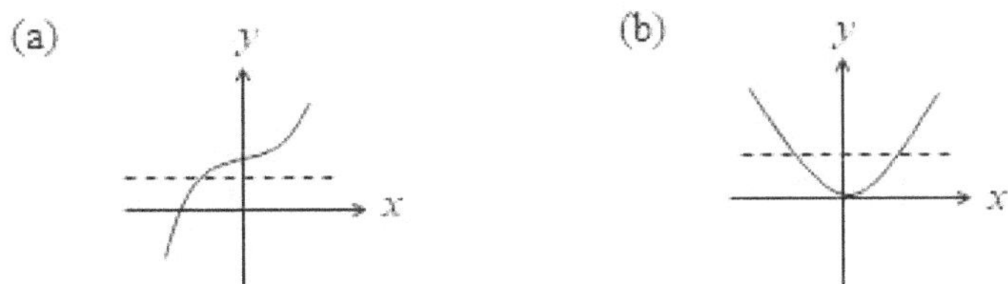

(a) A horizontal line drawn anywhere in the graph intersects the graph only once. The graph represents a one-to-one function.

(b) A horizontal line intersects the graph twice so the graph does not represent a one-to-one function.

Example 2

The function $f(x) = 5x - 3$ is defined on the set of real numbers. Show that the function is one-to-one.

Solution

Let $a, b \in$ R.

$f(x) = 5x - 3$ Write the original function

$f(a) = 5a - 3$ Substitute a for x

$f(b) = 5b - 3$ Substitute b for x

Equating the two expressions we get

$$5a - 3 = 5b - 3$$

$a = b$ Simplify

Since, $a = b$ the function f is one – to – one.

Example 3

The function g is defined on the set of real numbers by $g(x) = 3x^2 + 2$. Determine whether or not g is one-to-one.

Solution

Let $a, b \in$ R.

$g(x) = 3x^2 + 2$	Write the original function
$g(a) = 3a^2 + 2$	Substitute a for x
$g(b) = 3b^2 + 2$	Substitute b for x

Equating the two expressions, we get

$3a^2 + 2 = 3b^2 + 2$	
$a^2 = b^2$	Simplify
$a = \pm b$	Take square root of each side

Since a is equal to $-b$ or b the function g is not one – to – one.

Inverse Functions

Here are two illustrations of functions.

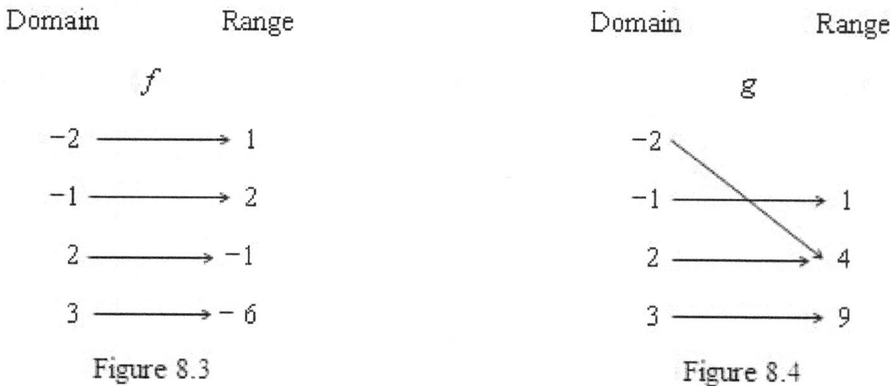

Figure 8.3 Figure 8.4

The function f (see Figure 8.3) maps each element in the domain to only one element in the co-domain. The reverse mapping also forms another function called the inverse function of f, denoted by f^{-1}. However, the reverse mapping of the function g (see Figure 8.4) does not represent a function because 4 is matched with – 2 and 2. The function g has no inverse function. You can see that, a function has an inverse function if and only if the function is one-to-one.

Notice from Figure 8.3 that the domain of the function f is the range of the inverse function f^{-1}, and the domain of f^{-1} is the range of f.

You can use the horizontal line test to verify that a function has an inverse function. A function has an inverse function if and only if no horizontal line intersects the graph of the function at more than one point.

Sometimes it may be possible to obtain an inverse of a function by restricting the domain of the function. For example, if we restrict the domain of the function $f(x) = x^2$ to only nonnegative values of x the function will have the inverse function $f^{-1}(x) = \sqrt{x}$.

Finding an Inverse Function Algebraically

The inverse of a function f can be represented by order pairs which are the coordinates of f reversed. This corresponds to interchanging the roles of x and y. To find the inverse function, follow the steps shown in Examples 4 and 5.

Example 4

Find the inverse function of the function $f(x) = 4x + 3$ defined on the set of real numbers.

Solution

$f(x) = 4x + 3$	Write the original function.
$y = 4x + 3$	Replace $f(x)$ with y.
$x = 4y + 3$	Interchange x and y.
$x - 3 = 4y$	Isolate the y term.
$\frac{x-3}{4} = y$	Solve for y.
$f^{-1}(x) = \frac{x-3}{4}$	Replace y with $f^{-1}(x)$

Example 5

Find the inverse of the function $g(x) = \frac{2x-3}{x+2}$, $x \neq -2$ defined on the set of real numbers.

Solution

$g(x) = \frac{2x-3}{x+2}$	Write the original function.
$y = \frac{2x-3}{x+2}$	Replace $g(x)$ with y
$x = \frac{2y-3}{y+2}$	Interchange x and y.

$$y = \frac{2x+3}{2-x}$$ Isolate the y term.

$$g^{-1}(x) = \frac{2x+3}{2-x}, x \neq 2$$ Replace y wiith $g^{-1}(x)$

Graphs of Inverse Functions

An inverse function interchanges the x- and y- coordinates in the ordered pair (x, y). It follows that, if a point (x, y) lies on the graph of a function f, then the point (y, x) must lie on the graph of its inverse function, f^{-1}. You should notice from your study of transformations in Geometry that the graph of f^{-1} is a reflection of the graph of f in the line $y = x$.

Example 6

Sketch the graphs of the function $f(x) = 2x + 6$ and its inverse function on the same rectangular coordinate system.

Solution

First, sketch the graph of $f(x) = 2x + 6$.

The graph of f has x- intercept at $(-3,0)$ and y- intercept at $(0,6)$. Obtain the graph of f by drawing a line through these points. To sketch the graph of f^{-1} reflect $(-3,0)$ and $(0,6)$ in the line $y = x$ to obtain the points $(0,-3)$ and $(6,0)$ respectively. Plot these points and draw a line through them, to obtain the graph of the inverse function.

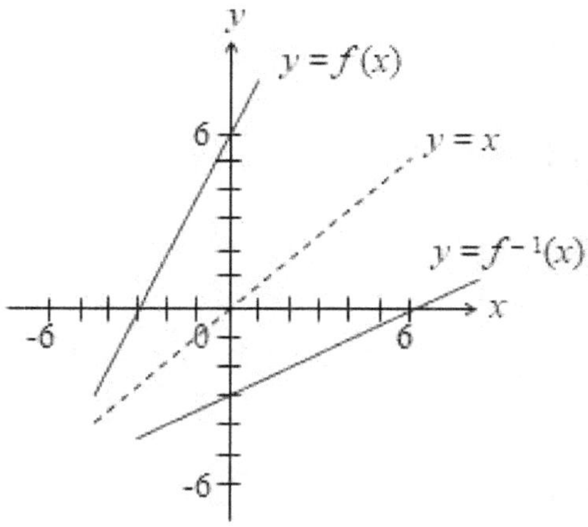

Exercise 8.2

In Exercises 1 – 4, determine whether each graph represents a one-to-one function.

1.
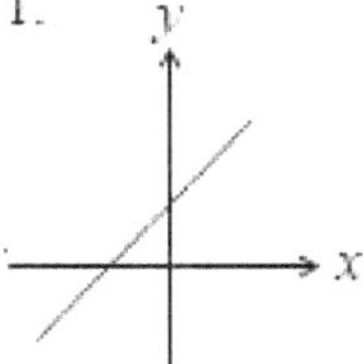

2.
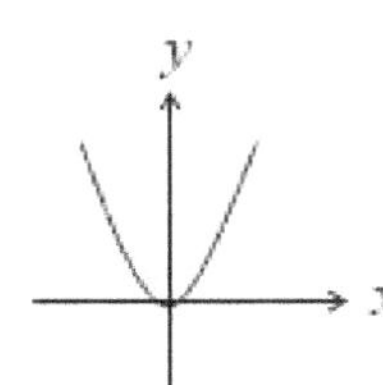

3.
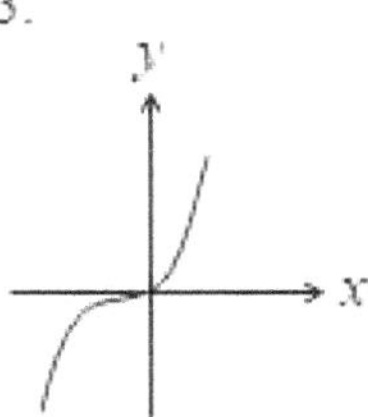

4.
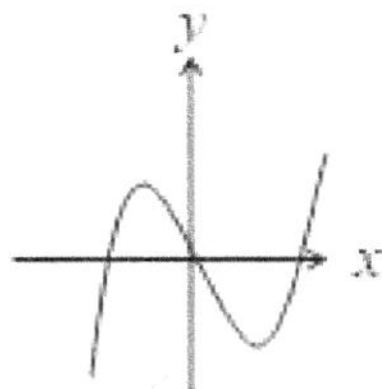

In Exercises 5 – 7, verify that the following functions are one-to-one.

5. $f(x) = 7x + 5$

6. $h(x) = \frac{3x}{2x-1}, x \neq \frac{1}{2}$

7. $g(x) = \frac{3x+2}{x+4}, x \neq -4$

In Exercises 8 – 11, determine whether each function is one- to-one.

8. $f(x) = 3x^2 - 2$

9. $g(x) = 5x + 3$

10. $h(x) = 5 - x^2$

11. $f(x) = \frac{2x+3}{x-5}, x \neq 5$

In Exercises 12 – 17, determine whether each function has an inverse function.

12.
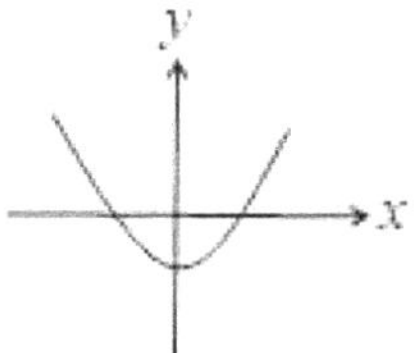

13.
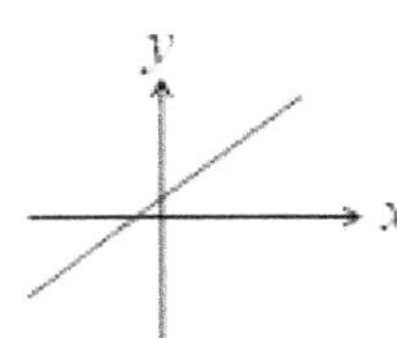

14.
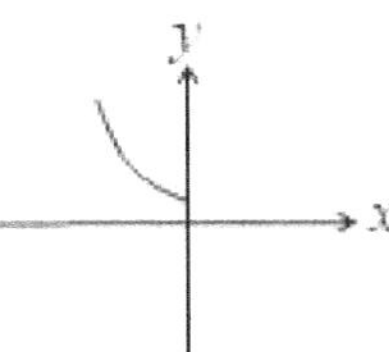

15.
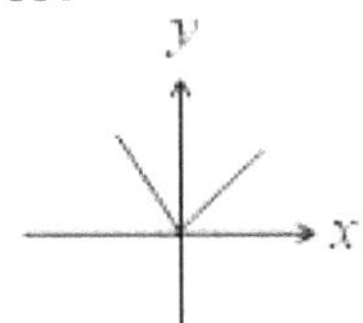

16.
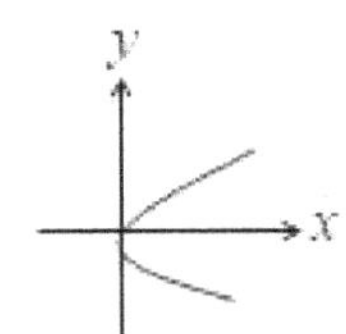

17.
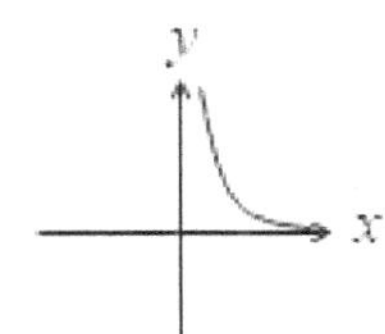

In Exercises 18 – 21, find the inverse function of each function.

18. $f(x) = 3x - 8$

19. $g(x) = 4 - x^2, x \geq 0$

20. $h(x) = \frac{5}{2x+1}, x \neq -\frac{1}{2}$

21. $f(x) = \frac{2x+3}{x-4}, x \neq 4$

22. Given that $g(x) = 3x - 12$ find $g^{-1}(3)$

23. Given that $h(x) = \frac{x+4}{2x-3}, x \neq \frac{3}{2}$ find $h^{-1}(-2)$.

24. Given that $f(x) = 3x - 5$ find x if $f^{-1}(x) = 2$

25. Given that $h(x) = \frac{2x+5}{x-3}, x \neq 3$ find x if $h^{-1}(x) = ¼$

8.3 Operations on Functions

You can form a new function by adding, subtracting, multiplying or dividing two functions. These operations can be performed using the following rules. If f and g are two functions then

1. $(f + g)(x) = f(x) + g(x)$ — Addition
2. $(f - g)(x) = f(x) - g(x)$ — Subtraction
3. $(f \cdot g)(x) = f(x) \cdot g(x)$ — Multiplication
4. $(f/g)(x) = f(x)/g(x), \; g(x) \neq 0$ — Division

Example 1

Let $f(x) = 3x - 2$ and $g(x) = 2x + 3$. Find

(a) $2f + g$

(b) $3f - 2g$

Solution

(a)

$$\begin{aligned}(2f + g)(x) &= 2f(x) + g(x)\\ &= 2(3x - 2) + (2x + 3)\\ &= 6x - 4 + 2x + 3\\ &= 8x - 1\end{aligned}$$

(b) $$(3f - 2g)(x) = 3f(x) - 2g(x)$$

$$= 3(3x - 2) - 2(2x + 3)$$

$$= 9x - 6 - 4x - 6$$

$$= 5x - 12$$

Example 2

Let $f(x) = x^2 + 2x$ and $g(x) = x^2 - 4$. Find

(a) $f \cdot g$ (b) f/g

Solution

(a) $$(f \cdot g)(x) = f(x) \cdot g(x)$$

$$= (x^2 + 2x)(x^2 - 4)$$

$$= x^4 + 2x^3 - 4x^2 - 8x$$

(b) $$\left(\frac{f}{g}\right)(x) = \frac{f(x)}{g(x)}$$

$$= \frac{x^2+2x}{x^2-4}$$

$$= \frac{x(x+2)}{(x-2)(x+2)}$$

$$= \frac{x}{x-2}, \quad x \neq 2$$

Composite Functions

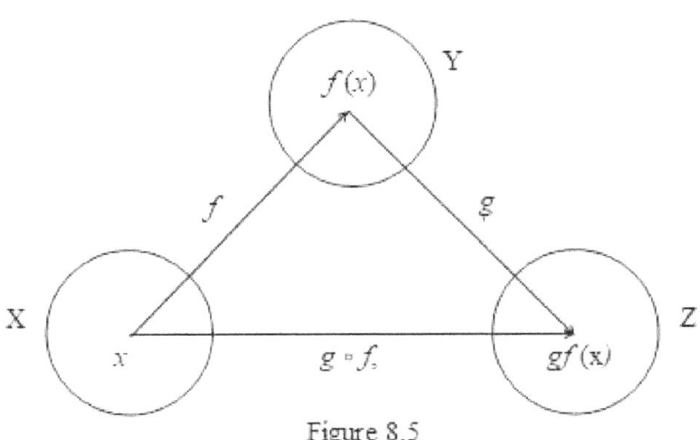

Figure 8.5

The function f assigns each element x in the set X an element $f(x)$ in the set Y. Then the function g takes $f(x)$ and assigns to it a value $gf(x)$ in the set Z, as illustrated in Figure 8.5. The two functions, f and g, can by combined to form a new function, denoted by $g \circ f$, which assigns each element x in the set X an element $gf(x)$ in the set Z. This new function is called " the composite function of g and f" or " the composition of g and f".

Definition of Composition of Two Functions

Let f and g be functions. The composition of the functions f and g is given by

$$(f \circ g)(x) = f(g(x))$$

The symbol $f(g(x))$, is read as "f of g of x." The domain of the composite function $f(g(x))$, is the set of all x in the domain of g such that $g(x)$ is in the domain of f.

Similarly, the composite function of g and f, denoted by $g \circ f$, is given by $g(f(x))$.

Example 3

If $f(x) = 2x - 3$ and $g(x) = 3x^2 + 2$ are defined on the set of real numbers, find

(a) $f \circ g$ (b) $g \circ f$

Solution

(a)

(a)	$(f \circ g)(x) = f(g(x))$	Definition of $f \circ g$.
	$= f(3x^2 + 2)$	Replace $g(x)$ by $3x^2 + 2$.
	$= 2(3x^2 + 2) - 3$	Substitute $3x^2 + 2$ for x.
	$= 6x^2 + 1$	Simplify.
(b)	$(g \circ f)(x) = g(f(x))$	Definition of $g \circ f$.
	$= g(2x - 3)$	Replace $f(x)$ by $2x - 3$.
	$= 3(2x - 3)^2 + 2$	Substitute $2x - 3$ for x.
	$= 12x^2 - 36x + 29$	Simplify.

Notice that $(f \circ g)(x) \neq (g \circ f)(x)$.

The composite function $f \circ g$ is generally not the same as the composite function $g \circ f$ .However, it is possible to find two functions such that $f(g(x)) = g(f(x))$.

Example 4

If $f(x) = 2x + 3$ and $f^{-1}(x) = (x - 3)/2$ are defined on the set of real numbers, find

(a) $f \circ f$ (b) $f \circ f^{-1}$

Solution

(a) $(f \circ f)(x) = f(f(x))$ — Definition of $f \circ f$.

$= f(2x + 3)$ — Replace $f(x)$ by $2x + 3$.

$= 2(2x + 3) + 3$ — Substitute $2x + 3$ for x.

$= 4x + 9$ — Simplify.

(b) $f \circ f^{-1}(x) = f(f^{-1}(x))$ — Definition of $f \circ f^{-1}$.

$= f\left(\frac{x-3}{2}\right)$ — Replace $f^{-1}(x)$ by $(x - 3)$

$= 2\left(\frac{x-3}{2}\right) + 3$ — Substitute $(x - 3)/2$ for x

$= x$ — Simplify.

You can verify that $(f^{-1} \circ f)(x) = x$ in the same manner. The functions $f^{-1} \circ f$ and $f \circ f^{-1}$ are called identity functions. Notice that two functions f and g are inverse functions of each other if $f(g(x)) = x$ and $g(f(x)) = x$.

Exercise 8.3

In Exercises 1 – 4, use the functions f and g to find:

(a) $f + g$ (b) $f - g$ (c) $f \cdot g$ (d) f/g

1. $f(x) = 3x - 4$, $g(x) = 2x + 3$
2. $f(x) = 2x^2 + 3x$, $g(x) = 2x + 3$
3. $f(x) = x + 3$, $g(x) = x^2 - 9$
4. $f(x) = (2x - 3)/(3x + 2)$, $g(x) = 4/(3x + 2)$

In Exercises 5 and 6, use the functions $f(x) = 5x + 3$ and $g(x) = x^2 - 2x + 1$ to find:

5. $2f + 3g$ 6. $f - 2g$

In Exercises 7 and 8, use the functions $f(x) = x^2 - 3x - 10$ and $g(x) = x^3 - 5x^2$ to find:

7. $f \cdot g$ 8. f/g

In Exercises 9 – 12, use the functions $f(x) = 5x + 3$ and $g(x) = 3x - 2$ to find:

9. $f^2(x)$ 10. $g^2(x)$ 11. $f(g(x))$ 12. $g(f(x))$

In Exercises 13 – 16, use the functions $f(x) = 2x + 3$ and $g(x) = x^2 - 1$ to find:

13. $f(g(x))$ 14. $g(f(x))$ 15. $f^2(x)$ 16. $g^2(x)$

In Exercises 17 and 18, use the functions $f(x) = 2x - 3$ and $g(x) = \frac{3x}{x+4}$, $x \neq -4$ to find:

17. $f \circ g$ 18. $g \circ f$

In Exercises 19 – 24, use the functions $f(x) = \frac{1}{2}x + 3$ and $g(x) = 3x - 2$ to find:

19. $f^{-1} \circ f$ 20. $f \circ f^{-1}$ 21. $g \circ g^{-1}$

22. $g^{-1} \circ g$ 23. $g^{-1} \circ f^{-1}$ 24. $f^{-1} \circ g^{-1}$

In Exercises 25 – 28, verify that the functions f and g are inverse functions of each other.

25. $f(x) = 3x + 4$, $g(x) = \frac{1}{3}(x - 4)$

26. $f(x) = \frac{1}{2}(x + 5)$, $g(x) = 2x - 5$

27. $f(x) = \frac{2x-7}{x}, x \neq 0,\ g(x) = \frac{7}{2-x}\ x \neq 2$

28. $f(x) = \frac{3x+5}{x-4}, x \neq 4,\ g(x) = \frac{4x+5}{x-3}, x \neq 3$

In Exercises 29 – 32, use the functions $f(x) = 3x^2 - 2$ and $g(x) = 2x + 3$ to find:

29. $f(g(x))$ 30. $f(g(-1))$

31. $g(f(x))$ 32. $g(f(2))$

In Exercises 33 – 36, use the functions $f(x) = 3x - 1$ and $g(x) = x^2$ to find:

33. $f(g(3))$ 34. $g(f(x))$

35. $g(f(5))$ 36. $f(g(-2))$

In Exercises 37 – 40, use the functions $f(x) = (2x - 1)/5$ and $g(x) = (3x + 2)/(x - 1)$, $x \neq 1$ to find

37. $g(f(2))$ 38. $f(g(2))$

39. $g^{-1}(f(-3))$ 40. $g(f^{-1}(-3))$

In Exercises 41 – 45, use the functions $f(x) = 2x + 3$ and $g(x) = x^3 - 1$ to find:

41. $f(g(-2))$ 42. $g(f(2))$ 43. $f^2(-1)$

44. $g^2(-1)$ 45. $f^{-1}(g^{-1}(7))$

8.4 Absolute Value Functions

A function that is written within an absolute value symbol is called absolute value function. For example, $f(x) = |3x + 2|$ and $g(x) = |x^2 - 1|$ are absolute value functions.

Consider the function $f(x) = |x|$ defined as

$$f(x) = \begin{cases} -x & if\ x < 0 \\ x & if\ x \geq 0 \end{cases}$$

The graph of $f(x) = |x|$ is shown in Figure 8.6.

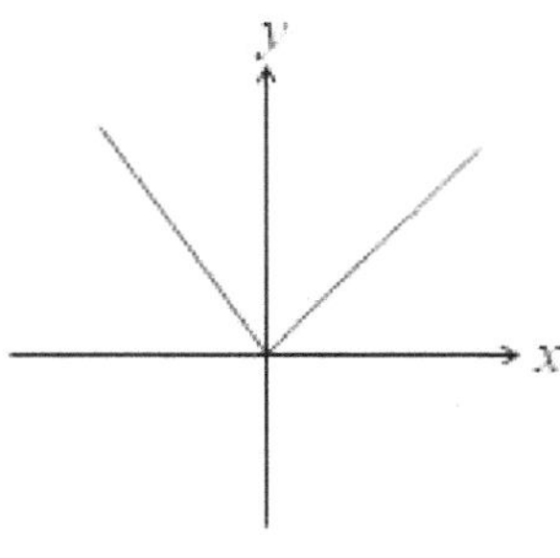

Figure 8.6

Observe that:

1. the domain of the function is the set of all real numbers.
2. the range is the set of all real numbers greater than or equal to zero, i.e. $y \geq 0$.
3. the vertex of the graph is (0, 0).

Absolute Value Equations

An absolute value equation is an equation that contains expressions in absolute value sign. Examples are $|2x + 1| = 9$ and $|5 - x| = 2|x - 3|$.

Solving Equations Involving Absolute Values

Consider the absolute value equation

$$|x| = 5$$

Remember that the expression inside the absolute value signs can be positive or negative. So, $x = 5$ or $-x = 5$, giving $x = -5$.

Thus , the solution of the equation is $x = 5$ and $x = -5$.

To solve an absolute value equation, isolate the absolute value and then write two separate equations and solve each of them, as shown in the following example.

Example

Solve: (a) $|2x + 5| = 13$ (b) $|3x + 2| - 6 = x$

Solution

(a) $|2x + 5| = 13$

First separate the equation into two equivalent form and then solve each equation separately.

$$2x + 5 = 13 \quad \text{or} \quad -(2x + 5) = 13$$

$$2x = 8 \qquad -2x = 18$$

$$x = 4 \qquad x = -9$$

You need to check your answers in the original equation. Note that it is possible that an absolute value equation may have no solution or one of the equivalent equations may not have a solution.

Check:

$$|2x + 5| = 13 \qquad |2x + 5| = 13$$

$$|2(-9) + 5| = 13 \qquad |2(4) + 5| = 13$$

$$|-13| = 13 \qquad |13| = 13$$

The solutions are $x = -9$ and $x = 4$.

(b) $|3x + 2| - 6 = x$

First isolate the absolute value

$$|3x + 2| = x + 6$$

Separate the equation into two equivalent equations and solve each equation separately.

$$3x + 2 = x + 6 \quad \text{or} \quad -(3x + 2) = x + 6$$

$$2x = 4 \qquad -4x = 8$$

$$x = 2 \qquad x = -2$$

The solutions are $x = -2$ and $x = 2$. Check these answers in the original equation

Example

Solve $|2x + 1| = |x + 8|$

Solution

$$|2x + 1| = |x + 8|$$

The equation is equivalent to

$$2x + 1 = -(x + 8) \quad \text{or} \quad 2x + 1 = x + 8$$

$2x + 1 = -x - 8$ $\qquad\qquad$ $x = 7$

$x = -3$

The solutions are $x = -3$ and $x = 7$. Check these in the original equation.

An alternate solution is given below:

$$|2x + 1| = |x + 8|$$

Squaring both sides we have

$$4x^2 + 4x + 1 = x^2 + 16x + 64$$

$x^2 - 4x - 21 = 0$ $\qquad$ Simplifying

$(x - 7)(x + 3) = 0$ $\qquad$ Factor the quadratic expression

$x = 7$ or $x = -3$

The solutions are $x = -3$ and $x = 7$.

Exercise 8.4

Sketch the graph of each function in the given domain. Find the domain of the inverse function from the graph.

1. $f(x) = |x - 5|, x \geq 5$
2. $f(x) = |3x - x^3|$
3. $f(x) = |x^2 - x - 6|$
4. $f(x) = |15 + 2x - x^2|$

Solve the following equations:

5. $|x + 2| = 6$
6. $|x + 5| = 7$
7. $|4x - 30| = 6$
8. $|9 - 7x| = 12$
9. $|2x + 5| = -16|$
10. $|21 - 7x| = 35$
11. $|3x - 1| = -5$
12. $|6x - 4| - 1 = -8$
13. $2|4 - 3x| - 6 = -2$
14. $|x + 8| = |2x + 1|$
15. $|2x - 3| = |x - 2|$
16. $|3x + 7| = |2x + 8|$
17. $|4x - 10| = 2|2x + 3|$
18. $3|3x - 2| = |9x - 12|$

8.5 Graphs and Transformation of Graphs of Functions

Cartesian Plane

You can represent ordered pairs of real numbers by points in a plane, called the rectangular coordinate system or the Cartesian plane, named after the French mathematician Rene' Descartes. The Cartesian plane consists of two real number lines intersecting at right angles. Each number line is called an axis (plural axes). The horizontal axis is usually called the x-axis, and the vertical axis is usually called the y -axis. The point of intersection of the two axes is called the origin, and the two axes separate the plane into four regions called quadrants, named as shown in Figure 8.7.

Each point in the plane corresponds to an ordered pair (x, y) of real numbers x and y, called coordinates of the point. For example, the coordinates of the origin are (0, 0). The x-coordinate represents the directed distance from the y-axis to the point, and the y-coordinate represents the directed distance from the x-axis to the point, as illustrated in Figure 8.7.

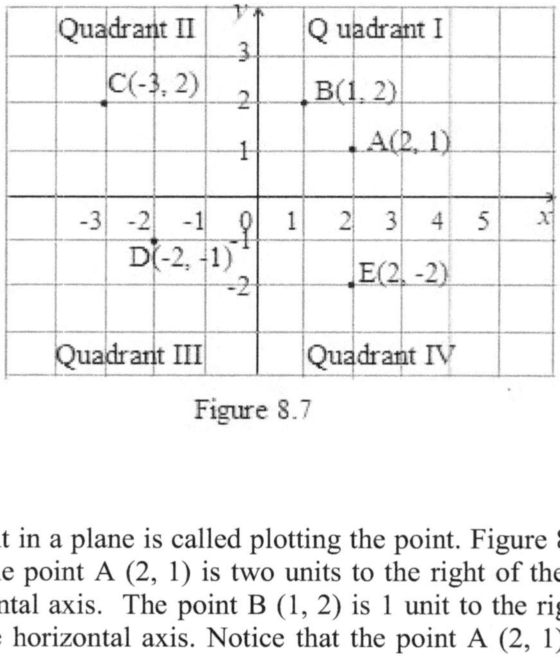

Figure 8.7

Locating a given point in a plane is called plotting the point. Figure 8.7 shows some points plotted in a plane. The point A (2, 1) is two units to the right of the vertical axis and one unit above the horizontal axis. The point B (1, 2) is 1 unit to the right of the vertical axis and 2 units above the horizontal axis. Notice that the point A (2, 1) and B (1, 2) are two different points. The order of the coordinates is important. The point C (− 3, 2) is 3 units to the left of the vertical axis and 2 units above the vertical line. The point D(− 2, −1) is 2 units to the left of the vertical axis and 1 unit below the horizontal axis, and the point E(2, −2) is 2 units to the right of the vertical axis and 2 units below the horizontal axis.

The Graph of an Equation

The graph of an equation is the set of all points in the Cartesian plane that are solutions of the equation.

Sketching the Graph of an Equation

To sketch the graph of an equation, find some ordered pairs (x, y) that satisfy the equation. Then plot the points and connect them with a smooth curve or line.

Example 1

Sketch the graph of $y = 3x + 2$.

Solution

Begin by finding some solutions of $y = 3x + 2$.

Choose convenient values of x. Then substitute these values into the equation to find the corresponding values of y. For example, if $x = -2$, then

$y = 3(-2) + 2 = -4$, and $(-2, -4)$ is a solution.

A similar calculation gives the following solution points:

$(-1, -1)$, $(0, 2)$ and $(1, 5)$.

The results are listed in a table such as shown below, called the table of values.

x	−2	−1	0	1
y	−4	−1	2	5

Now plot the points given in the table on a rectangular coordinate system and draw a line through them, as shown in Figure 8.8.

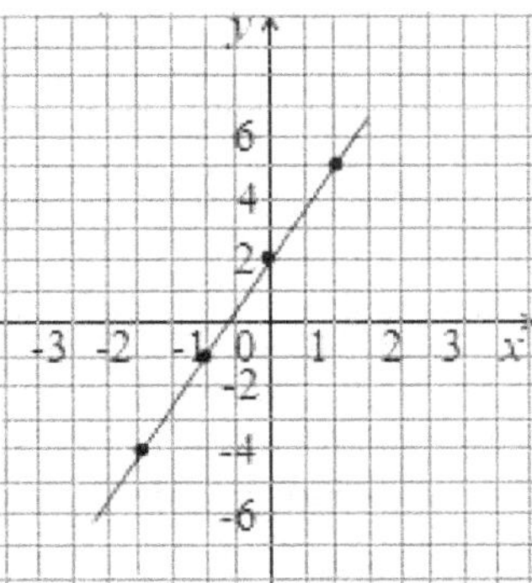

Figure 8.8

All the points of this graph appear to lie on a straight line. This straight line is the graph of $y = 3x + 2$. It represents all of the ordered pairs that are solutions of the equation.

You can graph a linear equation by plotting two points that are solutions of the equation. Although just two points are needed to determine a line, you can plot a third point to make it easier to detect an error.

Example 2

Sketch the graph of $y = x^2 - 5x + 3$ for $0 \leq x \leq 5$.

Solution

Begin by constructing a table of values.

Choose convenient x- values in the interval $0 \leq x \leq 5$, and then substitute these values into $y = x^2 - 5x + 3$ to find the corresponding y-values.

For example, if $x = 0$ then $y = 0^2 - 5(0) + 3 = 3$

So (0, 3) is a solution.

The results in the following table were found in a similar way

x	0	1	2	3	4	5
y	3	−1	−3	−3	−1	3

Next, plot the six points and then connect the points with a smooth curve as shown in Figure 8.9

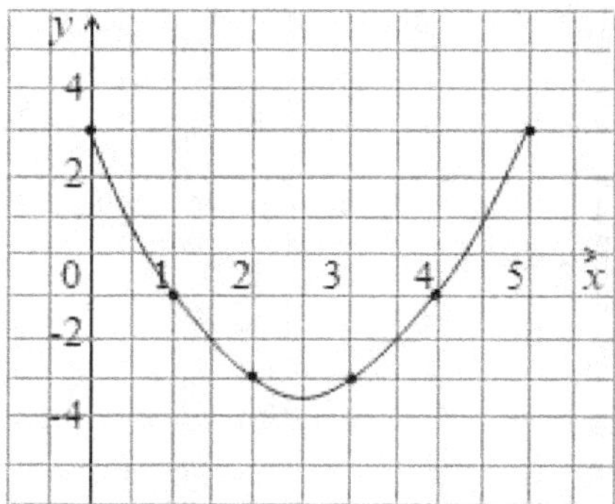

Figure 8.9

The smooth curve produced is called a parabola. Every quadratic function has a parabola as its graph. The minimum (or maximum) point on a parabola is its vertex. The graph of a quadratic function is symmetric with respect to the axis of the parabola. The axis of the parabola is a vertical line through the vertex.

Transformation of Graphs

Graph transformation is the process by which a change in the equation of a basic function affects the position and shape of the graph of that function.

Graphs of Basic Functions

It will be helpful to be familiar with the graphs of the basic functions shown in Figure 8.10.

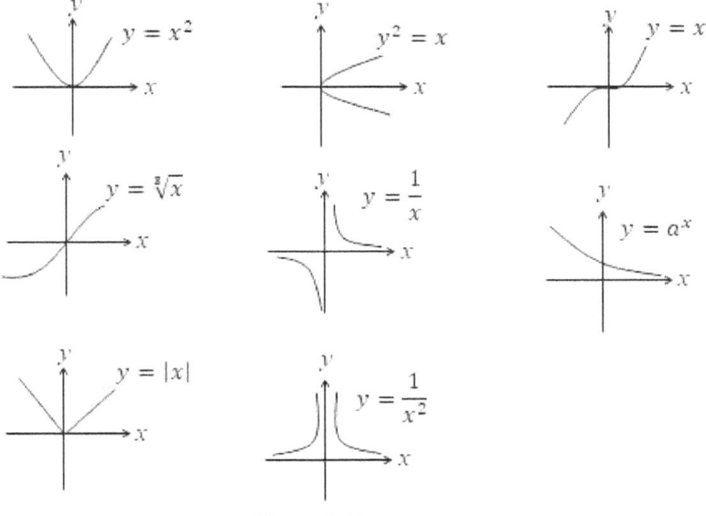

Figure 8.10

Many functions have graphs that are obtain by a simple transformation of the basic graphs. You need to understand what happens to the position or shape of a basic graph when the following changes are made to a function.

1. Translation in the y- direction

The graph of $y = f(x) + a$ is the graph of $y = f(x)$ moved a units up if a is positive (or down if a is negative).

Example 3

Use the graph of $y = x^2$ to sketch the graph of $y = x^2 + 3$.

Solution

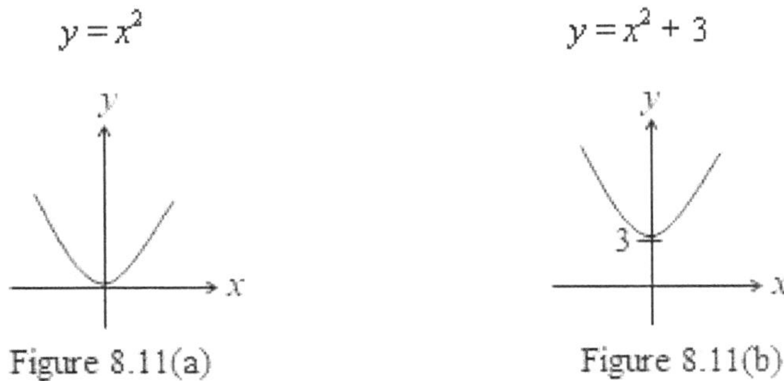

Figure 8.11(a) Figure 8.11(b)

The graph of $y = x^2$ is shown in Figure 8.11(a). The graph of $y = x^3 + 3$ has the same shape as that of $y = x^2$ but moved 3 units up, as shown in Figure 8.11(b).

2. Translation in the x-axis

The graph of $y = f(x + a)$ is the graph of $y = f(x)$ moved a units to the left if a is positive (or to the right if a is negative).

Example 4

Use the graph of $y = x^2$ to sketch the graph of $y = (x - 2)^2$.

Solution

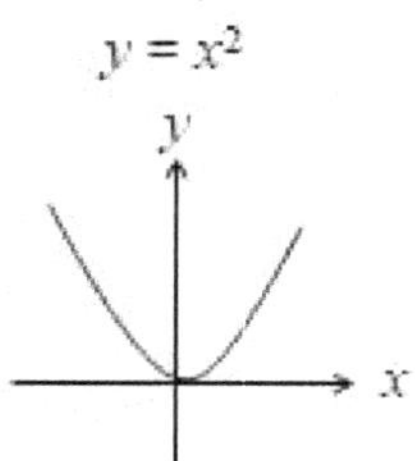

Figure 8.12(a)

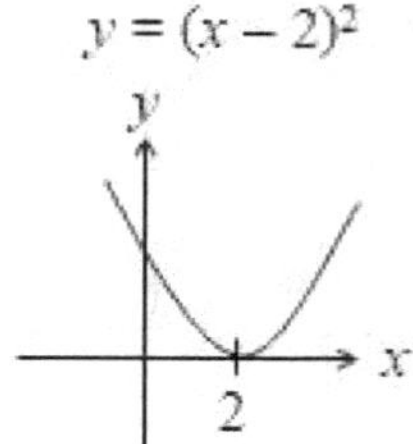

Figure 8.12(b)

The graph of $y = x^2$ is shown in Figure 8.12(a). The graph of $y = (x - 2)^2$ has the same shape as that of $y = x^2$ but moved 2 units to the right, as shown in Figure 8.12(b).

3. Reflection in the x-axis

Example 5

Use the graph of $y = x^2$ to sketch the graph of $y = -x^2$.

Solution

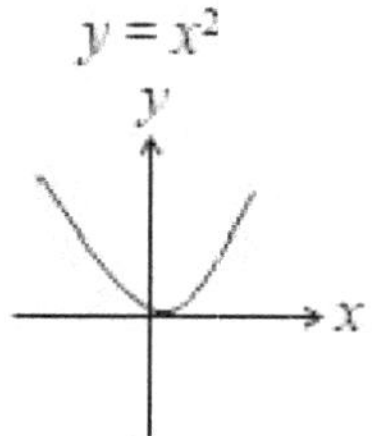

Figure 8.13(a)

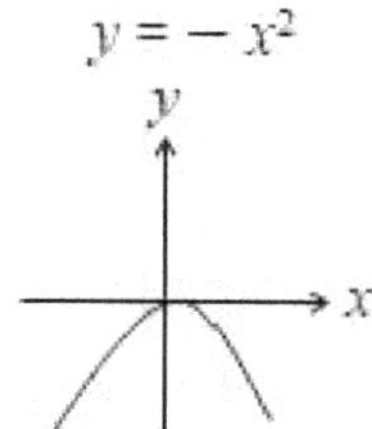

Figure 8.13(b)

Notice that multiplying $f(x)$ by a negative number reflects the graph of f in the x-axis.

4. Reflection in the y-axis

The graph of $y = f(-x)$ is the graph of $y = f(x)$ reflected in the y-axis.

Example 6

Use the graph of $f(x) = x^3$ to sketch $f(x) = -x^3$.

Solution

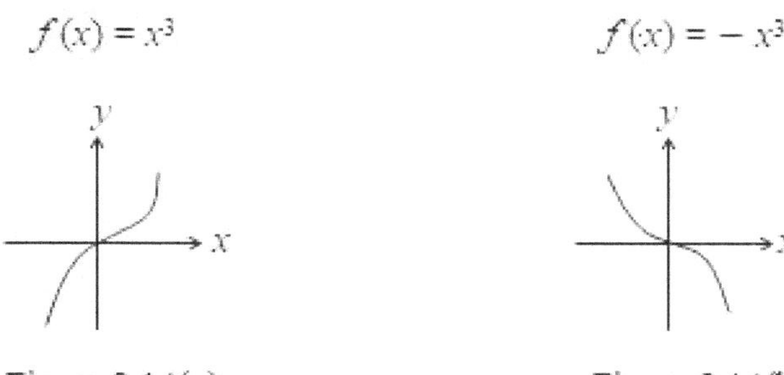

Figure 8.14(a) Figure 8.14(b)

Other transformations are:

5. The graph of $y = af(x)$ is obtained, if $a > 1$, by stretching the graph of $y = f(x)$ along the y-axis by a factor of a or compressing it if $0 < a < 1$.

6. The graph of $y = f(ax)$ is obtained, if $a > 1$, by compressing the graph of $y = f(x)$ along the x-axis by a factor of a or stretching it if $0 < a < 1$.

7. The graph of $y = f^{-1}(x)$ is obtained by reflecting the graph of $y = f(x)$ in the line $y = x$.

You can use a combination of these transformations to sketch the graph of the quadratic function $y = ax^2 + bx + c$ by rewriting $y = ax^2 + bx + c$ in the form $y = a(x - b)^2 + c$, using the method of completing the square. The process is illustrated in Example 7.

Example 7

Given the graph of $y = x^2$ sketch the graph of $y = x^2 - 6x + 8$.

Solution

First make $x^2 - 6x$ a perfect square by adding the square of one-half of the coefficient of x, which is $(-3)^2 = 9$. To make sure that the value of the function is not changed, we must subtract 9 from the function.

$y = x^2 - 6x + 9 - 9 + 8$ Add and subtract 9

$= x^2 - 6x + 9 - 1$

$= (x - 3)^2 - 1$ Factor the perfect square trinomial.

The graph of $y = x^2 - 6x + 8$ is obtained if the graph of $y = x^2$ is moved 3 units to the right, as shown in (b), and 1 unit downward, as shown in (c) (see Figure 8.15).

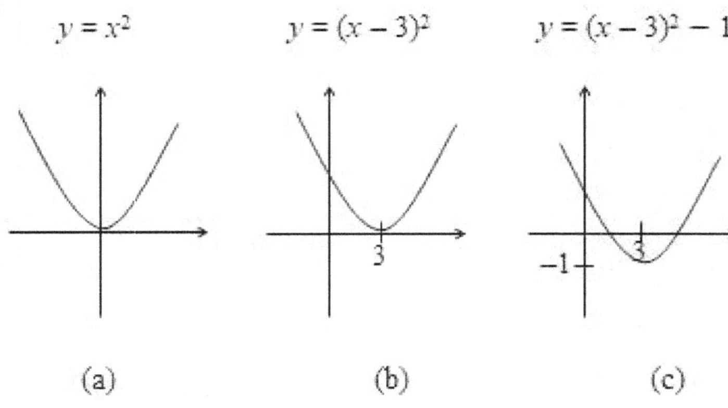

Figure 8.15

Exercise 8.5

For each of the following functions construct a table of values and sketch the graph.

1. $y = 2x + 5, \quad -4 \leq x \leq 3$
2. $y = 4 - 3x, \quad -3 \leq x \leq 5$
3. $3x + y = 7, \quad -1 \leq x \leq 5$
4. $y = x^2 - 4x + 3, \quad -1 \leq x \leq 5$
5. $y = 6 + 3x - 2x^2, \quad -2 \leq x \leq 4$
6. $y = (3 - x)(2 + x) \quad -4 \leq x \leq 5$

Use the graph of the applicable basic function to sketch the graph of each function:

7. $y = -x^3 + 1$
8. $y = x^2 + 4$
9. $y = -x^2 + 3$
10. $y = x^3 - 4$
11. $y = (x - 3)^2$
12. $y = (x + 2)^2 - 4$
13. $y = 2 - (x + 3)^2$
14. $y = x^2 - x - 6$
15. $y = x^2 - 8x + 10$
16. $y = 20 - x - x^2$
17. $y = -x^2 + 3x - 2$
18. $y^2 = x + 4$
19. $x + y^3 = 0$
20. $y = -\frac{1}{x} + 1$
21. $y = -\sqrt[3]{x} - 1$

Review Exercises

In Exercises 1 – 6, determine whether each of the following relations is a function.

1. Domain Range

2. Domain Range

2 → -2
2 → -1
3 → -1
4 → 0

3. Domain Range

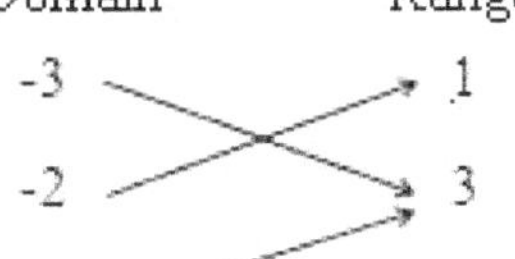

4. Domain Range

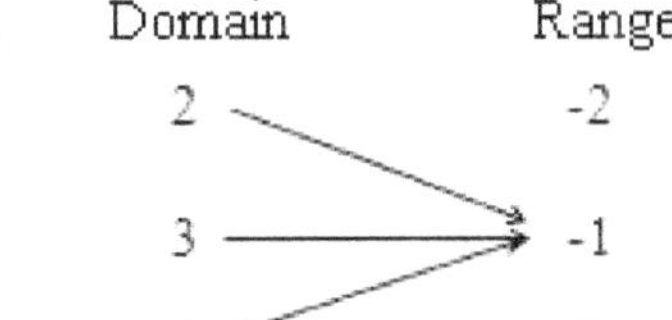

5. Domain Range

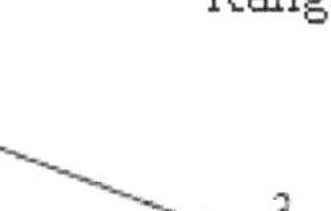

6. Domain Range

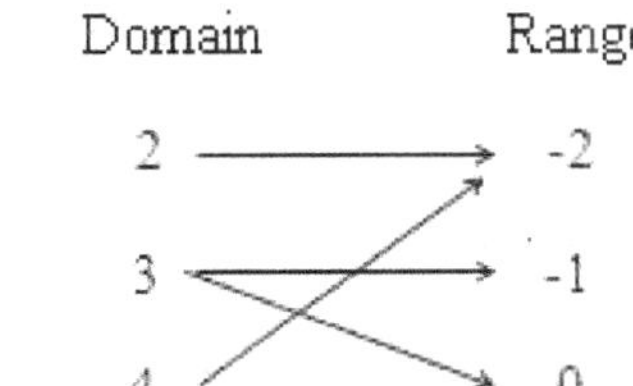

In Exercises 7 – 10, determine whether each graph represents a function.

7.

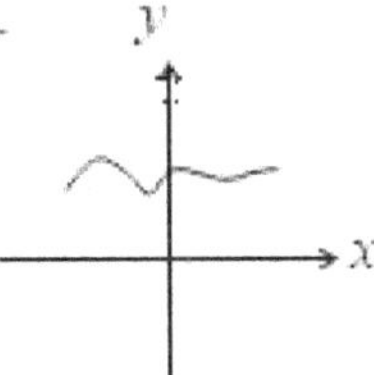

8.

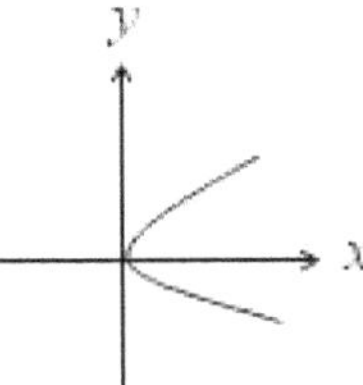

9.

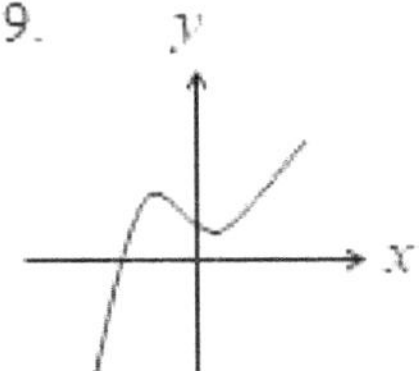

10.

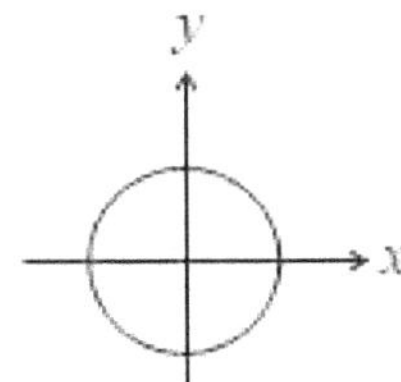

In Exercises 11 – 16, give the domain and the range of each function.

11.

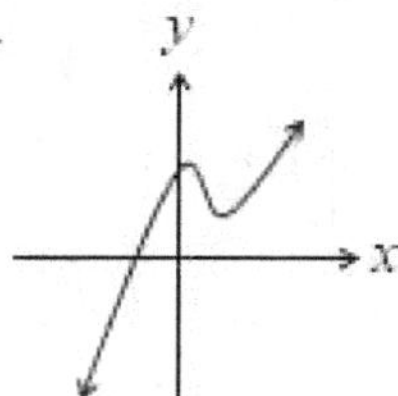

12.

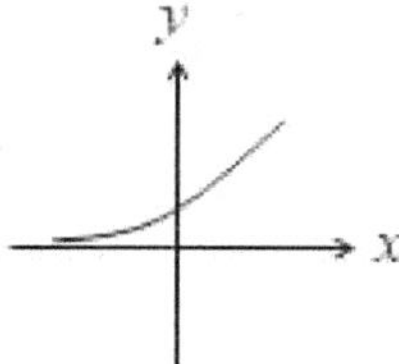

13.

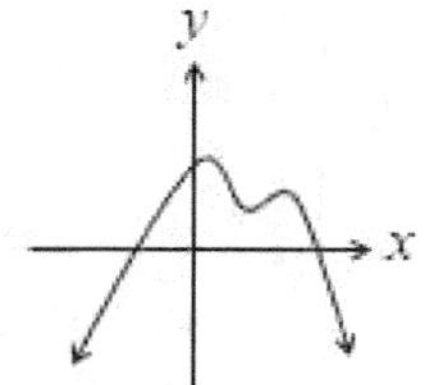

14.

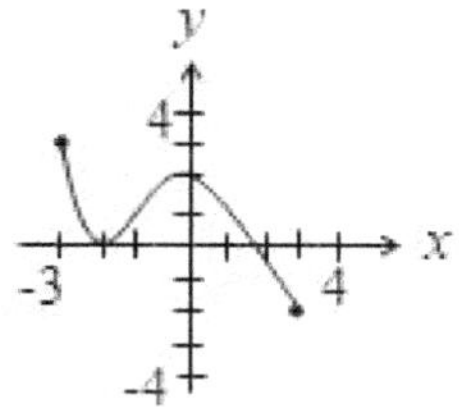

15.

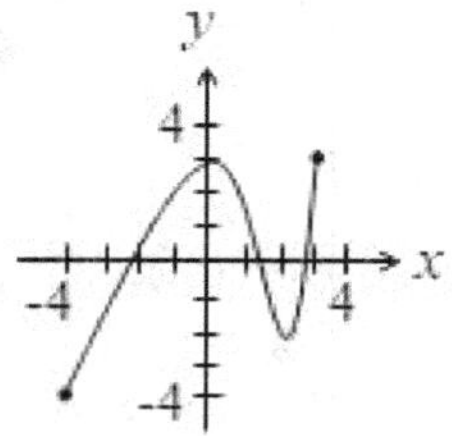

16.

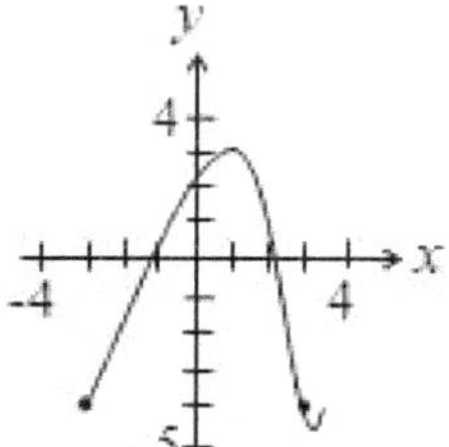

In Exercises 17 – 22, find the largest domain of each of the following functions defined on the set of real numbers.

17. $f(x) = 7 - 3x$

18. $f(x) = \frac{2}{3}x + 4$

19. $f(x) = \sqrt{x - 3}$

20. $f(x) = \sqrt{36 - x^2}$

21. $f(x) = \frac{2x-3}{x+5}$

22. $f(x)\frac{4x - 5}{x - 2}$

In Exercises 23 – 28, determine whether each graph is that of a one-to-one function.

23.
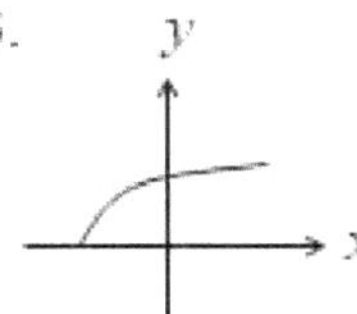

24.
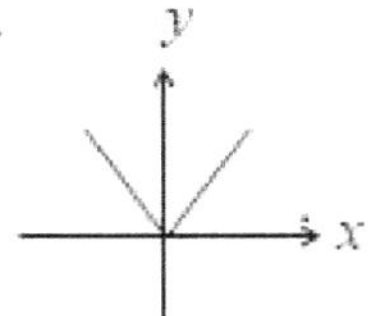

25.
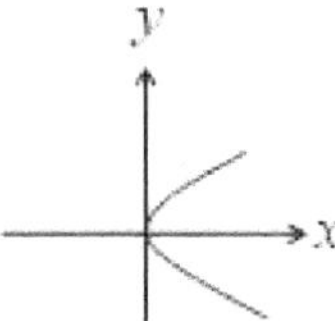

26.
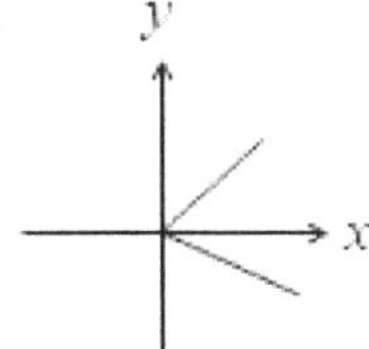

27.
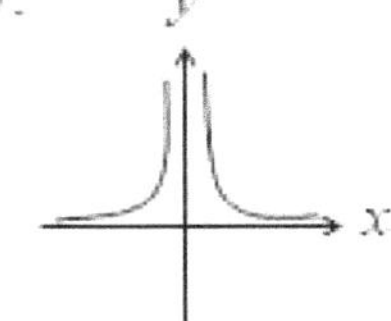

28.
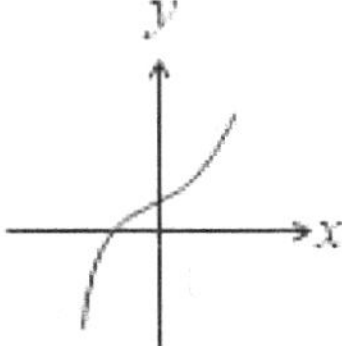

In Exercises 29 – 31, verify that each function is one-to-one.

29. $f(x) = 5x + 2$

30. $f(x) = 3 - 2x^3$

31. $h(x) = \frac{2x-3}{x+2}, x \neq -2$

In Exercises 32 – 37, determine whether each function has an inverse function.

32.
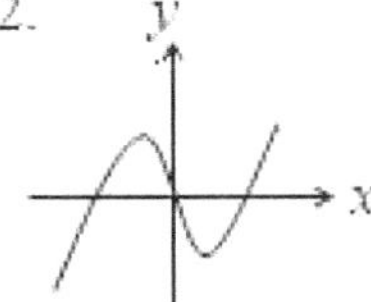

33.
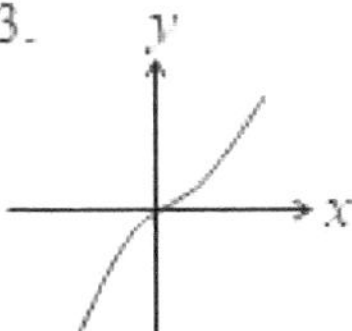

34.
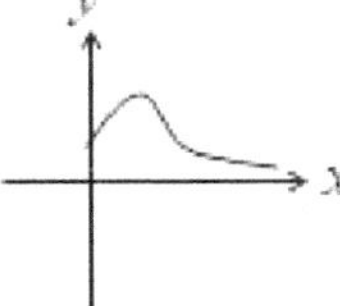

35.
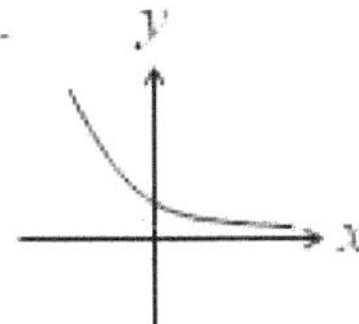

36.
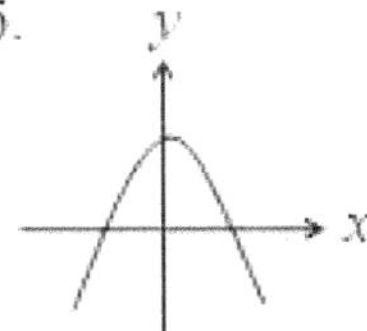

37.
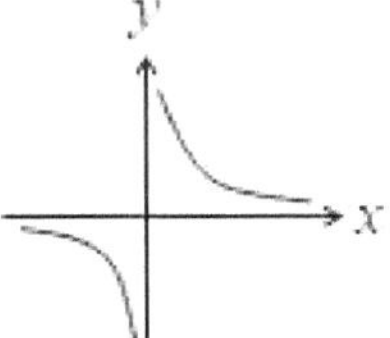

In Exercises 38 – 45, use $f(x) = 3x - 2$, $g(x) = 2x^2 - 3$ and $h(x) = 1 - 2x + x^2$ to find:

38. $f(2)$

39. $g(-2)$

40. $h(3)$

41. $g(2 - a)$

42. $h(2a)$

43. $f(a^2)$

44. $f(-3)$ 45. $h(-2)$

In Exercises 46 – 51, find the inverse function of each function.

46. $f(x) = 2x - 3$ 47. $f(x) = 9 - x^3$

48. $f(x) = (3/2)x + 4$ 49. $f(x) = 3x + 2$

50. $f(x) = 2x/(x - 3), x \neq 3$ 51. $f(x) = (3x - 2)/(x + 5), x \neq -5$

In Exercises 52 – 57, sketch the graph of each function and its inverse function on the same axes.

52. $f(x) = x + 2$ 53. $f(x) = x - 5$

54. $f(x) = 2x - 7$ 55. $f(x) = 3 - 3x$

56. $f(x) = x^2 - 1, x \geq 2$ 57. $f(x) = (x - 2)^2, x \geq 2$

In Exercises 58 – 61, use the given functions f and g to find:

(a) $f + g$ (b) $f - g$ (c) $f \cdot g$ (d) f/g

58. $f(x) = 3x + 4$, $g(x) = 2x - 3$

59. $f(x) = 2x^2 - 3x$, $g(x) = 2x - 3$

60. $f(x) = x - 3$, $g(x) = x^2 + 9$

61. $f(x) = (2x + 3)/(x - 3)$, $g(x) = (3x - 2)/(x - 2)$

In Exercises 62 – 65, use the functions $f(x) = 5x - 3$ and $g(x) = 3x + 2$ to find:

62. $2f - 3g$ 63. $f + 2g$

64. $3g - f$ 65. $-2f + g$

In Exercises 66 – 71, use $f(x) = 2x - 3$, $g(x) = x^2 + 1$ and $h(x) = \frac{1}{4}x - 1$ to find:

66. $f(g(x))$ 67. $g(f(x))$

68. $g(h(x))$ 69. $h(g(x))$

70. $f(h(x))$ 71. $h(f(x))$

In Exercises 72 – 77, use the functions $f(x) = 3x - 2$ and $g(x) = x + 2$ to find:

72. $f(g^{-1}(x))$ 73. $g(f^{-1}(x))$

74. $f^{-1}(g(x))$ 75. $g^{-1}(f(x))$

76. $f^{-1}(f^{-1}(x))$ 77. $g^{-1}(g^{-1}(x))$

In Exercises 78 – 81, verify that the functions f and g are inverse functions of each other.

78. $f(x) = 3x - 4$, $g(x) = ⅓(x + 4)$

79. $f(x) = ½(x - 5)$, $g(x) = 2x + 5$

80. $f(x) = (2x + 4)/x$, $x \neq 0$, $g(x) = 4/(x - 2)$, $x \neq 2$

81. $f(x) = (3x - 5)/(x + 4)$, $x \neq -4$, $g(x) = (4x + 5)/(3 - x)$, $x \neq -3$

In Exercises 82 – 87, use the functions $f(x) = 3x + 1$ and $g(x) = (2x - 3)/x$, $x \neq 0$ to find:

82. $f(g(2))$ 83. $g(f(-3))$

84. $f(g^{-1}(-2))$ 85. $g(f^{-1}(3))$

86. $f^{-1}(g^{-1}(1))$ 87. $g^{-1}(f^{-1}(1))$

88. The functions f and g are defined on the set of real numbers. If $f(x) = x^2$ and $f(g(x)) = x^2 + 6x + 9$, find $g(x)$.

89. If $f(x) = 2x + 1$ and $g(x) = 3x - 2$ find x when $g(f(x)) = 19$.

90. If $f(x) = 2x + 1$ and $g(x) = (3x + 2)/(x + 2)$, $x \neq -2$ find x when $f(g(x)) = x$.

In Exercises 91 – 102, solve each equation.

91. $|x - 1| = 5$ 92. $|x + 5| = 7$

93. $|2x + 3| = 9$ 94. $|7 + 2x| = 12$

95. $|5 - 2x| = 13$ 96. $|7 - 2x| + 5 = 9$

97. $-2|7 - 4x| + 16 = 0$ 98. $4|5x - 2| = 24$

99. $2|4 - 3x| + 2 = 6$ 100. $|3x - 1| = |x + 2|$

101. $|2x + 1| = |x + 8|$ 102. $|2 - x| = |3 + x|$

In Exercises 103 – 110, sketch the graph of each function.

103. $y = x^2 - 4$

104. $y = -x^2 + 5$

105. $y = -x^3 + 1$

106. $y = (x+3)^2$

107. $y = \sqrt[3]{x} - 1$

108. $y^2 = x - 3$

109. $y = |x| + 2$

110. $y = -x^2 + 3x - 2$

9 Linear Functions

A linear function is defined by the equation $y = mx + c$, where m and c are constants. Any equation of the form $y = mx + c$ is called a linear equation because the graph of the equation is a line with slope m and y-intercept at $(0, c)$.

9.1 The Distance Formula

We can develop a general formula for finding the distance between two points in a plane.

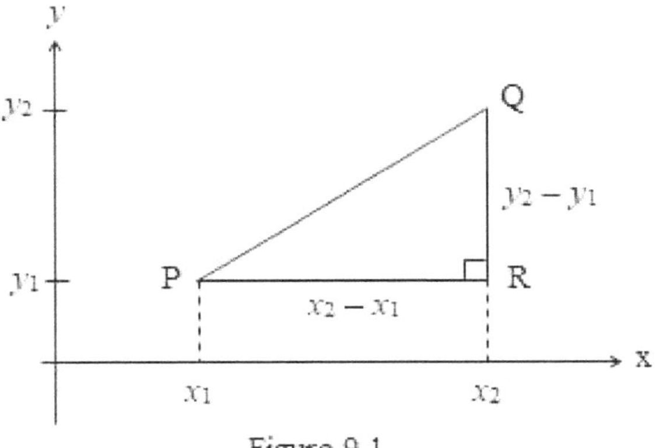

Figure 9.1

Figure 9.1 shows a straight line joining the points P (x_1, y_1) and Q (x_2, y_2). PR and RQ are drawn parallel to the x-axis and the y-axis respectively to form the right triangle PQR. By the Pythagorean Theorem the square of the distance between P and Q is

$$PQ^2 = PR^2 + RQ^2$$

$$= (x_2 - x_1)^2 + (y_2 - y_1)^2$$

We take the positive square root of each side to get

$$PQ = \sqrt{(x_2 - x_1)^2 + (y_2 - y_1)^2}$$

This result is the Distance Formula

The Distance Formula

The distance d between the points (x_1, y_1) and (x_2, y_2) is given by

$$d = \sqrt{(x_2 - x_1)^2 + (y_2 - y_1)^2}$$

Example

Find the distance between each pair of points.

(a) A(2, 3) and B(5, 7).

(b) P(-3, 4) and Q(2, 1)

Solution

(a) Let $(x_1, y_1) = (2, 3)$ and $(x_2, y_2) = (5, 7)$. Using the Distance Formula we get

$$AB = \sqrt{(5-2)^2 + (7-3)^2}$$

$$= \sqrt{9+16}$$

$$= \sqrt{25}$$

$$= 5$$

(b) Let $(x_1, y_1) = (-3, 4)$ and $(x_2, y_2) = (2, 1)$. Using the Distance Formula we get

$$PQ = \sqrt{(2+3)^2 + (1-4)^2}$$

$$= \sqrt{25+9}$$

$$= \sqrt{34}$$

$$\approx 5.83$$

Exercise 9.1

In Exercises 1 – 8, find the distance between the pair of points.

1. (2, 3); (14, 8)
2. (4, 3); (1, 7)
3. (4, 2); (10, 10)
4. (3, −2); (8, 3)
5. (2, 4); (5, 2)
6. (4, −1); (6, 2)
7. (−4, −1); (−2, −3)
8. (− 9, − 4); (6, 4)
9. A triangle has vertices A (1, 1), B (4, 4) and C (9, −1). Calculate the lengths of the sides of the triangle and verify that the triangle is a right triangle.
10. Verify that the triangle whose vertices are (3, −2), (−1, 1) and (2, 5) is isosceles.
11. The point (5, t) is equidistant from (2, −1) and (1, 6). Find the value of t.
12. The point (t, − 1) lies on a circle with a radius of 5 units and its centre at (−1, 3). Find the possible values of t.

13. Three points A, B and C have coordinates (9, 6), (3, −2) and (− 5, h). Given that h is positive and AB = BC, find the value of h.

.2 The Midpoint Formula

The point that divides a line segment joining two points into two equal line segments is called the midpoint of the line. .

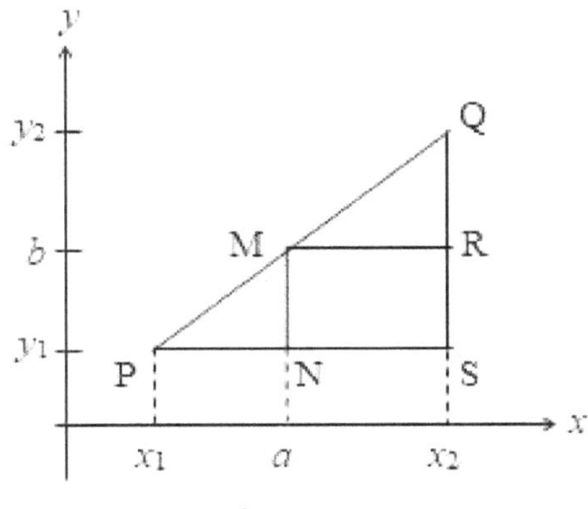

Figure 9.2

Figure 9.2 shows a line segment joining the points P(x_1, y_1) and Q(x_2, y_2). The midpoint M of PQ has coordinates (a, b). MN and MR are drawn parallel to QS and PS respectively. From Geometry, we know that the point N is the midpoint of the line segment PS.

Hence $a - x_1 = x_2 - a$, giving

$$a = \frac{x_1+x_2}{2}$$

Similarly, $b - y_1 = y_2 - b$, giving

$$b = \frac{y_1+y_2}{2}$$

Therefore, the midpoint of PQ has coordinates

$$\left(\frac{x_1+x_2}{2}, \frac{y_1+y_2}{2}\right).$$

This result gives us the following Midpoint Formula

The Midpoint Formula

The midpoint of the line segment joining the points (x_1, y_1) and (x_2, y_2) is given by

$$Midpoint = \left(\frac{x_1+x_2}{2}, \frac{y_1+y_2}{2}\right)$$

Example

Find the mid-point of the line joining the points

(a) (3, 2) and (5, − 8)

(b) (− 7, − 5) and (5, 9)

Solution

(a) Let $(x_1, y_1) = (3, 2)$ and $(x_2, y_2) = (5, -8)$. Using the Midpoint Formula we get

$$Midpoint = \left(\frac{3+5}{2}, \frac{2+(-8)}{2}\right) = (4, -3)$$

(b) Let $(x_1, y_1) = (-7, -5)$ and $(x_2, y_2) = (5, 9)$. Using the Midpoint Formula we get

$$Midpoint = \left(\frac{-7+5}{2}, \frac{-5+9}{2}\right)$$

$$= (-1, 2)$$

Exercise 9.2

In Exercises 1 – 12, find the midpoint of the line segment joining the pair of points.

1. (3, 2); (5, 4)
2. (4, 3); (0, 5)
3. (4, 2); (−10, −10)
4. (−1, −2); (−3, 4)
5. (3, −2); (−7, 8)
6. (4, 3); (−2, −5)
7. (7, −3); (−2, 9)
8. (−8, 3); (7, −5)
9. (−9, −4); (6, 4)
10. (5, −7); (−7, 9)
11. (1, 4); (7, 2)
12. (−4, 0); (−2, −4)

13 a. Find the coordinates of the midpoints of the sides of the triangle whose vertices are A (3, 2), B (7, 4) and C (−6, −5).

b. Verify that the length of the line segment joining the midpoint of two sides of triangle ABC in (a) is one-half of the length of the third side.

14 a. Find the coordinates of the midpoints of the sides of the triangle whose vertices are A (4, 3), B (4, 1) and C (2, 1).

b. Verify that the length of the line segment joining the midpoint of two sides of triangle ABC in (a) is one-half of the length of the third side.

15. The midpoint of a line segment AB is (3, 2). Given that the coordinates of A are (2, −4), find the coordinates of B.

16. If (6, 2) is the midpoint of the line segment connecting (3, −1) to P (x, y), find the values of x and y.

17. If (1, 2) is the midpoint of the line segment connecting (−3, 7) to P (x, y), find the values of x and y.

9.3 The Slope of a Line

The slope of a line is a measure of the steepness of the line.

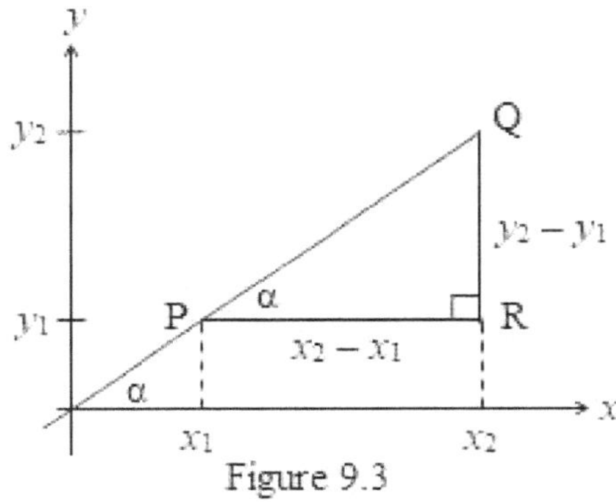

Figure 9.3

Because PR is parallel to the x-axis, ∠QPR = α. The slope of the line is defined as tan α. From triangle PQR

$$\tan\alpha = \frac{QR}{PR} = \frac{y_2 - y_1}{x_2 - x_1}$$

In general, the slope of a non-vertical line through (x_1, y_1) and (x_2, y_2) is given by

$$\text{Slope} = \frac{y_2 - y_1}{x_2 - x_1}$$

Notice that the slope is the ratio of the change in y-values to the change in x-values. Note that, you will obtain the same slope for any two different points on a line.

The slope of lines that go up from left to right is positive, (see Figure 9.4(a)), and the slope of lines that go down from left to right is negative, (see Figure 9.4(b)). The slope of a horizontal line is 0, and the slope of a vertical line is undefined.

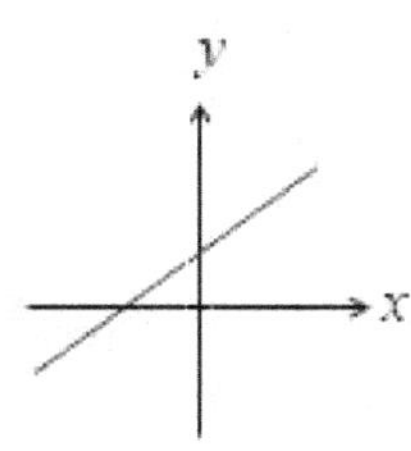

Figure 9.4(a)

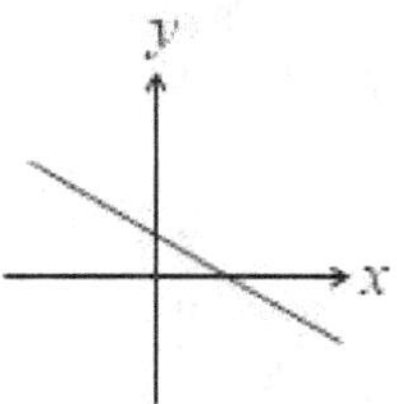

Figure 9.4(b)

Example

Find the slope of the line passing through each pair of points:

(a) A (3, 2) and B (5, 6)

(b) P (−3, 2) and Q (1, 5)

Solution

(a) Let $(x_1, y_1) = (3, 2)$ and $(x_2, y_2) = (5, 6)$. Using the definition of slope gives

$$Slope = \frac{y_2 - y_1}{x_2 - x_1}$$

$$= \frac{6-2}{5-3}$$

$$= 2$$

(b) Let $(x_1, y_1) = (-3, 2)$ and $(x_2, y_2) = (1, 5)$

$$slope = \frac{5-2}{1+3}$$

$$= \frac{3}{4}$$

Exercise 9.3

In Exercises 1 – 8, find the slope of the line passing through the pair of points.

1. (5, 4); (9, 11)
2. (−4, 9); (2, −3)
3. (4, −3); (−4, −7)
4. (−1, −2); (5, 7)
5. (6, 7); (11, 3)
6. (−3, 5); (9, −5)
7. (−1, 1); (5, 3)
8. (2, 3); (−4, − 6)

In Exercises 9 – 11, verify that the points are collinear.

9. (−1, −2); (4, 3); (5, 4)
10. (−4, 2); (−6, 5); (−8, 8)

11. (3, 5); (−2, −5); (1, 1)

12. If the points (−4, 3), (−1, 4) and (x, 9) are collinear find the value of x.

13. The points (−1, −1), (3, 11) and (1, t) lie on the same line. What is the value of t?

14. The points A (3, x) and B (7, 5) lie on a straight line. If the slope of the line is ¾, find the value of x.

15. The points (3, 2) and (x, 5) lie on the same line. If the slope of the line is −3/2, find the value of x.

16. Three points A (a, b), B (4/3, a) and C (b, 5/2) lie on a straight line. If the slope of the line is −3/2, find the values of a and b.

9.4 Equations of a Line

In this section we consider various forms of the equation of a line.

The Slope-Intercept Form

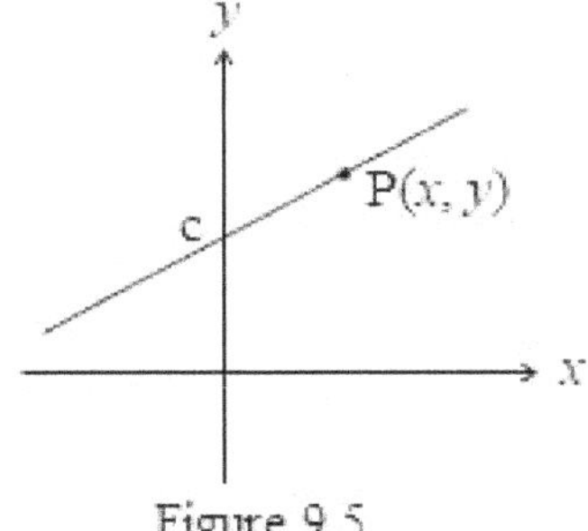

Figure 9.5

The line shown in Figure 9.5 has a slope m and y-intercept (0, c). If P (x, y) is any other point on the line, then using the definition of slope we have

$$\frac{y-c}{x-0} = m$$

Multiplying both sides by x and adding c to both sides gives $y = mx + c$. This equation is call the slope-intercept form of a line. The Slope-intercept form allows us to determine the slope of a line by simply inspecting its equation.

Example 1

Find an equation in slope –intercept form for a line:

(a) through (0, −2) with slope 3

Solution

Let (x, y) be any point on the line. Using the definition of the slope we have

$$\frac{y+2}{x} = 3$$

$$y = 3x - 2$$

(b) through $(-3, 2)$ with slope 2.

Solution

Let (x, y) be any point on the line. Using the definition of slope we have

$$\frac{y-2}{x+3} = 2$$

$$y - 2 = 2x + 6$$

$$y = 2x + 8$$

Example 2

Find the slope of the line whose equation is $3x + 2y - 6 = 0$.

Solution

To find the slope, solve the equation for y.

$$3x + 2y - 6 = 0$$

$$2y = -3x + 6 \qquad \text{Isolate } y$$

$$y = -\frac{3}{2}x + 6 \qquad \text{Divide both sides by 2}$$

The coefficient of x is $-3/2$, which is the slope of the line.

Example 3

Find the y-intercept of a line through $(4, -2)$ and has slope ¾ .

Solution

Substitute $x = 4$, $y = -2$ and $m = ¾$ into $y = mx + c$.

$$-2 = 4\left(\frac{3}{4}\right) + c$$

$$-2 = 3 + c$$

$$-5 = c$$

Hence, the y-intercept is (0, −5).

The Point-Slope Form

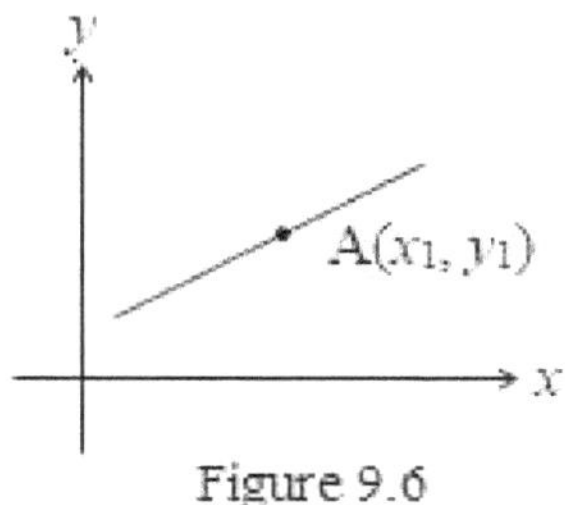

Figure 9.6

Let m represents a line passing through a point $A(x_1, y_1)$. If (x, y) represents any other point on the line then by the definition of slope

$$\frac{y-y_1}{x-x_1} = m$$

$$y - y_1 = m(x - x_1)$$

This is the point-slope form of the equation of a line.

Example 4

Find the equation of the line that passes through the point (3, 5) and has slope 2.

Solution

Substitute $x_1 = 3$, $y_1 = 5$ and $m = 2$ into the point-slope form $y - y_1 = m(x - x_1)$.

$$y - 5 = 2(x - 3)$$

$$y - 5 = 2x - 6$$

$$y = 2x - 1$$

Example 5

Find an equation of the line through (−1, 3) and (−3, 7).

Solution

Begin by finding the slope m of the line that passes through the given points.

$$m = \frac{7-3}{-3+1} = -2$$

You can use either (−1, 3) or (−3, 7) in the point-slope form with $m = -2$. If $(x_1, y_1) = (-1, 3)$, then

$$y - y_1 = m(x - x_1)$$

$$y - 3 = -2(x + 1)$$

$$y - 3 = -2x - 2$$

$$y = -2x + 1$$

The same result can be obtained if you use (− 3, 7).

General Form

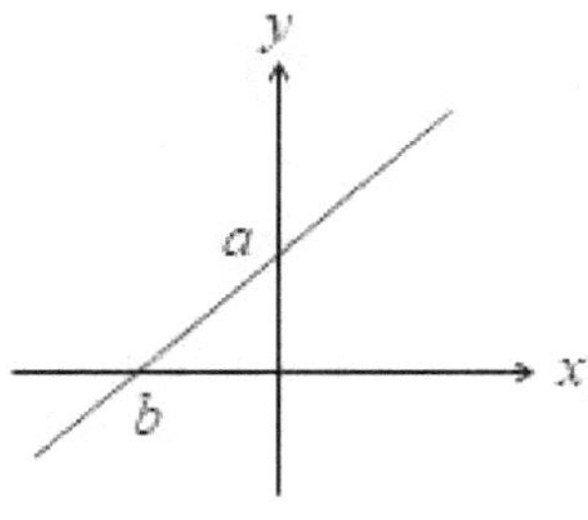

Figure 9.7

Figure 9.7 shows a line which crosses the x- axis at the point $(b, 0)$ and y-axis at the point $(0, a)$. If P(x, y) is any other point on the line, then using the definition of slope

$$m = \frac{0-a}{b-0} = -\frac{a}{b}$$

Using (0, a) and the slope $-a/b$ gives

$$\frac{y-a}{x} = -\frac{a}{b}$$

$$by - ab = -ax$$

$$ax + by - ab = 0$$

Replacing $-ab$ with c we get

$$ax + by + c = 0$$

This equation is called the general equation of a line.

Also, from $ax + by - ab = 0$ we get

$$ax + by = ab$$

Dividing both sides by ab gives

$$\frac{x}{b} + \frac{y}{a} = 1$$

This equation is called the intercept form of the equation of a line.

Notice that $(b, 0)$ and $(0, a)$ are the intercepts on the x- and y-axes respectively.

Example 6

Find the equation of the line that passes through $(-1, 2)$ and $(3, 5)$.

Solution

Begin by finding the slope of the line.

Let $(x_1, y_1) = (-1, 2)$ and $(x_2, y_2) = (3, 5)$.

$$m = \frac{5-2}{3+1} = \frac{3}{4}$$

Now substitute $x_1 = -1$, $y_1 = 2$ and $m = ¾$ into $y - y_1 = m(x - x_1)$.

$$y - 2 = \frac{3}{4}(x + 1)$$

$$4y - 8 = 3x + 3$$

$$3x - 4y + 11 = 0$$

Horizontal and Vertical Lines

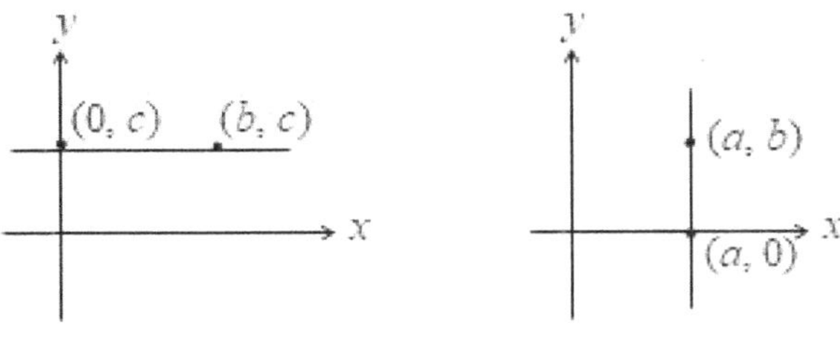

Figure 9.8(a) Figure 9.8(b)

The slope of the horizontal line, shown in Figure 9.8(a) is $\frac{c-c}{b-0} = 0$.

From the point-slope form of the equation of a line, you can see that the line has an equation $y - c = 0(x - b)$ or $y = c$. In general, every horizontal line has a slope of zero and equation of the form $y = c$, where c is the y-value of all ordered pairs on the line. The vertical line, shown in Figure 9.8(b) has slope

$$\frac{b-0}{a-a} = \frac{b}{0}.$$

Since division by 0 is undefined the equation of a vertical line cannot be written in the slope-intercept form. Every point on this line has an x-value of a, therefore, the equation of the line is $x = a$. In general, every vertical line has an equation of the form $x = a$, where a is the x-value of all the ordered pairs on the line.

Exercise 9.4

In Exercises 1 – 6, find the equation of the line that passes through the given y -intercept and has the indicated slope..

1. (0, 4); 3
2. (0, 5); −2
3. (0, 0); 4
4. (0, - 4); − ¾
5. (0, 1); 2/3
6. (0, − 3); − 3/2

In Exercises 7 – 12, find the slope and the y-intercept of each of the following lines.

7. $3x - 2y = 4$
8. $4y - 3x = 10$
9. $6x + 2y - 5 = 0$
10. $4x - 5y = 13$
11. $4x + 3y = 2$
12. $3x + 6y - 9 = 0$

In Exercises 13 – 18, find the y-intercept of the line that passes through the given point and has the indicated slope.

13. (1, 2); 3
14. (3, −8); − 4
15. (−3, −1); 2
16. (−4, 2); ½
17. (-4, -2); − ¾
18. (6, 5); 3/2

In Exercises 19 – 30, find the equation of the line that passes through the given point and has the indicated slope.

19. (2, 4); 3
20. (−1, −2); −3
21. (2, −3); −4
22. (4, −1); 2
23. (−2, 2); −7
24. (2, 3); 4
25. (6, 3); − 3/2
26. (−3, −2); 5/3
27. (4, −2); 2/5
28. (−3, −5); − 3/2
29. (2, 3); 4/3
30. (−4, 5); − 3/5

In Exercises 31 – 45, find the equation of the line passing through the pair of points.

31. (3, 4); (6, 7)
32. (4, −1); (7, 5)

33. (0, 4); (5, −1)
34. (−2, 3); (5, 3)
35. (4, −5); (6, 1)
36. (0, 3); (3, −3)
37. (−2, 4); (−4, −2)
38. (4, 5); (5, 2)
39. (−1, 6); (−2, 4)
40. (−2, 2); (−4, 5)
41. (−3, 6); (3, −2)
42. (2, −7); (6, −10)
43. (4, 3); (8, −2)
44. (4, 3); (8, 6)
45. (−1, −6); (2, −4)

9.5 Parallel and Perpendicular Lines

Parallel Lines

Two lines in a plane that never meet are called parallel lines. You can use the slope of a line to determine whether two lines are parallel.

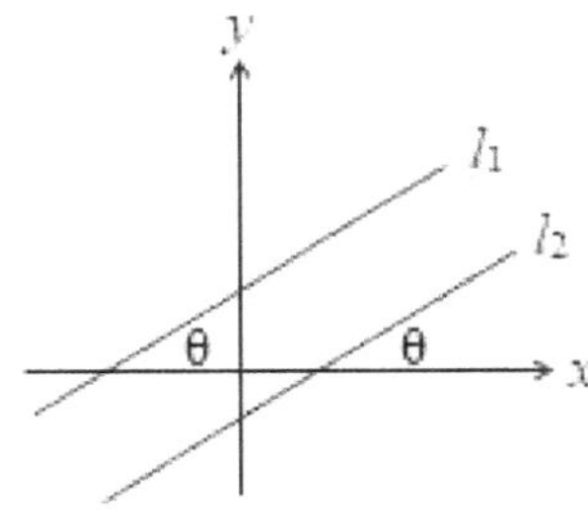

Figure 9.9

Figure 9.9 shows two parallel lines l_1 and l_2. Each line makes an angle θ with the x-axis and has the same slope tan θ. In general, two lines with the same slope are parallel.

Example 1

Determine whether the lines $2x - 3y - 12 = 0$ and $6x - 9y + 45 = 0$ are parallel.

Solution

Begin by rewriting each equation in the slope-intercept form.

	$2x - 3y - 12 = 0$	Original equation
	$-3y = -2x + 12$	Isolate term in y
	$y = \frac{2}{3}x - 4$	Divide each side by -3
Also	$6x - 9y + 45 = 0$	Original equation

$-9y = -6x - 45$ Isolate term in y

$y = \frac{2}{3}x + 5$ Divide each side by -9

Since both lines have slope 2/3, the lines are parallel.

Example 2

Find the equation of the line that passes through the point $(2, -3)$ and is parallel to the line $3x + 4y - 12 = 0$.

Solution

Begin by finding the slope of $3x + 4y - 12 = 0$ by writing the equation in the slope-intercept form.

$$3x + 4y - 12 = 0$$

$4y = -3x + 12$ Isolate term in y

$y = -\frac{3}{4}x + 3$ Divide each side by 4

The slope of this line is $-\frac{3}{4}$. Any line parallel to the given line must also have a slope of $-3/4$. So, you can find the equation of the line through $(2, -3)$ by substituting $x_1 = 2$, $y_1 = -3$ and $m = -3/4$ into

$$y - y_1 = m(x - x_1)$$

$$y + 3 = -\frac{3}{4}(x - 2)$$

$$4y + 12 = -3x + 6$$

$$3x + 4y + 6 = 0$$

Perpendicular Lines

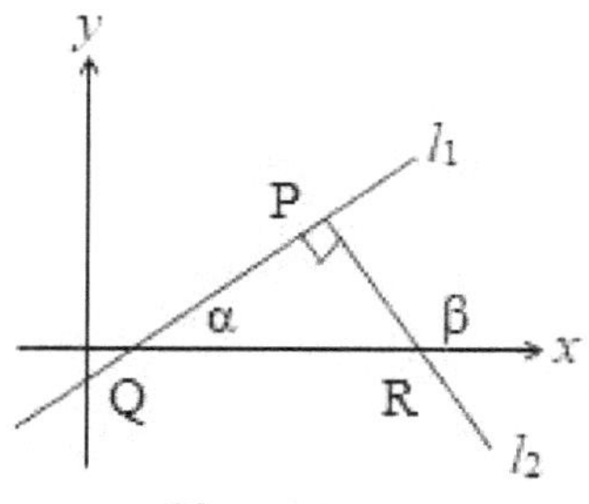

Figure 9.10

Figure 9.10 shows two perpendicular lines l_1 and l_2 which make angles α and β respectively with the x-axis. Let the gradients of the lines l_1 and l_2 be m_1 and m_2 respectively. From triangle QPR,

$$\tan\alpha = \frac{PR}{PQ}$$

$$\frac{PQ}{PR}\tan\alpha = 1$$

But $\frac{PQ}{PR} = \tan(180° - \beta)$, since $\angle QRP = 180° - \beta$

$= -\tan\beta$, since $\tan(180° - \beta) = -\tan\beta$

Hence, $\tan\alpha \tan\beta = -1$.

Replacing $\tan\alpha$ and $\tan\beta$ with m_1 and m_2 respectively we have $m_1 \times m_2 = -1$ or $m_2 = -1/m_1$. In general, two lines having slopes with a product of – 1 are perpendicular.

Example 3

Determine whether the lines $4x - 3y + 16 = 0$ and $3x + 4y - 29 = 0$ are perpendicular.

Solution

Begin by rewriting each equation in the slope-intercept form

$4x - 3y + 16 = 0$

$-3y = -4x - 16$ Isolate term in y

$y = \frac{4}{3}x + \frac{16}{3}$ Divide each side by 3

You can see that the slope of this line is 4/3.

Also $3x + 4y - 29 = 0$

$4y = -3x + 29$ Isolate term in y

$y = -\frac{3}{4}x + \frac{29}{3}$ Divide each side by 4

The slope of this line is – ¾.

Because – ¾ is a negative reciprocal of 4/3, the two lines are perpendicular.

Example 4

Find the equation of the line passing through (2, 3) and perpendicular to the line having the equation $3x - 2y - 9 = 0$.

Solution

Begin by writing $3x - 2y - 9 = 0$ in slope-intercept form.

$$3x - 2y - 9 = 0$$

$$-2y = -3x + 9 \qquad \text{Isolate term in } y$$

$$y = \frac{3}{2}x - \frac{9}{2} \qquad \text{Divide each side by 2}$$

The slope of this line is 3/2.

Any line perpendicular to this line must have a slope of − 2/3. So, the equation of the line perpendicular through (2, 3) is obtained by substituting $x_1 = 2$, $y_1 = 3$ and $m = -2/3$ into $y - y_1 = m(x - x_1)$.

$$y - 3 = -\tfrac{2}{3}(x - 2)$$

$$3y - 9 = -2x + 4$$

$$2x + 3y - 13 = 0$$

Exercise 9.5

In Exercises 1 – 4, determine whether the pair of lines represented by the equations are parallel.

1. $x + 2y = 4,\ 2x + 4y = 5$

2. $3x - 4y + 1 = 0,\ 3x + 4y - 10 = 0$

3. $2x + 3y + 2 = 0,\ 2x + 3y - 3 = 0$

4. $2x + 3y + 8 = 0,\ 4x + 6y - 21 = 0$

In Exercises 5 – 10, find the equation of the line passing through the given point and parallel to the given line.

5. (4, 3), $4x - 3y + 12 = 0$

6. (−1, 4), $5x - 2y - 1 = 0$

7. (−3, 2), $2x + 3y - 5 = 0$

8. (−2, −1), $3x - 2y + 7 = 0$

9. (−2, 5), x-axis

10. (5, 3), y-axis

11. A line passing through (−2, 1) and (3, y) is parallel to a line with slope of 2. Find the value of y.

12. A line passing through (5, 5) and (x, 8) is parallel to the line $3x + 2y + 8 = 0$, find the value of x.

13. Find the equation of the line which passes through the mid- point of the line joining the points A (2, 1) and B (5, 3) and parallel to the line $3x + 2y - 8 = 0$.

In Exercises 14 – 19, state the slope of the line perpendicular to the line with slope:

14. 3 15. -2 16. ½

17. $-1/3$ 18. 3/4 19. $-3/2$

In Exercises 20 – 23, determine whether the pair of lines represented by the equations are perpendicular.

20. $x - 3y = 6,\ 3x + y = 3$

21. $4x + 3y - 6 = 0,\ 3x - 4y + 4 = 0$

22. $2x - 3y + 12 = 0,\ 4x + 3y - 6 = 0$

23. $3x - 4y + 1 = 0,\ 3x + 4y - 10 = 0$

In Exercises 24 – 29, find the equation of the line passing through the given point and perpendicular to the given line.

24. (4, 3), $5x + 3y - 10 = 0$

25. (−2, 5), $x + 3y + 2 = 0$

26. (−3, 2), $3x + 2y - 7 = 0$

27. (3, −4), $2x - 3y + 8 = 0$

28. (1, 2), $2y + 3x - 1 = 0$

29. (2, 3), $2x - 3y + 5 = 0$

30. Find the equation of the perpendicular bisector of the line joining the points A (2, 1) and B (6, 5).

31. A line passing through (3, 2) and (y, 5) is perpendicular to a line with slope $\frac{4}{3}$. Find the value of y.

32. A line passing through (x, −2) and (3, −1) is perpendicular to the line $3x + 4y - 8 = 0$. Find the value of x.

33. Find the equation of the perpendicular bisector of the line joining the points A (−4, −1) and B (8, 7).

34. Three points A (1, 1), B (4, 3) and C (−1, 4) are the vertices of a triangle. Verify that triangle ABC is a right triangle. What is its area?

9.6 Sketching Graphs of Linear Functions

Every equation of the form $ax + by = c$ has a straight line as its graph. Two points are needed to sketch the graph of a linear function. Knowing where the line crosses the x-axis and the y-axis gives us another way of sketching the graph of a linear equation.

Example 1

Sketch the graph of $3x + 2y = 6$.

Solution

Begin by finding the x- and y-intercepts. To find the x-intercept, let $y = 0$ and solve for x.

$$3x + 2(0) = 6$$

$$3x = 6$$

$$x = 2$$

The x-intercept is (2. 0).

Similarly, to find the y-intercept let $x = 0$ and solve for y.

$$3(0) + 2y = 6$$

$$2y = 6$$

$$y = 3$$

The y-intercept is (0, 3).

Plot these points and draw a line through the points, as shown in Figure 9.11.

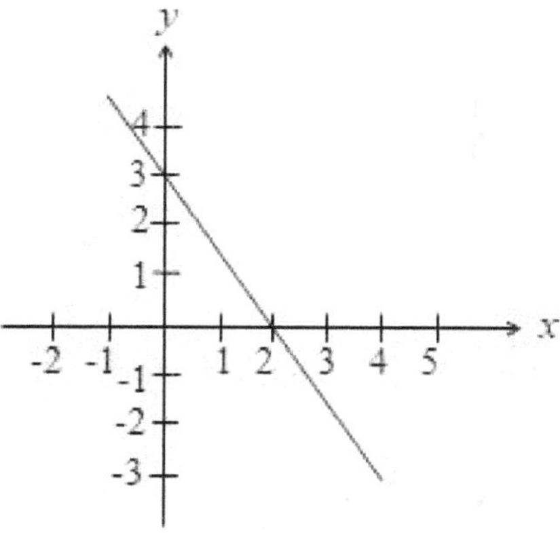

Figure 9.11

Using the Slope and *y*-intercept

Example 2

Sketch the graph of $3x - 2y = -4$.

Solution

Begin by writing the equation in the slope-intercept form.

$$3x - 2y = -4$$

$$-2y = -3x - 4$$

$$y = \frac{3}{2}x + 2$$

You can see that the slope is 3/2 and the *y*-intercept is (0, 2).

Plot the point (0, 2). Because the slope is 3/2 the line rises three units for each two units the line moves to the right. So, stating from (0, 2) move three units up and then two units to the right. This locate the point (2, 5). Plot this point and draw a line through (0, 2) and (2, 5), as shown in Figure 9.12.

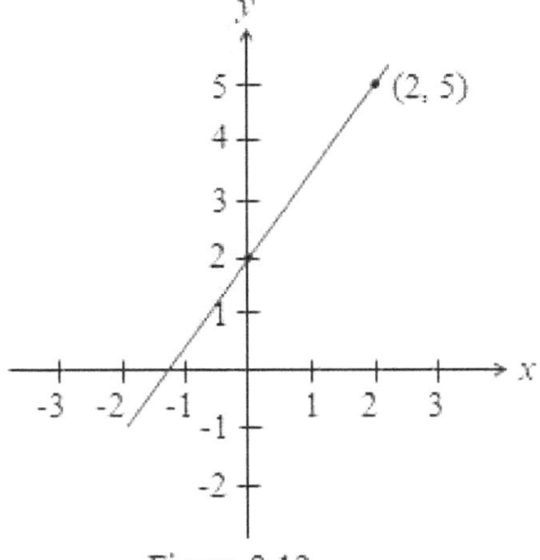

Figure 9.12

Sketching a Graph of a Line through the Origin

Example 3

Sketch the graph of $3x + 2y = 0$.

Solution

Begin by solving the equation of $3x + 2y = 0$ for y.

$$3x + 2y = 0$$

$$2y = -3x \qquad \text{Subtract } 3x \text{ from each side}$$

$$y = -\frac{3}{2}x \qquad \text{Divide by 2}$$

The line passes through the origin (0, 0), so the line has only one intercept. To get a second point you can choose a convenient value of x and find the corresponding value of y, or you can use the slope of the line.

If $x = 2$, we have

$$3(2) + 2y = 0$$

$$2y = -6$$

$$y = -3$$

giving the point (2, −3). Draw a line through this point and the origin, as shown in Figure 9.13.

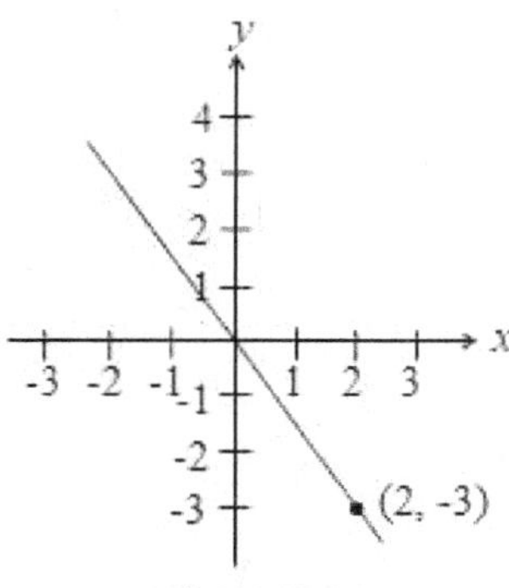

Figure 9.13

Exercise 9.6

Sketch the graph of each of the following linear functions:

1. $x + 3y = 6$
2. $2x - y = 6$
3. $2x - 3y = 6$
4. $x - 4y + 8 = 0$
5. $3x + 2y = 12$
6. $4x + 5y = 20$
7. $8x - 12y = 24$
8. $3x - 4y - 48 = 0$
9. $3x - 4y - 12 = 0$
10. $5x - 6y = 0$
11. $2x + 7y = 0$
12. $8x = 12y$

Review Exercises

In Exercises 1 – 12, find the distance between the following pairs of points:

1. (7, 3); (5, 8) 2. (3, 1); (6, 5)

3. (10, 3); (5, 15) 4. (−5, 0); (−8, 2)

5. (−2, 1); (3, 4) 6. (−1, −4); (−8, −4)

7. (− 6, 3); (− 6, −5) 8. (3, −1); (−2, −3)

9. (−3, 4); (4, −6) 10. (3, 3); (9, 7)

11. (7, 4); (−2, − 8) 12. (−3, 1); (4, −5)

In Exercises 13 – 24, find the mid-point of the line segment joining the points.

13. (0, −2); (8, 4) 14. (−2, −3); (2, 7)

15. (6, 1); (3, 6) 16. ((5,6); (−3, −4)

17. (1, 1); (7, 5) 18. (2, 7); (5, 5)

19. (4, − 5); (1, 5) 20. (− 8, −5); (3, 1)

21. (−1, 2); (5, 4) 22. (3, 6); (6, 3)

23. (−7, 3); (3, − 5) 24. (8, −3); (−4, 7)

In Exercises 25- 32, find the slope of the line passing through the pair of points.

25. (4, 3); (7, −3) 26. (5, −8); (2, −4)

27. (−2, −3); (4, 6) 28. (5, 6); (−3, −4)

29. (−2, 1); (1, 5) 30. (2, −1); (−2, − 4)

31. (5, −3); (3, 4) 32. (−3, −2); (− 6, −4)

In Exercises 33 – 35, determine whether the points are collinear.

33. (−2, − 3); (4, 6); (6, 9)

34. (2, −1); (−2, − 4); (6, 2)

35. (5, − 3); (3, 4); (7, −10)

In Exercises 36 – 43, find the equation of the line that passes through the given point and has the indicated slope.

36. (0, −2); 3 37. (2, −3); 4

38. (3, 0); −2 39. (−2,3); ¾

40. (2, 0); −3 41. (0, 3); 2

42.(−2, 0); 2/3 43. (−2, −3); ¼

In Exercises 44 – 51, find the equation of the line that passes through the pair of points.

44. (1, 4); (5, 6) 45. (2, 6); (4, 1)

46. (2, 3); (5, 6) 47. (1, -3); (4, 2)

48. (-3, 4); (2, 4) 49. (4, −5); (2, 1)

50. (−2, −3); (−5, 1) 51. (5, 3); (8, 5)

In Exercises 52 – 57, find the slope and the y-intercept of each of the following lines.

52. $5x - 3y = 15$ 53. $4y + 3x = 10$

54. $6x - 2y + 5 = 0$ 55. $4x + 5y = 12$

56. $4x - 3y = 8$ 57. $3x - 6y + 9 = 0$

In Exercises 58 – 61, find the equation of the line passing through the given point and parallel to the given line.

58. (4, −3); $4x - 3y + 12 = 0$ 59. (−1, 2); $5x - 2y - 1 = 0$

60. (−3, 1); $2x + 3y - 5 = 0$ 61. (2, −2); $3x - 2y + 7 = 0$

In Exercises 62 – 67, find the equation of the line passing through the given point and perpendicular to the given line.

62. (4, 3); $5x - 3y - 10 = 0$ 63. (−2, −5); $x + 3y + 2 = 0$

64. (−3, 2); $3x - 2y + 7 = 0$ 65. (3, −2); $2x - 3y + 8 = 0$

66. (−1, −2); $2y + 3x - 1 = 0$ 67. (2, 3); $2x + 3y - 5 = 0$

In Exercises 68 – 76, sketch the graph of the linear function.

68. $2x + 3y = 6$ 69. $5x - 4y = 20$ 70. $2y - 3x = 0$

71. $3x + 2y = 0$ 72. $x - 3y = 6$ 73. $x/3 - y/2 = 4$

74. $3y + 5 = 0$ 75. $5x - 2 = 0$ 76. $2x + 3y = 9$

10 Quadratic Functions

The quadratic function is defined by the equation of the form $y = ax^2 + bx + c$ where a, b and c are real numbers and $a \neq 0$.

Zeros of Quadratic Functions

The zeros of a quadratic function f are the x-values for which $f(x) = 0$. To find the zeros of a function, set the function equal to zero and solve for x.

10.1The Quadratic Formula

In Chapter 5, you learned to apply the Quadratic Formula to solve quadratic equations. In this section, we derive the Quadratic Formula by using the method of completing the square.

$$ax^2 + bx + c = 0$$ Original equation

$$ax^2 + bx = -c$$ Isolate c.

$$x^2 + \frac{b}{a}x = -\frac{c}{a}$$ Divide each side by a.

$$x^2 + \frac{b}{a}x + \frac{b^2}{4a^2} = \frac{b^2}{4a^2} - \frac{c}{a}$$ Add the square of $b/2a$ to each side.

$$\left(x + \frac{b}{2a}\right)^2 = \frac{b^2-4ac}{4a^2}$$ Factor the expression on the left

$$x + \frac{b}{2a} = \pm\sqrt{\frac{b^2-4ac}{4a^2}}$$ Take square root of each side.

$$x = -\frac{b}{2a} \pm \sqrt{\frac{b^2-4ac}{4a^2}}$$ Solve for x.

$$x = \frac{-b \pm \sqrt{b^2-4ac}}{2a}$$ Simplify.

So, if $ax^2 + bx + c = 0$, where $a \neq 0$, then

$$x = \frac{-b \pm \sqrt{b^2-4ac}}{2a}.$$

This solution of the general quadratic equation is called the Quadratic Formula. The solutions are also called roots of the quadrant equation.

Discriminant

The expression $b^2 - 4ac$ inside the radical is called the discriminant, and it gives us useful information about the number and type of solutions of .a quadratic equation.

Let $ax^2 + bx + c = 0$, where a, b and c are real numbers and $a \neq 0$.

1. If $b^2 - 4ac > 0$, there are two distinct real solutions.

2. If $b^2 - 4ac = 0$, there is one (repeated) real solution.

3. If $b^2 - 4ac < 0$, there are no real solutions.

The x-intercepts of the graph of the quadratic function $f(x) = ax^2 + bx + c$ are the solutions of the equation $ax^2 + bx + c = 0$. The graph of a quadratic function may intersect the x-axis at two points or intersect at one point or may not intersect at all. The number of the x-intersect can be determined using the value of the denominator.

The graph of the quadratic equation $ax^2 + bx + c = 0$ intersects the x-axis at two points if $b^2 - 4ac > 0$.

The graph of the quadratic equation $ax^2 + bx + c = 0$ intersects the x-axis at one point if $b^2 - 4ac = 0$.

The graph of the quadratic equation $ax^2 + bx + c = 0$ would not intersect the x-axis if $b^2 - 4ac < 0$.

The graphs shown in Figure 10.1 illustrate the three cases.

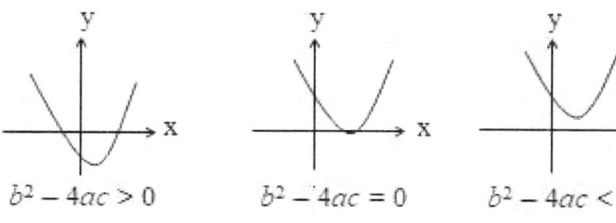

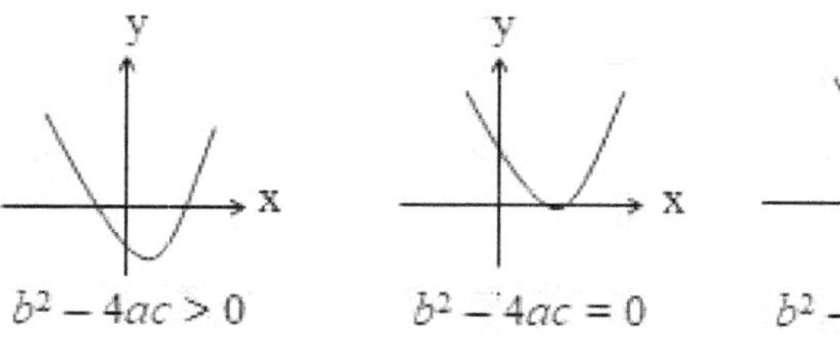

Figure 10.1

Example 1

Determine the type of solution(s) for each quadratic equation.

(a) $x^2 + 2x - 3 = 0$

(b) $9x^2 + 6x + 1 = 0$

(c) $2x^2 - 3x + 5 = 0$

Solution

(a) $x^2 + 2x - 3 = 0$ has two distinct real solutions

since $2^2 - 4(2)(-3) = 28$

(b) $9x^2 + 6x + 1 = 0$ has one real solution since

$6^2 - 4(9)(1) = 0$

(c) $2x^2 - 3x + 5 = 0$ has no real solutions since

$(-3)^2 - 4(2)(5) = -31$

Example 2

Determine all real values of k for which the equation $x^2 + 2x - k = 0$ has real roots.

Solution

Since the equation has real roots $b^2 - 4ac \geq 0$.

$$2^2 - 4(1)(-k) \geq 0$$

$$4 + 4k \geq 0$$

$$k \geq -1$$

Exercise 10.1

In Exercises 1 – 6, determine the type of solution(s) for each quadratic equation.

1. $x^2 - 4x + 3 = 0$
2. $x^2 - 3x + 5 = 0$
3. $x^2 - 4x + 4 = 0$
4. $2x^2 - 3x + 4 = 0$
5. $3x^2 + 7x - 6 = 0$
6. $4x^2 + 4x + 1 = 0$

In Exercises 7 – 15, find the value of the discriminant and give the number of real solutions.

7. $3x^2 + 8x = 0$
8. $2x^2 - 5x = 0$
9. $x^2 - 8x + 16 = 0$
10. $4x^2 + 12x + 9 = 0$
11. $3x^2 - 7x + 1 = 0$
12. $2x^2 - x + 5 = 0$
13. $3x^2 - x - 2 = 0$
14. $2x^2 - 5x + 11 = 0$
15. $9x^2 + 12x + 4 = 0$
16. Find the values of m for which the equation $x^2 + (m + 3)x + 4m = 0$ has equal roots.
17. If the equation $px^2 + (p + 1)x + p = 0$ has equal roots, find the value(s) of p.
18. Find the range of values of k for which $kx^2 + 4x + 2 = 0$ has real roots.
19. Find the range of values of k for which $x^2 + x - k = 0$ has two distinct roots.
20. Find the smallest possible integral value of k such that $2x^2 + 3x + k = 0$ will have imaginary roots.
21. Find the range of values of k for which the equation $x^2 + (3 - k)x + 1 = 0$ has real roots.

10.2 Finding Maximum and Minimum Values

You can find the maximum or minimum value of a quadratic function by a process of completing the square. The following examples illustrate the process.

Example 1

Find the minimum value of:

(a) $f(x) = x^2 + 6x + 14$ (b) $f(x) = 2x^2 - 5x + 7$

Solution

(a) $f(x) = x^2 + 6x + 14$

$= x^2 + 6x + 9 - 9 + 14$ Complete the square

$= x^2 + 6x + 9 + 5$

$= (x + 3)^2 + 5$ Factor the quadratic expression

Because $(x + 3)^2$ is a square it is always positive or zero. The value of the function increases whenever x is replaced with any real number except – 3. Thus, the minimum value of the function is $f(-3) = 5$.

(b) $f(x) = 2x^2 - 5x + 7$

$= 2\left[x^2 - \frac{5}{2}x + \frac{7}{2}\right]$ Factor 2

$= 2\left[x^2 - \frac{5}{2}x + \frac{25}{16} - \frac{25}{16} + \frac{7}{2}\right]$ Complete the square

$= 2\left[\left(x - \frac{5}{4}\right)^2 + \frac{31}{16}\right]$ Factor the quadratic expression.

$= 2\left(x - \frac{5}{4}\right)^2 + \frac{31}{8}$ Remove the bracket.

Because $(x - 5/4)^2$ is a square it is always positive or zero. The value of the function increases whenever x is replaced with any real number except 5/4.Thus, the minimum value of the function is $f(5/4) = 31/8$.

Example 2

Find the maximum value of:

(a) $f(x) = 5 - 3x - x^2$ (b) $f(x) = 5 + 2x - 3x^2$

Solution

(a) $f(x) = 5 - 3x - x^2$

$= -x^2 - 3x + 5$

$= -(x^2 + 3x - 5)$ Factor – 1

$= -\left(x^2 + 3x + \frac{9}{4} - \frac{9}{4} - 5\right)$ Complete the square

$= -\left[\left(x + \frac{3}{2}\right)^2 - \frac{29}{4}\right]$ Factor the quadratic expression

$= \frac{29}{4} - \left(x + \frac{3}{2}\right)^2$ Use the distributive property

Because $(x + 3/2)^2$ is a square it is always positive or zero. The value of the function decreases whenever x is replaced with any real number except $-3/2$. Thus, the maximum value of the function is $f(-3/2) = 29/5$.

(b) $f(x) = 5 + 2x - 3x^2$

$= -3x^2 + 2x + 5$

$= -3\left[x^2 - \frac{2}{3}x - \frac{5}{3}\right]$ Factor – 3

$= -3\left[x^2 - \frac{2}{3}x + \frac{1}{9} - \frac{1}{9} - \frac{5}{3}\right]$ Complete the square

$= -3\left[\left(x - \frac{1}{3}\right)^2 - \frac{16}{9}\right]$ Factor the quadratic expression

$= \frac{16}{3} - 3\left(x - \frac{1}{3}\right)^2$ Use the distributive property

Because $(x - ⅓)^2$ is a square it is always positive or zero. The value of the function decreases whenever x is replaced with any real number except 1/3. Thus, the maximum value of the function is $f(1/3) = 16/3$.

Example 3

Find the minimum point of the parabola $f(x) = x^2 - x - 6$.

Solution

$$f(x) = x^2 - x - 6$$

$$= x^2 - x + \frac{1}{4} - \frac{1}{4} - 6$$

$$= x^2 - x + \frac{1}{4} - \frac{25}{4}$$

$$= \left(x - \frac{1}{2}\right)^2 - \frac{25}{4}$$

The minimum point is (½, – 25/4).

Exercise 10.2

In Exercises 1 – 6, find the maximum value of each of the following functions:

1. $y = 2 + x - x^2$
2. $y = 5 - 2x - x^2$
3. $y = 6 + 5x - x^2$
4. $y = 3 - 5x - 2x^2$
5. $y = 2 + 4x - 3x^2$
6. $y = 1 - 6x - 3x^2$

In Exercises 7 – 12, find the minimum value of each of the following functions:

7. $y = x^2 + 3x - 2$
8. $y = x^2 - 8x + 6$
9. $y = x^2 + 2x + 3$
10. $y = 2x^2 + 3x - 4$
11. $y = 3x^2 - 6x + 4$
12. $y = 2x^2 + 6x + 5$

In Exercises 13 – 18, find the minimum or maximum point of the parabola of each function.

13. $y = x^2 - 6x + 10$
14. $y = 6 - x - x^2$
15. $y = 2 - 2x - x^2$
16. $y = 2x^2 + 8x + 9$
17. $y = 3x^2 + 6x + 7$
18. $y = -7 + 12x - 3x^2$

10.3 Sketching Graphs of Quadratic Functions

The graph of the quadratic function $f(x) = ax^2 + bx + c$, $a \neq 0$ is a parabola which opens either upward or downward, as shown in Figure 10.2.

$a > 0$

$a < 0$

Figure 10.2

If a is positive, the parabola opens upward and has a minimum (or lowest) point. If a is negative, the parabola opens downward and has a maximum (or highest) point. The highest (or lowest) point on a parabola is called the vertex (or the turning point) of the parabola.

The y-coordinate of the vertex is the minimum (or the maximum) value of the function. The graph of a quadratic function is symmetric with respect to a vertical line through the vertex, as shown in Figure 10.3. This line, called the axis of symmetry, is midway between any pair of symmetric points on the parabola.

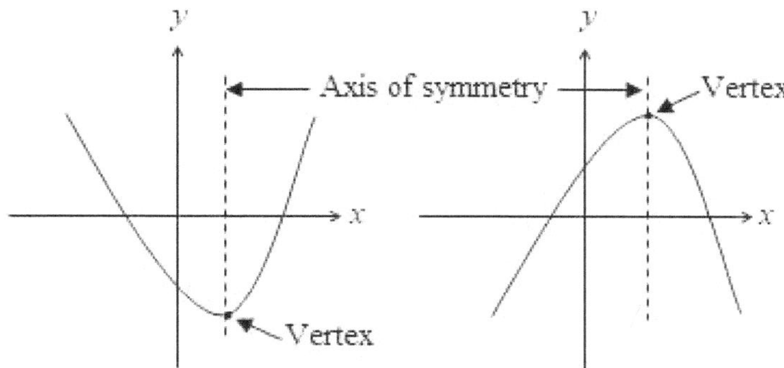

Figure 10.3

The graph of the quadratic function $y = ax^2 + bx + c$ intersects the y-axis at $(0, c)$. This point lies on the line $y = c$. Substituting c for y we have

$$c = ax^2 + bx + c$$

$$0 = ax^2 + bx$$

$$0 = x(ax + b)$$

Setting each factor to 0 and solving for x, gives

$$x = 0 \quad \text{and} \quad x = -\,b/a$$

Hence, the points $(0, c)$ and $(-\,b/a, c)$ are two symmetric points. The x- coordinate of the midpoint of the line joining these points is $[0 + (-\,b/a)]/2 = -\,b/2a$.

So, the x-coordinate of the vertex is $-$ b/2a. You can find the y-coordinate of the vertex by substituting $-\,b\,/2a$ into the original equation. The equation of the axis of symmetry is $x = -\,b/2a$.

Example 1

Sketch the graph of $y = x^2 - 4x - 12$.

Solution

Begin by finding the x- and y-intercepts. To find the x-intercept, let $y = 0$ and solve for x.

$$x^2 - 4x - 12 = 0$$

$$(x - 6)(x + 2) = 0$$

$$x - 6 = 0 \quad \text{or} \quad x + 2 = 0$$

$$x = 6 \qquad\qquad x = -\,2$$

The x-intercepts are (6, 0) and (− 2, 0).

Letting $x = 0$, we get $y = -12$. So, the y-intercept is (0,−12)

Now we find the vertex of the graph. Identify the values of a and b in $y = x^2 - 4x - 12$, and use $x = -b/2a$ to find the x-coordinate of the vertex.

Here $a = 1$, $b = -4$ and $c = -12$.

So $x = -(-4)/(2 \cdot 1) = 2$

Find the y-coordinate of the vertex by substituting 2 for x in $y = x^2 - 4x - 12$.

$$y = 2^2 - 4 \cdot 2 - 12$$

$$= 4 - 8 - 12$$

$$= -16$$

The vertex of this parabola is (2, −16).

Since the coefficient of x^2 is positive, the curve opens upward.

Plotting the vertex, the x- and y-intercepts gives the graph shown in Figure 10.4.

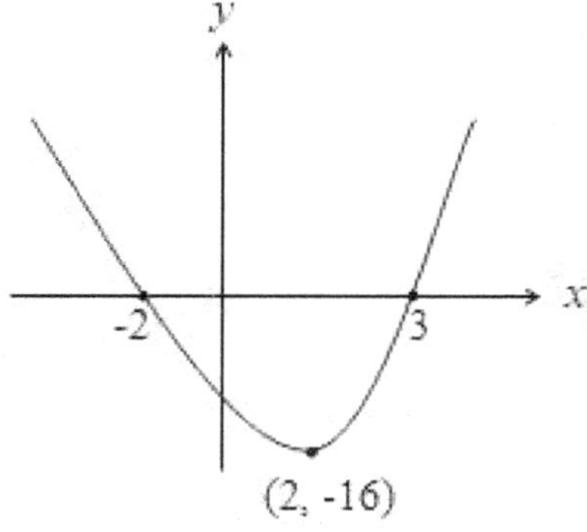

Figure 10.4

Example 2

Sketch the graph of $y = 5 + 6x - x^2$.

Solution

The y-intercept of the curve is (0, 5).

The quadratic expression $5 + 6x - x^2$ cannot be factored, so we find two symmetric points.

Substituting 5 for y in $y = 5 + 6x - x^2$ gives

$$5 = 5 + 6x - x^2$$

$$0 = x(6 - x)$$

So, $x = 0$ and $x = 6$.

The points (0, 5) and (6, 5) are symmetric points.

Next, find the vertex of the curve. The x-coordinate of the vertex is

$$x = -\frac{b}{2a} = -\frac{6}{2 \cdot -1} = 3$$

The y-coordinate of the vertex is

$$y = 5 + 6 \cdot 3 - 3^2$$

$$= 5 + 18 - 9$$

$$= 14$$

The vertex of this parabola is (3, 14).

Because the coefficient of x^2 is negative the curve opens downward. Plotting the vertex, the x- and y-intercepts gives the graph shown in Figure 10.5.

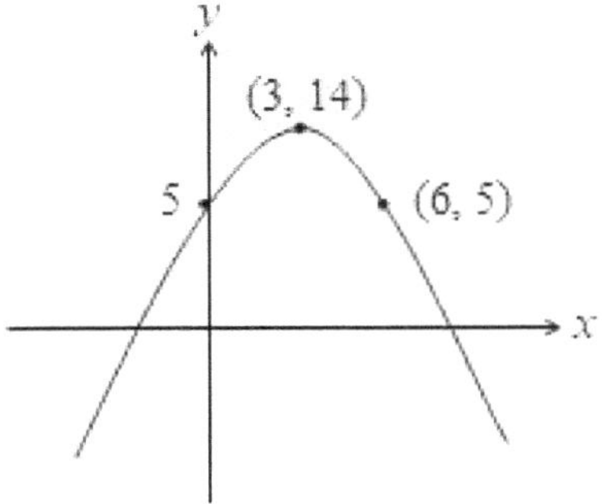

Figure 10.5

Example 3

Sketch the graph of $y = 12 - 5x - 2x^2$.

Solution

By substituting $x = 0$ into $y = 12 - 5x - 2x^2$ we get $y = 12$.

Also, substituting 0 for y, we get

$$12 - 5x - 2x^2 = 0$$

$$(3 - 2x)(4 + x) = 0$$

So $x = 3/2$ and $x = -4$.

The x-intercepts are $(3/2, 0)$ and $(-4, 0)$.

One way to find the vertex of a parabola is to use the technique described in Section 10.2.

Now
$$12 - 5x - 2x^2 = -2\left[x^2 + \frac{5}{2}x - 6\right]$$
$$= -2\left[x^2 + \frac{5}{2}x + \frac{25}{16} - \frac{25}{16} - 6\right]$$
$$= -2\left[\left(x + \frac{5}{4}\right)^2 - \frac{121}{16}\right]$$
$$= \frac{121}{8} - 2\left(x + \frac{5}{4}\right)^2$$

You can see that the vertex of the parabola is $(-5/4, 121/8)$.

The curve opens downward since the coefficient of x^2 is negative. Plotting the vertex, the x- and y-intercepts gives the graph shown in Figure 10.6.

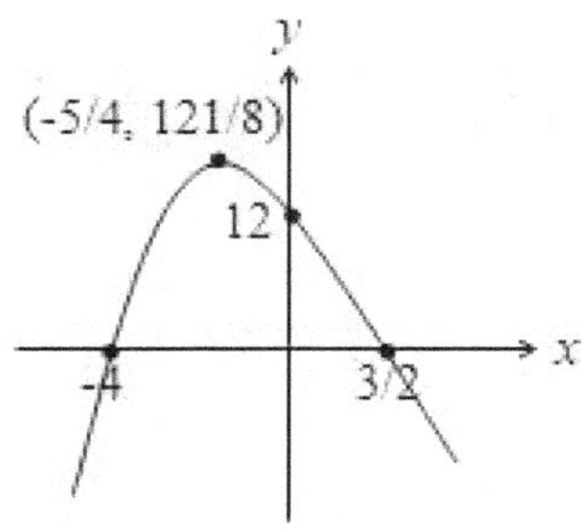

Figure 10.6

The process for sketching graphs of quadratic functions illustrated in the preceding examples is summarized below.

1. Find the vertex by completing the square or by using the formula $x = -b/2a$.

2. Find the x- and y- intercepts (or two symmetric points).

3. Plot the vertex of the graph.

4. Plot either the x- and y-intercepts (or the symmetric points).

5. Complete the sketch, using the fact that the curve opens upward if a is positive and opens downward if a is negative.

Exercise 10.3

Sketch the graph of each of the following quadratic functions:

1. $y = x^2 + 6x + 5$ 2. $y = 4 - 3x - x^2$ 3. $y = x^2 + 2x - 10$

4. $y = x^2 - 2x - 15$ 5. $y = -x^2 + 6x - 8$ 6. $y = x^2 + 6x + 8$

7. $y = 12 + x - x^2$ 8. $y = 6 + x - 2x^2$ 9. $y = 3x^2 + 12x + 3$

10. $y = -2x^2 - 4x - 1$ 11. $y = 2x^2 + 4x - 1$ 12. $y = 1 + 6x - 3x^2$

13. $y = 2x^2 - 6x + 7$ 14. $y = 3x^2 - 6x + 4$ 15. $y = |x^2 + x - 6|$

16. $y = |2x^2 - x - 6|$ 17. $y = |4 - 3x - x^2|$ 18. $y = |3 - 5x - 2x^2$

10.4 Quadratic Inequalities

An expression such as $ax^2 + bx + c < 0$, where $a \neq 0$ is called a quadratic inequality. Note that the inequality symbol $<$ can be replaced by the symbol $>$, $\leq$ or $\geq$.

Solving Quadratic Inequalities

The solutions of a quadratic inequality can be found graphically or by using algebraic methods.

Graphical Solution of Quadratic Inequalities

To solve the quadratic inequality $ax^2 + bx + c < 0$ graphically, use the technique in Section 10.3 to graph $y = ax^2 + bx + c$. Then determine from the graph the values of x for which y is negative.

Example 1

Solve the inequality $x^2 + 2x - 15 < 0$.

Solution

Sketch the graph of $y = x^2 + 2x - 15$. The graph is shown in Figure 10.7.

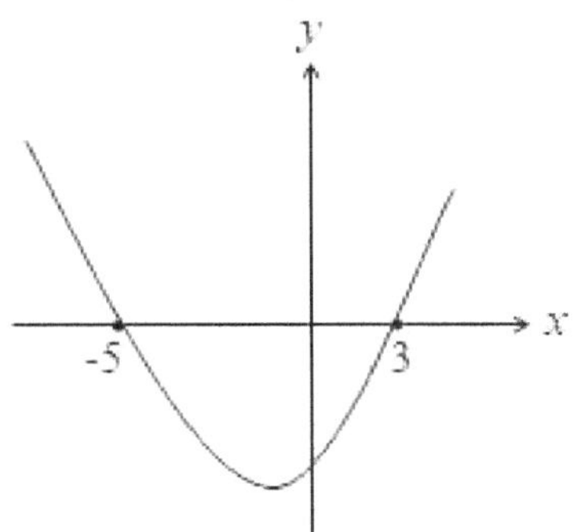

Figure 10.7

The graph intercepts the x axis when x is -5 or 3. From the graph determine the values of x that make $x^2 + 2x - 15 < 0$ a true statement. You can see that the graph is below the x axis for values of x between -5 and 3. So the solution set of the inequality is $(-5, 3)$.

Example 2

Solve the inequality $x^2 - 3x - 20 \geq 8$.

Solution

First, rewrite the inequality in the general form.

$$x^2 - 3x - 28 \geq 0$$

Next, sketch the graph of $y = x^2 - 3x - 28$.

The graph is shown in Figure 10.8.

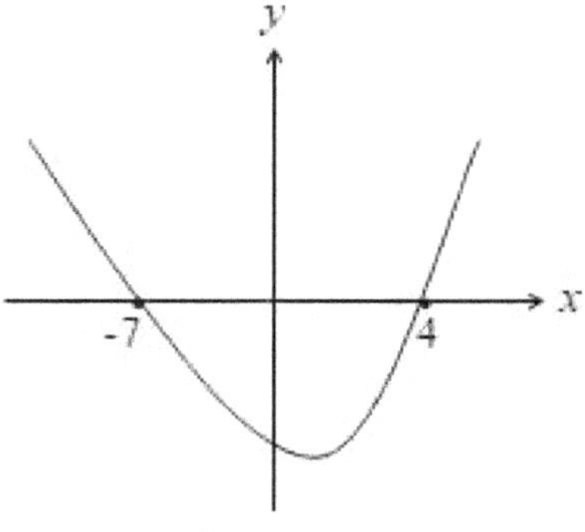

Figure 10.8

In this case, look for all values of x for which $y > 0$ or $y = 0$. You can see that y is positive or zero in the interval $(-\infty, -7]$ and in the interval $[4, \infty)$. Hence, the solutions set of the inequality $2x^2 - 3x - 20 \geq 8$ consists of the union of the intervals $(-\infty, -7]$ and $[4, \infty)$, which is written as $(-\infty, -7] \cup [4, \infty)$.

Algebraic Solution of Quadratic Inequalities

Example 3

Solve the inequality $x^2 - 4x - 32 < 0$.

Solution

First, find the solutions of the corresponding quadratic equation.

$$x^2 - 4x - 32 = 0$$

$$(x - 8)(x + 4) = 0$$

$x = -4$ or $x = 8$

The quadratic equation has solutions – 4 and 8 called the critical numbers of $x^2 - 4x - 32 < 0$. The critical numbers divide a number line into three regions called the test regions, as shown in Figure 10.9.

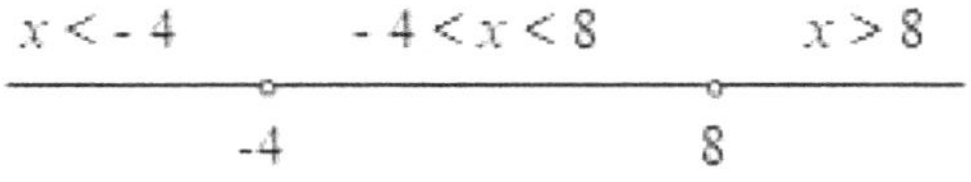

$x < -4$ $\quad -4 < x < 8 \quad$ $x > 8$

-4 $\quad$ 8

Figure 10.9

Next, choose a number in each of these test regions and substitute this number into the quadratic expression to determine whether your answer is positive or negative. If you get a negative answer, it means that the value of the expression is negative for all values of x in that region. Similarly if you get a positive answer it means that all values of the expression are positive for all values of x in that region.

Test interval	Chosen x value	$x^2 - 4x - 32$
$x < -4$	$x = -5$	$25 + 20 - 32 > 0$
$-4 < x < 8$	$x = 1$	$1 - 4 - 32 < 0$
$x > 8$	$x = 9$	$81 - 36 - 32 > 0$

Note that, we need not find the actual value of the expression. All that we need is the sign of the value of the expression. We add the information from the table to the number line, as shown in Figure 10.10.

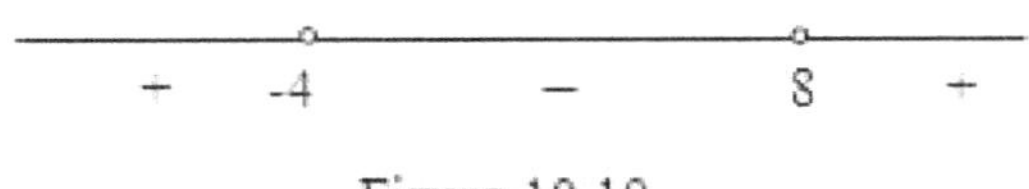

Figure 10.10

The inequality $x^2 - 4x - 32 < 0$ is satisfied by all x-values in the interval $(-4, 8)$, so the solution set of the inequality is the interval $(-4, 8)$.

Notice that the critical numbers were found by factoring the quadratic expression. If the quadratic expression cannot be factored, use the Quadratic Formula to find the critical numbers.

Example 4

Solve the inequality $(x - 4)(x + 8) \geq 13$.

Solution

First rewrite the inequality in the general form

$$(x-4)(x+8) \geq 13 \qquad \text{Original inequality}$$

$$x^2 + 4x - 45 \geq 0 \qquad \text{Standard form}$$

Next find the critical numbers by solving the quadratic equation.

$$x^2 + 4x - 45 = 0$$

$$(x-5)(x+9) = 0$$

$$x = 5 \text{ or } x = -9$$

Choose a number in each test region to determine if the inequality is positive or negative, as shown the following table.

Test interval	Chosen x value	$x^2 + 4x - 45$
$x < -9$	$x = -10$	$100 - 40 - 45 > 0$
$-9 < x < 5$	$x = 1$	$1 + 4 - 45 < 0$
$x > 5$	$x = 6$	$36 + 24 - 45 > 0$

The information is added to the number line shown in Figure10.11.

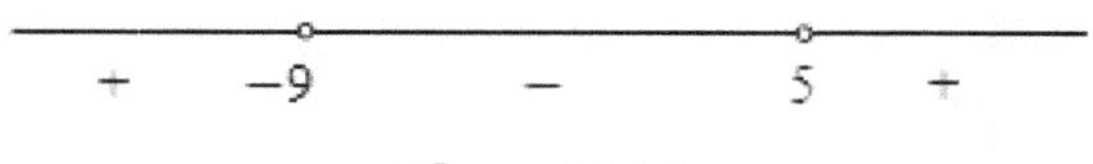

Figure 10.11

From the graph you can see that the inequality $(x-4)(x+8) \geq 13$ is satisfied by all x-values in the intervals $(-\infty, -9]$ and $[5, \infty)$. So, the solution set of the inequality is $(-\infty, -5] \cup [5, \infty)$.

The following alternative method can be used to find the solution set of a quadratic inequality if the quadratic expression can be factored.

Example 5

Solve the inequality $x^2 + 2x - 15 < 0$.

Solution

First, factor the quadratic expression.

$$(x-3)(x+5) < 0$$

Next, find the critical numbers.

$$(x-3)(x+5)=0$$

$$x=3 \text{ or } x=-5$$

The value of $x-3$ is positive when $x>3$ and negative when $x<3$. Also, the value of $x+5$ is positive when $x>-5$ and negative when $x<-5$. The information is illustrated in Figure 10.12.

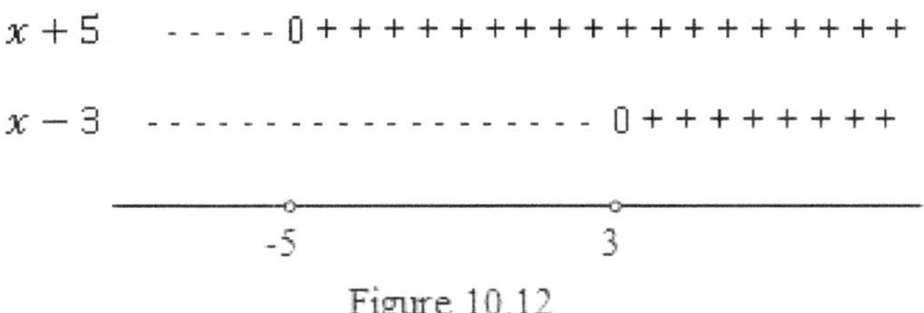

Figure 10.12

Here, we want those values of x for which the product $(x-3)(x+5)$ is negative. You can see from the sign graph above that the solution is $(-5, 3)$.

Example 6

Solve the inequality $x^2 > 4x + 21$.

Solution

First, rewrite the inequality in standard form.

$$x^2-4x-21>0$$

Next, factor the quadratic expression.

$$(x-7)(x+3)>0$$

The sign graph is shown in Figure 10.13.

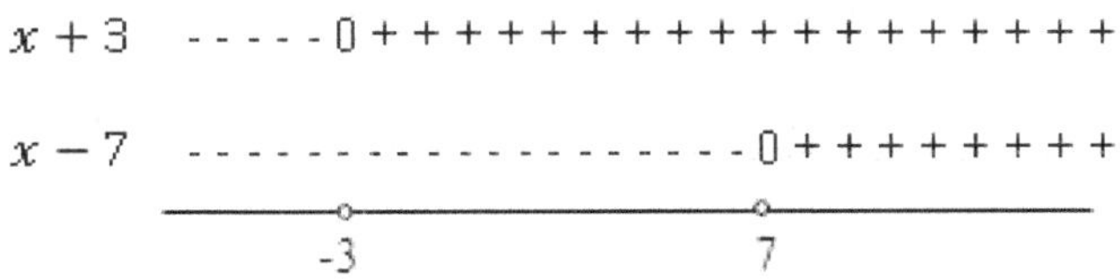

x + 3 - - - - - 0 + + + + + + + + + + + + + + + + + + +

x − 7 - - - - - - - - - - - - - - - - - - - 0 + + + + + + + + +

-3 7

Figure 10.13

We are looking for a region where $(x-7)(x+3)$ is positive. From the graph we see that the solution is $(-\infty, -3) \cup (7, \infty)$.

Exercise 10.4

In Exercises 1 – 10, use graphs to solve each of the following quadratic inequalities:

1. $x^2 + x - 12 < 0$
2. $x^2 + 3x - 10 > 0$
3. $x^2 + 4x - 12 > 0$
4. $x^2 + 2x - 15 < 0$
5. $x^2 - 5x + 6 \geq 0$
6. $x^2 + 7x + 10 \leq 0$
7. $2x^2 + x - 6 \leq 0$
8. $3x^2 - 10x - 8 \leq 0$
9. $4x^2 + x - 3 < 0$
10. $x(x + 5) \geq 24$

In Exercises 11 – 22, use an algebraic method to solve each of the following inequalities:

11. $x^2 - 3x - 4 > 0$
12. $x^2 - 2x - 8 < 0$
13. $x^2 - x - 12 \leq 0$
14. $x^2 + 7x + 12 \geq 0$
15. $x^2 - 3x - 10 < 0$
16. $x^2 + 6x - 27 > 0$
17. $(x + 3)(x - 4) \leq 0$
18. $x^2 - 3x > 18$
19. $2x^2 + 5x + 3 < 0$
20. $3x^2 + 8x + 5 \geq 0$
21. $1 - x - 2x^2 \leq 0$
22. $6 + 7x \geq 3x^2$

Review Exercises

In Exercises 1 – 6, find the nature of the roots of each of the following equations:

1. $x^2 - 4x + 3 = 0$
2. $x^2 - 3x + 5 = 0$
3. $x^2 - 4x + 4 = 0$
4. $x^2 + 6x + 8 = 0$
5. $2x^2 + 3x + 3 = 0$
6. $3x^2 - 2x - 1 = 0$

In Exercises 7 – 12, for each quadratic equation find the value of the discriminant and give the number of real solutions:

7. $x^2 - 11x + 28 = 0$
8. $3x^2 - 7x = 0$
9. $4x^2 - 4x + 1 = 0$
10. $2x^2 + 4x - 3 = 0$
11. $9x^2 + 24x + 16 = 0$
12. $3 - 6x - 4x^2 = 0$
13. Find the range of values of k for which the equation $x^2 + (3 - k)x + 1 = 0$ has real roots.
14. Find the largest possible value of k such that $kx^2 + 4x + 2 = 0$ has real roots.

15. Find the value of k for which $x^2 - (k - 2)x + 2k + 1 = 0$ has equal roots.

16. Find the smallest possible integral value of k such that $2x^2 - 3x + k = 0$ will have imaginary roots.

17. Find the range of values of k for which the equation $x^2 + (k + 2)x + 3k - 2 = 0$ has an imaginary roots.

18. Find the range of values for k for which the expression $3x^2 + 2x + k$ is always positive for all real values of x.

19. Find the range of values of k for which the expression $3x^2 + 6x + k$ is always positive for all real values of x.

In Exercises 20 – 25, find the maximum value of each of the following functions:

20. $y = 3 - 6x - 4x^2$
21. $y = 10 + 3x - x^2$
22. $y = 6 + 4x - x^2$
23. $y = 2 - 3x - x^2$
24. $y = 5 - 6x - 3x^2$
25. $y = 3 + 8x - 2x^2$

In Exercises 26 – 31, find the minimum value of each of the following functions:

26. $y = x^2 + 6x + 8$
27. $y = x^2 - 2x - 4$
28. $y = 3x^2 - 2x - 1$
29. $y = 2x^2 + 4x - 3$
30. $y = 2x^2 - 5x + 6$
31. $y = 4x^2 + 8x - 3$

In Exercises 32 – 37, find the vertex of the parabola of each function:

32. $y = x^2 + 6x + 5$
33. $y = x^2 - 8x + 15$
34. $y = 1 - 2x - x^2$
35. $y = x^2 - 4x + 7$
36. $y = 2x^2 + 3x - 4$
37. $y = 3x^2 - 4x - 1$

In Exercises 38 – 43, sketch the graph of each of the following quadratic functions:

38. $y = x^2 - 3x + 5$
39. $y = 1 + x - x^2$
40. $y = 2x^2 - 5x + 3$
41. $y = 5 - 2x - 4x^2$
42. $y = x^2 + 2$
43. $y = 3x^2 + 2x$

In Exercises 44 – 55, solve the quadratic inequality.

44. $x^2 + 4x - 5 \geq 0$
45. $x^2 - 6x - 7 < 0$
46. $x^2 + 3x - 10 \leq 0$
47. $8x - x^2 > 12$

48. $3x^2 + 2x - 8 \leq 0$

49. $2x^2 - 3x \geq 20$

50. $x^2 - 11x + 28 < 0$

51. $x^2 + 6x + 8 > 0$

52. $3x^2 - 2x - 1 \leq 0$

53. $6x^2 - x - 2 \geq 0$

54. $x^2 + 9x + 14 < 0$

55. $12 - 5x - 2x^2 > 0$

11 Polynomial Functions

You are already familiar with the polynomial function. In earlier chapters, we discussed linear and quadratic functions. These functions are special types of polynomial functions.

An expression of the form $a_n x^n + a_{n-1}x^{n-1} + \ldots + a_1 x + a_0$, where $a_n \neq 0$ and $n \geq 0$ is an integer, is called a polynomial function of degree n. The coefficients of the polynomial function, a_n, a_{n-1}, …, a_1 and a_0, are real numbers.

The degree of a polynomial function is the largest power of x occurring in the expression. The linear and the quadratic functions are polynomial functions of degree 1 and 2 respectively. The function $f(x) = a_0$, where a_0 is a real number, is a polynomial of degree 0. The function $f(x) = a_0$ is called the constant function. Polynomials of degree 3 and 4 are called cubic and quartic polynomials respectively. Examples of polynomial functions include

$f(x) = 3$	Constant function
$f(x) = 2x + 1$	Linear function
$f(x) = 3x^2 + 5x - 5$	Quadratic function
$f(x) = x^3 - 2x^2 + x + 3$	Cubic function
$f(x) = 2x^4 - x^3 + 3x^2 + 5x - 2$	Quartic function

Zeros and Domain of Polynomial Functions

A polynomial function has at least one real value of x for which $f(x) = 0$. Such vales of x are called the zeros of the function. The zeros are the x-intercept of the graph of the function. The domain of every polynomial function is the set of all real numbers.

11.1 Dividing Polynomials

A polynomial can be divided by another polynomial by a process similar to the process you would have used when you divide positive integers. The following examples, illustrate the process of long division of a polynomial.

Example 1

Divide $x^2 + 2x - 8$ by $x - 2$.

Solution

Begin by dividing the first term in the dividend x^2 by the first term in the divisor x. Then multiply the divisor by your answer, i.e. x, to get $x^2 - 2x$. Next subtract $x^2 - 2x$, from $x^2 + 2x$, which gives $4x$. Bring down $-$ 8 and repeat the process until the degree of the remainder is less than that of the divisor, as illustrated below.

$$\begin{array}{r} x+4 \\ x-2\overline{)x^2+2x-8} \end{array}$$

$\underline{x^2 - 2x}$	Multiply the divisor by x
$4x - 8$	Subtract and bring down -8
$\underline{4x - 8}$	Multiply $x - 2$ by 4
0	

So the quotient is $x + 4$ and the remainder is 0.

Example 2

Divide $3 + 5x^2 - 6x + 2x^3$ by $3 + x$.

Solution

First write the terms of the divisor and dividend in descending powers of x before you divide.

$$\begin{array}{r} 2x^2-x-3 \\ x+3\overline{)2x^3+5x^2-6x+3} \end{array}$$

$\underline{2x^3 + 6x^2}$	Multiply $x + 3$ by $2x^2$
$-x^2 - 6x$	Subtract and bring down $-6x$
$\underline{-x^2 - 3x}$	Multiply $x + 3$ by $-x$
$-3x + 3$	Subtract and bring down 3
$\underline{-3x - 9}$	Multiply x + 3 by -3
12	Subtract

The quotient is $2x^2 - x - 3$ and the remainder is 12.

Example 3 Divide $x^3 - 8$ by $x + 2$.

Solution

Any missing term of the dividend can be written in using 0 for the coefficients.

$$\begin{array}{r} x^2+2x+4 \\ x-2\overline{)x^3+0x^2+0x-8} \end{array}$$

$\underline{x^3-2x^2}$	Multiply $x-2$ by x^2
$2x^2+0x$	Subtract and bring down $0x$
$\underline{2x^2-4x}$	Multiply $x-2$ by $2x$
$4x-8$	Subtract and bring down -8
$\underline{4x-8}$	Multiply $x-2$ by 4
0	Subtract

The quotient is x^2+2x+4 and the remainder is 0.

Exercise 11.1

In Exercises 1 – 10, use long division to find the remainder for each division.

1. $(x^3-3x^2-5x+2)\div(x-3)$
2. $(x^3+2x^2-3x-4)\div(x+2)$
3. $(x^3-4x^2+5x-2)\div(x-2)$
4. $(3x^3+4x^2+x-2)\div(x+3)$
5. $(2x^3+5x^2-11x+4)\div(2x-1)$
6. $(x^3+15)\div(x+2)$
7. $(2x^2+x^2-13x+6)\div(x-2)$
8. $(2x^3-3x^2+4x+4)\div(2x+1)$
9. $(3x^3-2x^2+x-2)\div(x+3)$
10. $(3x^3-4x+1)\div(x-3)$

In Exercises 11 – 20, use long division to find the quotient and the remainder.

11. $(x^3+2x^2+3x+1)\div(x-1)$
12. $(x^3-3x^2+4x-2)\div(x-2)$
13. $(x^3+5x^2+3x+1)\div(x+3)$
14. $(2x^3+5x^2-2x+2)\div(x+2)$
15. $(3x^3-2x^2+5x-2)\div(x-1)$
16. $(4x^3+3x-5)\div(x-2)$
17. $(2x^3-7x^2+11x-4)\div(2x-1)$
18. $(2x^3+5x^2+6x-8)\div(2x+1)$
19. $(2x^3-5x^2+2x-2)\div(x+2)$
20. $(2x^3-4x^2-7x+5)\div(x-3)$

11.2 The Remainder and Factor Theorems

Consider the division of $x^3 - 2x^2 - 5x + 8$ by $x - 3$.

The remainder can be found without performing the division.

Suppose that when $x^3 - 2x^2 - 5x + 8$ is divided by $x - 3$ the quotient is Q(x) and the remainder is R.

Then $x^3 - 2x^2 - 5x + 8 = (x - 3)Q(x) + R$

Substituting 3 for x gives

$$3^3 - 2(3^2) - 5(3) + 8 = 0 \cdot Q(3) + R$$

$$27 - 18 - 15 + 8 = R$$

$$2 = R$$

The remainder when $x^3 - 2x^2 - 5x + 8$ is divided by $x - 3$ is 2. You can verify this by performing the long division.

This example shows that if a polynomial P(x) is divided by $x - a$, the remainder is P(a). This result is known as the Remainder Theorem.

Observe that you can obtain the remainder by simply substituting x = a.

Example 1

Find the remainder when $3x^3 + x^2 - 8x + 6$ is divided by $3x - 2$.

Solution

Let $P(x) = 3x^3 + x^2 - 8x + 6$.

Solve $3x - 2 = 0$ to get $x = 2/3$. Substitute 2/3 for x.

$$P\left(\frac{2}{3}\right) = 3\left(\frac{2}{3}\right)^3 + \left(\frac{2}{3}\right)^2 - 8\left(\frac{2}{3}\right) + 6$$

$$= \frac{8}{9} + \frac{4}{9} - \frac{16}{3} + 6$$

$$= 2$$

The remainder is 2.

If the remainder in a long division is zero then the divisor divides exactly into the dividend. This suggests the Factor Theorem.

Factor Theorem

If $x - \alpha$ is a factor of the polynomial $P(x)$, then $P(\alpha) = 0$. The number α is called the zero of the polynomial function.

Factoring a Polynomial

To factor a polynomial is to write it as a product of two or more simpler polynomials which are factors of the original polynomial.

Example 2

Factor $2x^3 + 3x^2 - 11x - 6$ and find the zeros.

Solution

You can use the try and error method to find one factor of this expression, and then find another factor by division.

Let $P(x) = 2x^3 + 3x^2 - 11x - 6$

Begin by substituting values for x until you find a value that makes the expression zero.

We will start with 1.

$$P(1) = 2(1)^3 + 3(1)^2 - 11(1) - 6$$

$$= 2 + 3 - 11 - 6$$

$$= -12$$

Because the remainder is -12, $(x - 1)$ is not a factor of this polynomial.

Next, we substitute 2.

$$P(2) = 2(2)^3 + 3(2)^2 - 11(2) - 6$$

$$= 16 + 12 - 22 - 6$$

$$= 0$$

Because the remainder is 0, $(x - 2)$ is a factor of the polynomial.

Dividing $2x^3 + 3x^2 - 11x - 6$ by $x - 2$ gives $2x^2 + 7x + 3$.

Finally, factoring $2x^2 + 7x + 3$ we get $(2x + 1)(x + 3)$.

So, $2x^3 + 3x^2 - 11x - 6 = (x - 2)(2x + 1)(x + 3)$.

You can see that, the zeros are -3, $-\frac{1}{2}$ and 2.

Example 3

Find the value of a if $x - 2$ is a factor of $x^3 + ax^2 + x + 6$ and factor the expression completely.

Solution

Let $P(x) = x^3 + ax^2 + x + 6$

$P(2) = 2^3 + a(2^2) + 2 + 6$ Substitute 2 for x.

$= 8 + 4a + 2 + 6$

$= 16 + 4a$

Because $x - 2$ is a factor of $x^3 + ax^2 + x + 6$ the remainder is 0.

So, $16 + 4a = 0$

$a = -4$

Hence, $P(x) = x^3 - 4x^2 + x + 6$.

Dividing $x^3 - 4x^2 + x + 6$ by $x - 2$ gives $x^2 - 2x - 3$.

So, $x^3 - 4x^2 + x + 6 = (x - 2)(x^2 - 2x - 3)$

$= (x - 2)(x - 3)(x + 1)$

Example 4

Find the values of a and b if $x^2 - 5x + 6$ is a factor of $2x^3 + ax^2 + bx + 6$ and state the third factor of the expression.

Solution

Let $P(x) = 2x^3 + ax^2 + bx + 6$.

Because $x^2 - 5x + 6$ is a factor of the expression then both $x - 2$ and $x - 3$ are factors of $P(x)$.

$P(2) = 16 + 4a + 2b + 6$ Substitute 2 for x.

$= 22 + 4a + 2b$

$22 + 4a + 2b = 0$ Equate to 0

$2a + b = -11$ (1) Simplify

$P(3) = 54 + 9a + 3b + 6$ Substitute 3 for x

$= 60 + 9a + 3b$

$60 + 9a + 3b = 0$ Equate to 0

$3a + b = -20$ (2) Simplify

Subtracting Equation (2) from Equation (1) gives

$a = -9$

Substitute $a = -9$ into Equation (1)

$-18 + b = -11$

$b = 7$

So $P(x) = 2x^3 - 9x^2 + 7x + 6$.

Dividing $2x^3 - 9x^2 + 7x + 6$ by $x^2 - 5x + 6$ gives $2x + 1$.

The third factor is $2x + 1$.

Exercise 11.2

In Exercises 1 – 6, use the remainder theorem to find the remainder when:

1. $x^3 + 3x^2 - 4x + 2$ is divided by $x - 1$
2. $x^3 + x^2 - 3x - 2$ is divided by $x + 2$
3. $x^3 + 3x^2 - 8x + 1$ is divided by $x - 3$
4. $4x^2 - 3x^2 + 2x - 7$ is divided by $x + 1$
5. $x^3 - 5x^2 + 11x - 6$ is divided by $x - 2$
6. $4x^3 - 6x^2 + 5$ is divided by $2x - 1$

In Exercises 7 – 12, factor completely:

7. $x^3 - 3x^2 - x + 3$ given that $x - 1$ is a factor
8. $x^3 - 5x^2 - x + 5$ given that $x + 1$ is a factor
9. $x^3 + 2x^2 - 5x - 6$ given that $x - 2$ is a factor
10. $2x^3 + 3x^2 - 5x - 6$ given that $x + 2$ is a factor
11. $2x^3 - 3x^2 - 3x + 2$ given that $2x - 1$ is a factor
12. $2x^3 - 17x - 9$ given that $x + 3$ is a factor

In Exercises 13 – 18, completely factor the polynomial.

13. $x^3 - 2x^2 + 3x - 6$

14. $x^3 + 6x^2 + 11x + 6$

15. $x^3 + 5x^2 - 2x - 24$

16. $2x^3 + 3x^2 - 11x - 6$

17. $5x^2 - 8x^2 - 11x + 14$

18. $2x^3 + 9x^2 + 7x - 6$

19. If $x - 1$ is a factor of $3x^3 + kx^2 - 4x - 3$, find the value of k.

20. If $3x^3 - kx + 7$ is divisible by $x + 1$, find the value of k.

21. If $2x^3 + x^2 + px + 2p^2$ is divisible by $x + 1$, find the value of p.

22. If $x^4 + px^3 + qx^2 - x + 2$ is divisible by $x^2 - 1$, find the values of p and q.

23. If $x - 2$ and $x + 1$ are both factors of $ax^3 + 3x^2 - 9x + b$ find the values of a and b. State the third factor of the expression.

24. When the polynomial $x^3 - x^2 + ax + b$ is divided by $x - 1$ the remainder is -8. If $x + 1$ is a factor of the polynomial, find the values of a and b and factor the polynomial completely.

25. If the polynomial $x^3 + ax^2 + bx - 2$ is divided by $x - 1$ the remainder is 2 and when divided by $x + 1$ the remainder is 4. Find the values of a and b.

26. When the expression $x^3 + x^2 + px + q$ is divided by $x^2 - 1$, the remainder is $2x + 3$. Find the values of p and q.

27. The expression $ax^4 + bx^3 + 3x^2 - 2x + 3$ has remainder $x + 1$ when divided by $x^2 - 3x + 2$. Find the values of a and b.

Review Exercises

In Exercises 1 – 4, divide:

1. $3x^3 + 4x^2 + 7x + 2$ by $3x + 1$

2. $3x^3 + 2x^2 - 3x - 2$ by $3x + 2$

3. $2x^3 + 7x^2 - x + 12$ by $x + 4$

4. $2x^3 - x^2 - 2x + 1$ by $2x - 1$

In Exercises 5 – 8, find the remainder when:

5. $x^3 - 2x^2 + 5x + 8$ is divided by $x - 2$

6. $12x^3 + 16x^2 - 5x - 3$ is divided by $2x - 1$

7. $2x^3 + x^2 - 13x + 6$ is divided by $x - 2$

8. $x^3 - 4x^2 + 2x + 3$ is divided by $x - 1$

In Exercises 9 – 12, factor fully:

9. $x^3 - 5x^2 + 2x + 8$

10. $3x^3 - 8x^2 - x + 10$

11. $2x^3 - x^2 - 8x + 4$

12. $x^3 - 7x + 6$

13. Verify that $x - 3$ is a factor of $x^3 - 6x^2 + 11x - 6$. Find the other factors.

14. Verify that $x + 1$ is a factor of $2x^3 - 3x^2 - 2x + 3$. Find the other factors.

In Exercises 15 – 20, factor the polynomial. State the zeros of each of the polynomials.

15. $x^3 - 2x^2 - 5x + 6$

16. $x^3 - 4x^2 + x + 6$

17. $x^3 - 6x^2 + 11x - 6$

18. $2x^3 - x^2 - 8x + 4$

19. $2x^3 + x^2 - 15x - 18$

20. $x^4 - 8x^3 + 14x^2 + 8x - 15$

21. If $x^4 + 3x^3 + ax^2 + bx - 18$ is divisible by $x^2 - 9$, find the values of a and b.

22. The polynomial $f(x) = 2x^3 - x^2 + px + 1$ has $(2x - 1)$ as a factor. Find the value of p and the zeros of $f(x)$.

23. The polynomial $f(x) = x^3 + ax^2 + bx - 12$ has $(x - 2)$ as a factor. When it is divided by $(x - 1)$ the remainder is -12. Find:

a. the values of a and b

b. the other factors of $f(x)$

c. the remainder when $f(x)$ is divided by $(x - 3)$

24. When the polynomial $f(x) = x^3 + ax^2 + bx + 6$, where a and b are constants, is divided by $(x - 2)$ the remainder is -4. When divided by $(x + 1)$ the remainder is 8. Find:

a. the values of a and b

b. the zeros of the function $f(x)$

25. The polynomial $f(x) = x^3 - x^2 - 4kx + 3k^2$, where k is a constant, has $(x - 2)$ as a factor.

a. Find the possible values of k.

b. For the integral value of k find the remainder when $f(x)$ is divided by $(x - 3)$.

26. If the polynomial $f(x) = ax^2 + bx + c$, where a, b and c are constants, is divided by $(x - 1)$ the remainder is -5 and if it is divided by $(x + 2)$ the remainder is 4. If $(x - 2)$ is a factor of $f(x)$, find the values of a, b and c. Hence find the zeros of $f(x)$.

27. The function $f(x) = ax^3 - 3x^2 + bx + 2$, where a and b are constants, is such that $2x - 1$ is a factor. Given that the sum of the remainder when $f(x)$ is divided by $x + 2$ and the remainder when $f(x)$ is divided by $x - 3$ is 0, find the values of a and b.

12 Rational Functions

If $p(x)$ and $q(x)$ are polynomials, any function written in the form

$$f(x) = \frac{p(x)}{q(x)},$$

where $q(x) \neq 0$, is called a rational function.

12.1 Sketching Graphs of Rational Functions

To sketch graphs of rational functions, you need to find the intercepts on the axes and the domain of the function, and note where the function is undefined. Any values of x that make $q(x) = 0$ are excluded from the domain.

Finding the Zeros of a Rational Function

The zeros of a rational function f are those values of x that make $f(x) = 0$. Recall that the zeros are the x-intercepts of the graph of the function. To find the zeros of a rational function equate the numerator to 0.

Example 1

Find the zeros of

$$f(x) = \frac{x^2-2x-3}{(x-6)(x^2-4)}$$

Solution

Set the numerator equal to 0 and solve the resulting equation by factoring.

$x^2 - 2x - 3 = 0$

$(x - 3)(x + 1) = 0$ Factoring

$x - 3 = 0$ or $x + 1 = 0$ Set each factor to 0

$x = 3$ $x = -1$ Solving each equation

Thus, the zeros are $x = 3$ and $x = -1$.

Finding the Domain of a Rational Function

Recall that the denominator of a rational expression cannot be zero because division by zero is undefined. To find the domain of a rational function set the denominator to zero and solve. Exclude from the domain any values of x that makes the denominator zero.

Example 2

For what values of x is

$$\frac{3x+2}{x^2-x-6}$$

undefined?

Solution

Set the denominator $x^2 - x - 6$ to 0.

$x^2 - x - 6 = 0$	Original equation
$(x - 3)(x + 2) = 0$	Factoring
$x - 3 = 0$ or $x + 2 = 0$	Set each factor to 0
$x = 3$ $\quad$ $x = -2$	Solving each equation

So, $\frac{3x+2}{x^2-x-6}$ is undefined when $x = 3$ and $x = -2$.

Example 3

Find the domain of the rational function

$$f(x) = \frac{5x-3}{x^2-x-2}$$

Solution

Set the denominator to 0 and solve.

$x^2 - x - 2 = 0$	
$(x - 2)(x + 1) = 0$	Factoring
$x - 2 = 0$ or $x + 1 = 0$	Set each factor to 0
$x = 2$ $\quad$ $x = -1$	Solving each equation

The domain consists of all real values of x except $x = -1$ and $x = 2$.

The graph of a rational function f in x often has one or more breaks because the function is undefined when the denominator is zero.

Asymptotes

An asymptote of a graph is a straight line to which the graph approaches but does not intersects as x increases or decreases without bound.

Vertical Asymptote

If a function gets larger and larger in magnitude without bound as x approaches the number c, then the line $x = c$ is a vertical asymptote.

Finding Vertical Asymptotes

A quick way to find any vertical asymptotes of a rational function in x, is to find all values of x for which the denominator is zero. If a number c makes the denominator 0 but does not make the numerator 0, then the line $x = c$ is a vertical asymptote. If, however, a number c makes both the denominator and the numerator 0, then the line $x = c$ is not a vertical asymptote.

Horizontal Asymptotes

If the values of a rational function f approach a number c as $|x|$ gets larger and larger, then the line $f(x) = c$ is a horizontal asymptote.

Finding Horizontal Asymptotes

To find the horizontal asymptotes of a rational function in x, begin by comparing the degree of the numerator of the function and the degree of the denominator.

1. If the degree of the numerator is the same as that of the denominator, divide the coefficient of the highest power of x in the numerator by the coefficient of the highest power of x in the denominator.

2. If the degree of the denominator is greater than that of the numerator then the x- axis will be the horizontal asymptote.

3. If the degree of the numerator is greater than that of the denominator then you will have a slant asymptote which you can find by doing long division.

Example 4

Find any possible vertical or horizontal asymptotes for each rational function.

(a) $y = \frac{2x}{x+3}$ (b) $y = \frac{3x^2+2x-1}{x^2-x-6}$

Solution

(a) $y = \frac{2x}{x+3}$

The denominator is zero when $x + 3 = 0$. Solving this equation, you find the denominator is zero when $x = -3$. So, the line $x = -3$ is a vertical asymptote.

Because the degree of the numerator and denominator are the same, divide coefficient of $2x$ by the coefficient of x to get $2/1 = 2$. This means that the line $y = 2$ is a horizontal asymptote.

(b) $y = \frac{1+2x-3x^2}{2x^2-3x+1}$

To find the vertical asymptote, we set the denominator to 0.

$2x^2 - 3x + 1 = 0$

$(2x - 1)(x - 1) = 0$ Factor

$2x - 1 = 0$ or $x - 1 = 0$ Set each factor to 0

$x = ½$ $x = 1$

Notice that the value $x = 1$ also makes the numerator 0, so $x = 1$ is not a vertical asymptote of this function. The only vertical asymptote is the line $x = ½$.

To find the horizontal asymptote, divide the coefficient of $-3x^2$ by the coefficient of $2x^2$, which gives $-3/2$. So, the line $y = -3/2$ is a horizontal asymptote.

A general method for sketching graphs of rational functions is illustrated in Examples 5 and 6.

Example 5

Sketch a graph of $y = \frac{3x+2}{x-2}$

Solution

Begin by finding the x- and y-intercepts.

To find the x-intercepts set the numerator to 0, and solve $3x + 2 = 0$.

$3x + 2 = 0$

$x = -⅔$

The x-intercept is $(-2/3, 0)$.

To find the y-intercept, we substitute 0 for x to get $y = -1$. The y-intercept is $(0, -1)$.

The denominator is zero when $x = 2$. So, the line $x = 2$ is a vertical asymptote. It is common to draw a vertical dashed line at $x = 2$ to indicate this, as shown in Figure 12.1. The only horizontal asymptote is at $y = 3$. Use the intercepts and the asymptotes to get the graph shown in Figure 12.1.

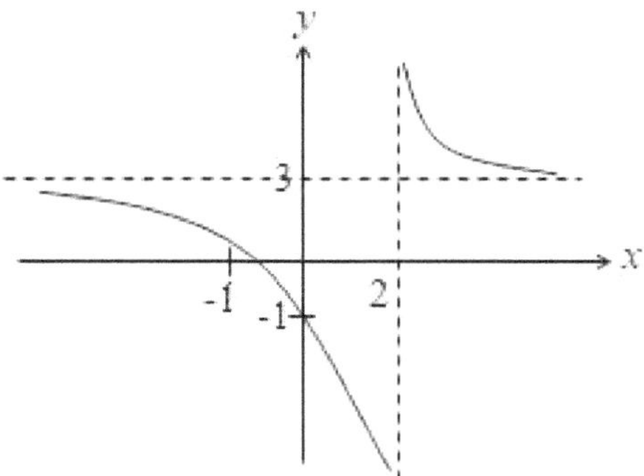

Figure 12.1

Note: As x approaches 2 from the left the value of the function becomes very large and negative, so the value of the function tends to $-\infty$. However, as x approaches 2 from the right the value of the function becomes very large and positive, so the value of the function tends to ∞.

Example 6

Sketch a graph of $y = \frac{x^2+2x-3}{x^2+x-12}$.

Solution

Begin by finding the x- and y-intercepts.

To find the x-intercepts, we solve $x^2 + 2x - 3 = 0$.

$$x^2 + 2x - 3 = 0$$

$$(x - 1)(x + 3) = 0$$

$$x - 1 = 0 \text{ or } x + 3 = 0$$

$$x = 1 \qquad x = -3$$

The x-intercepts are $(1, 0)$ and $(-3, 0)$.

To find the y-intercept substitute 0 for x to get

$y = \frac{-3}{-12} = \frac{1}{4}$.

The y-intercept is $(0, \frac{1}{4})$.

The denominator is zero when $x^2 + x - 12 = 0$. Solve this equation by factoring to get $x = -4$ and $x = 3$. So, the lines $x = -4$ and $x = 3$ are vertical asymptotes. You can see that the horizontal asymptote is the line $y = 1$. The graph will look as shown in Figure 12.2

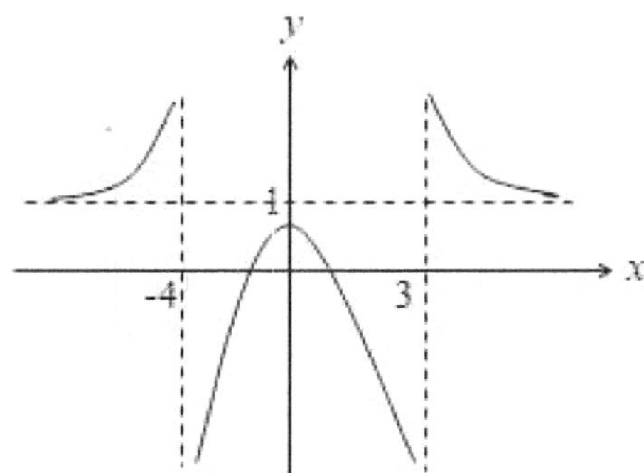

Figure 12.2

Exercise 12.1

Find the zeros of the following rational functions:

1. $f(x) = \frac{2x+1}{x-1}$
2. $f(x) = \frac{x+4}{(x-2)(x+3)}$
3. $f(x) = \frac{x-1}{(x-4)(x+2)}$
4. $f(x) = \frac{(x-2)(x+2)}{x+1}$
5. $f(x) = \frac{2x^2+x-1}{3x^2+4x+3}$
6. $f(x) = \frac{x^2-9}{2x^2+3}$

State the value(s) of x for which the following rational functions are undefined:

7. $\frac{2}{2x-3}$
8. $\frac{x+5}{(x-2)(x-3)}$
9. $\frac{2x+3}{(x+4)(x-2)}$
10. $\frac{x^2+3x-2}{x+3}$
11. $\frac{3x^2+5x-2}{x^2-4}$
12. $\frac{x^2-2x-3}{2x^2-x-6}$

Sketch the graph of each of the following rational functions:

13. $y = \frac{2x-3}{x-2}$
14. $y = \frac{x-3}{3x+2}$
15. $y = \frac{2x+5}{x-1}$
16. $y = \frac{x^2+5x-6}{x^2-x-12}$
17. $y = \frac{x^2-x-6}{x^2-4x}$
18. $y = \frac{x^2-1}{x^2+x-6}$

12.2 Partial Fractions

In Section 3.4, you learned to combine two or more rational expressions into a single rational expression. There may be situations in which it will be useful to reverse this process, that is, to express a rational expression as the sum of two or more simpler rational expressions, called partial fractions. For example,

$$\frac{x+5}{(x-1)(x+2)}$$

can be expressed in the form

$\frac{2}{x-1} - \frac{1}{x+2}$.

The technique used in expressing a rational expression as a sum of two or more simpler fractions is called decomposition of a rational expression into partial fractions.

To decompose a rational expression $p(x)/q(x)$, where $p(x)$ and $q(x)$ are polynomials, and $q(x) \neq 0$, we assume that the degree of $p(x)$ is less than the degree of $q(x)$. If the degree of $p(x)$ is greater than or equal to that of $q(x)$, you have to divide $p(x)$ by $q(x)$ to obtain

$$\frac{p(x)}{q(x)} = A(x) + \frac{B(x)}{q(x)}$$

where the degree of B(x) is less than that of $q(x)$.

Partial Fraction Decomposition

The following information would be useful in decomposing the rational expression $p(x)/q(x)$, where $q(x) \neq 0$, and the degree of $p(x)$ is less than $q(x)$.

1. If $q(x)$ has a non-repeating linear factor of the form $ax + b$, then the partial fraction decomposition of $p(x)/q(x)$ contains a term of the form

$$\frac{A}{ax+b},$$

where A is constant.

2. If $q(x)$ has a repeating linear factor such as $(ax + b)^2$, then the partial fraction decomposition of $p(x)/q(x)$ contain two terms of the form

$$\frac{A}{ax+b} + \frac{B}{(ax+b)^2}.$$

3. If $q(x)$ has a non-repeating quadratic factor of the form $ax^2 + bx + c$, which is not factorable, then the partial fraction decomposition of $p(x)/q(x)$ contains a term of the form

$$\frac{Ax+B}{ax^2+bx+c}.$$

Note: If $q(x)$ has n repeating linear or quadratic factors, the partial fraction decomposition of $p(x)/q(x)$ will contain n terms.

Example 1

Decompose $\frac{5x-8}{(x-1)(x-2)}$ into partial fractions.

Solution

First, factor any quadratic expression in the denominator, if possible. In this case, the denominator is already factored. Rewrite the fraction in the form

$$\frac{5x-8}{(x-1)(x-2)} = \frac{A}{x-1} + \frac{B}{x-2}$$

where A and B are constants.

Multiplying both sides of this equation by the least common denominator (LCD) and simplifying, we obtain

$$5x - 8 = (A + B)x - 2A - B$$

Because the expression on the left hand side equals the expression on the right hand side, the corresponding coefficients of x are equal, and the constant terms are equal. So, we have

$$A + B = 5 \qquad (1)$$

$$2A + B = 8 \qquad (2)$$

Solving the system of equations we get A = 3 and B = 2.

Hence, $\frac{5x-8}{(x-1)(x-2)} = \frac{3}{x-1} + \frac{2}{x-2}$

The same result is obtained as follows.

$$\frac{5x-8}{(x-1)(x-2)} = \frac{A}{x-1} + \frac{B}{x-2}$$

Then $5x - 8 = A(x - 2) + B(x - 1)$

This equation is an identity, so it must hold for all values x. If we let $x = 1$, then the second term of the right side will be 0, as shown below.

$$5(1) - 8 = A(1 - 2) + B(1 - 1)$$

$$-3 = -A + B \cdot 0$$

$$3 = A$$

Similarly, if we let $x = 2$, the first term will be 0, giving 2 = B.

Example 2

Decompose $\frac{5x-3}{(x+1)(x^2+1)}$. into partial fractions.

Solution

The quadratic in the denominator cannot be factored. So, we write the expression as

$$\frac{5x-3}{(x+1)(x^2+1)} = \frac{A}{x+1} + \frac{Bx+C}{x^2+1}$$

$$5x - 3 = A(x^2 + 1) + (Bx + C)(x + 1)$$

Substituting – 1 for x we get – 8 = 2A, giving A = – 4.

We complete the solution by the method of equating coefficients. Equating the coefficients of x^2 gives

A + B = 0.

Substituting – 4 for A we get B = 4.

Also equating the constant terms gives

A + C = – 3.

Substituting – 4 for A gives C = 1.

Hence, $\frac{5x-3}{(x+1)(x^2+1)} = -\frac{4}{x+1} + \frac{4x+1}{x^2+1}$

Example 3

Express $\frac{x^2+8x+9}{(x+1)(x+2)}$ in partial fractions

Solution

Here the numerator has the same degree as the denominator, so we perform long division to obtain

$$\frac{x^2+8x+9}{(x+1)(x+2)} = 1 + \frac{5x+7}{(x+1)(x+2)}$$

We then apply partial fraction decomposition to

$\frac{5x+7}{(x+1)(x+2)}$.

$$\frac{5x+7}{(x+1)(x+2)} = \frac{A}{x+1} + \frac{B}{x+2}$$

Then $5x + 7 = A(x + 2) + B(x + 1)$

Substituting – 1for x we get A = 2. Also substituting – 2 for x we get B = 3. Hence,

$$\frac{x^2+8x+9}{(x+1)(x+2)} = 1 + \frac{2}{x+1} + \frac{3}{x+2}.$$

Exercise 12.2

Determine the partial fraction decomposition of each of the following.

1. $\frac{5x+7}{(x+1)(x+2)}$ 2. $\frac{x+1}{(x-1)(x-2)}$ 3. $\frac{x}{(x-4)(x-1)}$ 4. $\frac{3}{(x+1)(x-1)}$

5. $\frac{x}{x^2-4}$ 6. $\frac{3x+2}{x^2+x-2}$ 7. $\frac{2x+3}{(x-2)(x-3)}$ 8. $\frac{3-2x}{(x-2)(x+3)}$

9. $\frac{2}{(x-1)(x^2+1)}$ 10. $\frac{2x+1}{(x^2+1)(x-2)}$ 11. $\frac{x^2}{(x-1)(x+1)}$ 12. $\frac{x^2-2}{(x+3)(x-1)}$

13. $\frac{x^3}{x^2-1}$ 14. $\frac{x+3}{(x+1)^2}$ 15. $\frac{x-1}{(x+2)^2}$ 16. $\frac{3x^2-8}{(x-2)^2}$

Review Exercises

In Exercises 1 – 4, find the zeros of the following rational functions.

1. $f(x) = \frac{3x-2}{x+1}$ 2. $f(x) = \frac{x^2-6x+5}{x+3}$ 3. $f(x) = \frac{2x^2-x-1}{x^2-5x+6}$ 4. $f(x) = \frac{x^2-4}{3x^2-2}$

In Exercises 5 – 10, state the value(s) of x for which the following rational functions are undefined.

5. $\frac{x}{x^2-1}$ 6. $\frac{3x+2}{(x^2-4)(x+1)}$ 7. $\frac{2x-3}{x^2-x-6}$

8. $\frac{(3x-2)(x+1)}{x^2-5x+6}$ 9. $\frac{2x^2+5x+2}{x^2-9}$ 10. $\frac{x^2-7x+12}{2x^2+x-6}$

In Exercises 11 – 16, sketch the graph of each of the following rational functions.

11. $y = \frac{3x+2}{2x+4}$ 12. $y = \frac{-4}{x+2}$ 13. $y = \frac{2}{3+2x}$

14. $y = \frac{x^2-x-2}{(x-3)(x+2)}$ 15. $y = \frac{2x}{x-3}$ 16. $y = \frac{x^2-4}{x^2-9}$

In Exercises 17 – 26, determine the partial fraction decomposition of each of the following.

17. $\frac{x+7}{x^2-7x+10}$ 18. $\frac{5x+2}{3x^2+x-4}$ 19. $\frac{5x-2}{x^2-3x-28}$ 20. $\frac{2x^2+6x-35}{x^2-x-12}$

21. $\frac{7x-21}{x^2+3x-10}$ 22. $\frac{3x+1}{x^2-4x+3}$ 23. $\frac{x^2}{x^2-4}$ 24. $\frac{2x^2-4x-9}{x^2-x-2}$

25. $\frac{2x+5}{(1+x)^2}$ 26. $\frac{x^2}{(x-1)(x^2+1)}$

13 Exponential and Logarithmic Functions

13.1 Exponential Functions

An exponential function is defined by an equation of the form $y = a^x$, where $a > 0$, $a \neq 1$ and x is any real number. Examples of exponential functions include $y = 2^x$ and $y = 3^{-x}$.

Consider the function defined by $y = 2^x$.The table below list some values of the function.

x	−2	−1	0	1	2	3
y	¼	½	1	2	4	8

Table 1

Plotting these points and then drawing a smooth curve through them we obtain the graph shown in Figure 13.1.

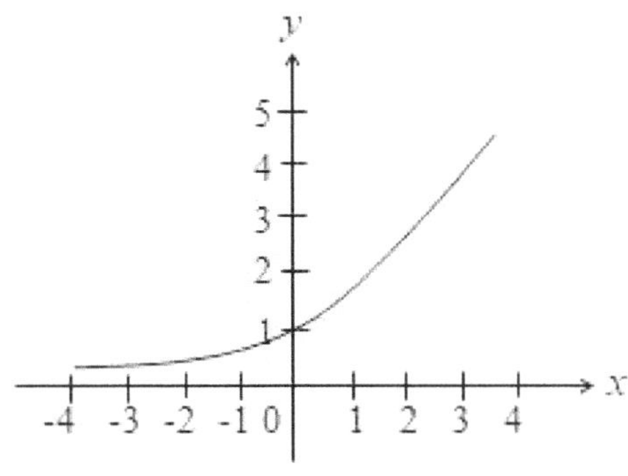

Figure 13.1

The graph has a y-intercept at (0, 1). You can see that as x get larger and larger, the value of the function also get lager. As x gets more and more negative, the value of the function becomes smaller and smaller, approaching but never reaching 0. Therefore, the x-axis is a horizontal asymptote. The domain of the function is the set of all real numbers, and the range is the set of all positive numbers. Because no horizontal line would intersect the graph of the function more than once; the function must have an inverse, and is one-to-one.

Next, we consider the exponential function defined by $y = 2^{-x}$.The values of the function are shown in Table 2.

x	−3	−2	−1	0	1	2
y	8	4	2	1	½	¼

Table 2

The graph of $y = 2^{-x}$is shown in Figure 13.2.

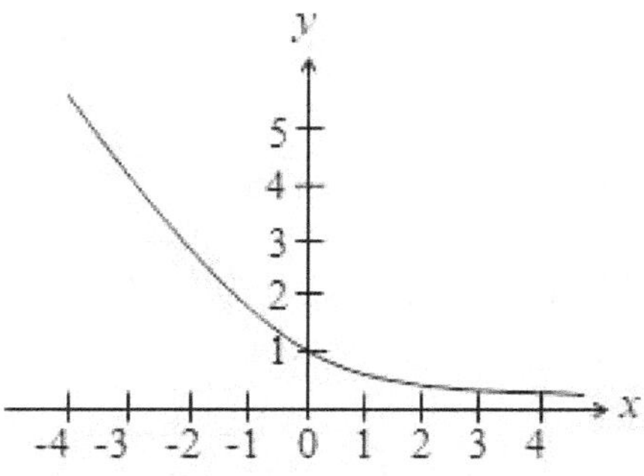

Figure 13.2

You can see that as x gets larger and larger, the value of the function becomes smaller and smaller, approaching but never reaching 0. The x-axis is a horizontal asymptote, but the function approaches the right side of the asymptote. The domain of the function includes all real numbers, and the range includes all positive numbers. The y-intercept is (0, 1).

The graph of $y = 2^{-x}$ is a reflection of the graph of $y = 2^x$ in the y-axis, as shown in Figure 13.3.

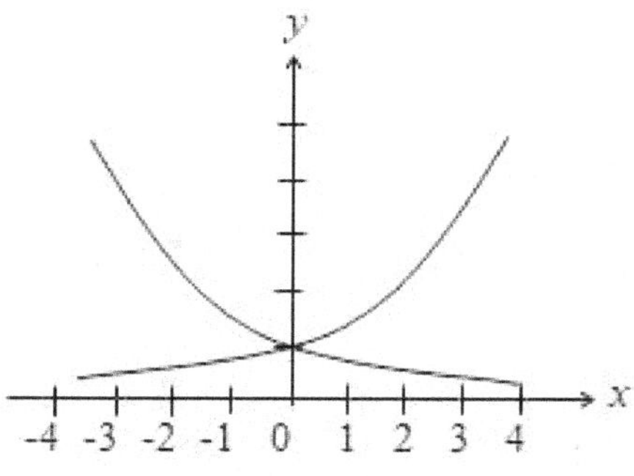

Figure 13.3

The Natural Exponential Function

Often the base of an exponential function is the irrational number, denoted by the letter "e", called the Euler's number. The number e is the constant number 2.71828 . . . .. The exponential function defined by $y = e^x$ is called the natural exponential function. The graph $y = e^x$ is shown in Figure 13.4.

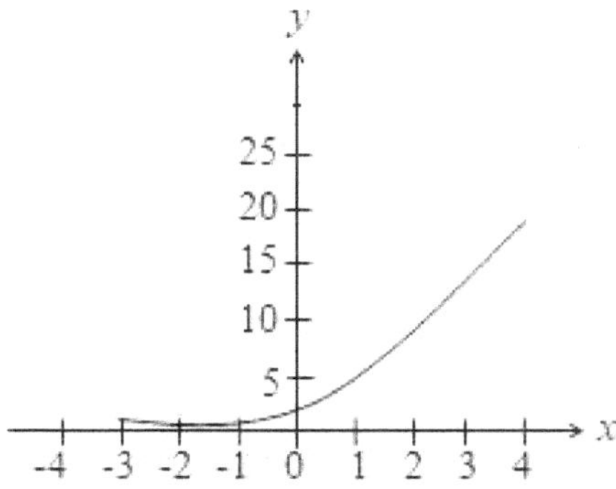

Figure 13.4

From the graph, you may notice that the characteristics of the natural exponential function are the same as that of the exponential function $y = a^x$, as summarized below.

1. If $a > 1$, the values of the function increase as x increases.

2. If $0 < a < 1$, the values of the function decrease as x increases.

3. The graph intersects the y-axis at (0, 1)

4. The x-axis is a horizontal asymptote of the graph

5. The function is one-to-one

6. The domain is the set of all real numbers

7. The range is the set of all positive numbers

The rules of exponents discussed in Chapter 2 can be extended to cover exponential functions.

Exponential Equations

An equation that has terms with constant bases and variable exponents is called an exponential equation.

Solving an Exponential Equation

To solve an exponential equation, we apply the following property.

If x and y are real numbers, and a is a positive number such that $a \neq 1$, then

$$a^x = a^y \qquad \text{if and only if } x = y.$$

Example

Solve;

(a) $3^{2x+1} = 243$ (b) $32^x = 8^{1-x}$

Solution

(a) First rewrite both sides of the equation so the bases are the same.

$3^{2x+1} = 243$	Original equation
$3^{2x+1} = 3^5$	Rewrite with the same base
$2x + 1 = 5$	Equate exponents
$2x = 4$	Subtract 1 from each side
$x = 2$	Divide each side by 2

(b) First, rewrite both sides of the equation so the bases are the same.

$32^x = 8^{1-x}$	Original equation
$(2^5)^x = 2^{3(1-x)}$	Rewrite with the same base
$2^{5x} = 2^{3-3x}$	Multiply exponents
$5x = 3 - 3x$	Equate exponents
$8x = 3$	Subtract $3x$ from each side
$x = 3/8$	Divide each side by 2

Exercise 13.1

Solve each of the following exponential equations

1. $8^x = 64$ 2. $9^x = 27$ 3. $128^x = 32$

4. $3^{2x-1} = 27$ 5. $2^{x-1} = \frac{1}{8}$ 6. $8^{x-1} = 16$

7. $(x - 1)^3 = 64$ 8. $x^{4/3} = 81$ 9. $(9/25)^{-x} = (5/3)^{3-x}$

10. $9^{x-2} = 3^{3x-1}$ 11. $4^{2x} = ½ (8)^x$ 12. $3^{\sqrt{x}} = 27^x$

13.2 Logarithmic Functions

Recall from Section 13.1 that every function of the form $y = a^x$ has an inverse function. By the definition of inverse functions, the inverse function of $y = 2^x$ can be obtained by interchanging x and y. That is, the inverse function of $y = 2^x$ is $x = 2^y$ .This inverse function define the logarithmic function with base 2, written briefly as $y = \log_2 x$, and read "y is the logarithm of x to the base 2."

In general, if a and x are positive real numbers such that $a \neq 1$, then the logarithmic function of x to the base a is defined by $y = \log_a x$ if and only if $x = a^y$.

Notice that $\log_a x$ is the exponent needed on the base to get x. For example, if $8 = 2^3$, then $\log_2 8 = 3$.

The values of $\log_2 x$ shown in Table 3 is obtained from Table 1.

x	¼	½	1	2	3	4
$x = 2^y$	−2	−1	0	1	2	3

Table 3

Plotting these points and then drawing a smooth curve through them gives us the graph shown in Figure 13.5

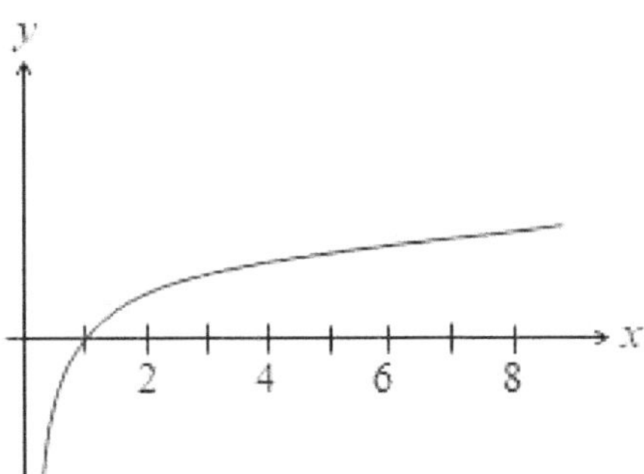

Figure 13.5

The basic characteristics of the graph of a logarithmic function are summarized below.

1. The graph has x-intercept at (1, 0).
2. The y-axis is a vertical asymptote of the graph.
3. The domain of the function is the set of positive numbers.
4. The range of the function is the set of all real numbers.
5. The value of the function increases as x increases.

Figure 13.6, shows the graphs of $f(x) = 2^x$ and $g(x) = \log_2 x$ on the same rectangular coordinate system.

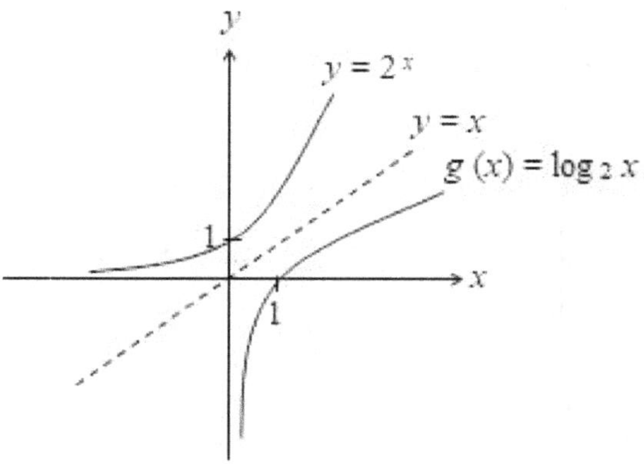

Figure 13.6

Figure 13.6 shows that, the functions are inverse functions of each other. The graphs are mirror images about the line $y = x$. Again notice that, the domain of the exponential function is the range of the logarithmic function. Also, the range of the exponential function is the domain of the logarithmic function.

The Natural Logarithmic Function

We often used the number *e* as a base. A logarithm to the base *e* is called natural logarithm, and $\log_e x$ is abbreviated In *x*. The natural logarithmic function has the same characteristics as that of the logarithmic function.

Common Logarithms

The logarithm to the base 10 is called the common logarithm. It is customary to omit the base in writing the common logarithm. If no other base is shown the base is assumed to be 10. That is log N means $\log_{10}$ N.

Evaluating Logarithms

You can evaluate a logarithm by rewriting it in an exponential notation using the definition of logarithm.

Example

Evaluate each logarithm

(a) $\log_2 32$ (b) $\log_3 \frac{1}{81}$

Solution

(a) Let $x = \log_2 32$

$$2^x = 32 \quad \text{Use definition of logarithm}$$

$$2^x = 2^5 \quad \text{Rewrite with the same base}$$

$$x = 5 \quad \text{Equate exponents}$$

So, $\log_2 32 = 5$

(b) Let $x = \log_3 \frac{1}{81}$

$$3^x = \frac{1}{81} \quad \text{Use definition of logarithm}$$

$$3^x = 3^{-4} \quad \text{Rewrite with the same base}$$

$$x = -4 \quad \text{Equate exponents}$$

So, $\log_3 \frac{1}{81} = -4$

Exercise 13.2

In Exercises 1 – 6, express in logarithmic form:

1. $5^3 = 125$ 2. $2^2 = 4$

3. $32^{1/5} = 2$ 4. $e^0 = 1$

5. $27^{-1/3} = ⅓$ 6. $10^3 = 1000$

In Exercises 7 – 11, express in exponent form:

7. $\log_2 8 = 3$ 8. $\log_4 64 = 3$

9. $\log_9 9 = 1$ 10. $\log_4 (1/16) = -2$

11. $\log_9 (1/27) = -3/2$

In Exercises 12 – 23, evaluate each logarithm:

12. $\log_2 32$ 13. $\log_3 243$

14. $\log_4 64$ 15. $\log_2 (1/128)$

16. $\log 1000$ 17. $\log_6 (1/36)$

18. $\log_{25} 5$ 19. $\log_{\sqrt{2}} 8$

20. $\log_{0.1} 100$ 21. $\log_{\sqrt[3]{2}} (1/64)$

22. $\log_{1/3} 27$ 23. $\log_{3/4} 1(7/9)$

13.3 Properties of Logarithms

The proof of following properties of logarithms depends on the definition of logarithm.

For any real number $a > 0$ and $a \neq 1$, $a^0 = 1$ and $a^1 = a$, it follows that $\log_a 1 = 0$ and $\log_a a = 1$.

Further properties of logarithm can be proofed using the rules of exponents.

Product Rule

Let $\log_a x = p$ and $\log_a y = q$. Then by the definition of logarithm $x = a^p$ and $y = a^q$.

Hence $xy = a^p \cdot a^q = a^{p+q}$

Now use the definition of logarithm to write

$$\log_a xy = p + q$$

Since $p = \log_a x$ and $q = \log_a y$

$$\log_a xy = \log_a x + \log_a y$$

We summarized the properties of logarithms below. Proofs of properties 2 and 3 are left for exercises. You can easily prove them by arguments similar to those used in the preceding proof.

Properties of Logarithms

1. $\log_a xy = \log_a x + \log_a y$
2. $\log_a \frac{x}{y} = \log_a x - \log_a y$
3. $\log_a x^n = n \log_a x$
4. $\log_a a = 1$
5. $\log_a 1 = 0$

Using Properties of Logarithms

Rewriting a Logarithmic Expression as Sums of logarithms

One use of the properties of logarithms is to write a logarithmic expression as the sum or difference of logarithmic expressions.

Example 1

Use the properties of logarithms to rewrite each expression.

(a) $log_a \frac{a^3b^4}{c^2}$ (b) $log\sqrt[4]{\frac{x^2}{y^6z^4}}$

Solution

(a) $log_a \frac{a^3b^4}{c^2}$

$$log_a \frac{a^3b^4}{c^2} = log_a a^3b^4 - log_a c^2 \quad \text{Use Property 2}$$

$$= \log_a a^3 + \log_a b^4 - \log_a c^2 \quad \text{Use Property 1}$$

$$= 3\log_a a + 4\log_a b - 2\log_a c \quad \text{Use Property 3}$$

$$= 3 + 4\log_a b - 2\log_a c \quad \text{Use Property 4}$$

(b) $log\sqrt[4]{\frac{x^2}{y^6z^4}}$

$$log\sqrt[4]{\frac{x^2}{y^6z^4}} = log\left(\frac{x^2}{y^6z^4}\right)^{\frac{1}{4}} \quad \text{Use properties of exponents}$$

$$= \frac{1}{4} log\left(\frac{x^2}{y^6z^4}\right) \quad \text{Use Property 3}$$

$$= \frac{1}{4}(\log x^2 - \log y^6z^4) \quad \text{Use Property 2}$$

$$= \frac{1}{4}(\log x^2 - \log y^6 - \log z^4) \quad \text{Use Property 1}$$

$$= \frac{1}{4}(2\log x - 6\log y - 4\log x) \quad \text{Use Property 3}$$

$$= \frac{1}{2} logx - \frac{3}{2} logy - logz \quad \text{Simplify}$$

When you rewrite a logarithmic expression as sums and difference of logarithms, you are expanding the expression.

Rewriting the Sums of Logarithms as a single Logarithm

We can reverse the process discussed above by using the properties of logarithm to rewrite the sum or difference of logarithmic expressions as a single logarithm.

Example 2

Use the properties of logarithm to rewrite each expression

(a) $2\log_a x + 3\log_a y - 5\log_a z$

(b) $5\log a + 3 - \frac{1}{2}\log b$

Solution

(a) $2\log_a x + 3\log_a y - 5\log_a z = \log_a x^2 + \log_a y^3 - \log_a z^5$

$= \log_a x^2y^3 - \log_a z^5$

$= log_a \frac{x^2y^3}{z^5}$

(b) $5loga + 3 - \frac{1}{2}logb = loga^5 + log1000 - log\sqrt{b}$

$= log1000a^5 - log\sqrt{b}$

$= log\left(\frac{1000a^5}{\sqrt{b}}\right)$

Using Logarithm Properties to Simplify Logarithms

Example 3

Simplify each of the following logarithms.

(a) $\log_2 64$ (b) $log_3\sqrt[4]{27}$ (c) $\frac{log\,32}{log\,8}$

Solution

(a) Rewriting 64 in exponent form.

$\log_2 64 = \log_2 2^6$

$= 6\log_2 2$ Use Property 3

$= 6$ Use Property 4

(b) $log_3\sqrt[4]{27} = log_3 27^{\frac{1}{4}}$

$= log_3 3^{\frac{3}{4}}$ Rewrite 27 as3^3

$= \frac{3}{4}log_3 3$ Use Property 3

$= \frac{3}{4}$ Use Property 4

(c) $\frac{log\ 27}{log\ 9} = \frac{log\,3^3}{log\,3^2} = \frac{3log\,3}{2log\,3} = \frac{3}{2}$

Note that $\frac{log\,27}{log\,9} \neq log27 - log9$

Solving Exponential Equations using Logarithms

Recall in Section 13.1, exponential equations like $8^{2x} = 16$ were solved by writing each side of the equation as a power of 2. However, this method cannot be used always. For example, an equation such as $2^x = 27$, cannot be solved using this method, since 27 cannot be written as a power of 2. Example 4 shows how to solve these types of equations.

Example 4

Solve each of the following equations.

(a) $3^x = 7$ (b) $\left(\frac{1}{4}\right)^{x-2} = 5^{3x}$

Solution

(a) $3^x = 7$ Original equation

Taking the logarithm to base 10 on both sides gives

$$\log 3^x = \log 7$$

$$x \log 3 = \log 7 \quad \text{Use Property 3}$$

$$x = \frac{log\,7}{log\,3} \quad \text{Divide each side by } log3$$

$$= \frac{0.8451}{0.4771}$$

$$= 1.7713$$

(b) $\left(\frac{1}{4}\right)^{x-2} = 5^{3x}$

$$\log\left(\frac{1}{4}\right)^{x-2} = \log 5^{3x} \quad \text{Take logarithm to base 10}$$

$$\log 2^{-2(x-2)} = \log 5^{3x}$$

$$(-2x + 4)\log 2 = 3x \log 5$$

$$x = \frac{4log\,2}{2log\,2 + 3log\,5}$$

$$= \frac{1.2041}{2.6990}$$

$$= 0.4461$$

Exercise 13.3

In Exercises 1 – 12, expand the following:

1. $log_a x^3 y^2$ 2. $log_a \frac{x^2}{y}$ 3. $log x^2 \cdot \sqrt[3]{y}$ 4. $log_2 \frac{1}{x^5}$

5. $\ln x^3 \sqrt{y}$ 6. $log_3 \frac{x^2 y^3}{z^2}$ 7. $log_5 \frac{x^2 y}{\sqrt[3]{z}}$ 8. $log_{10} \frac{x^4 . \sqrt[3]{y}}{z^3}$

9. $log \sqrt[4]{\frac{xy^3}{z^5}}$ 10. $log \sqrt[3]{\frac{xy^2}{z^4}}$ 11. $log_2 \frac{x^2 y^3}{\sqrt{z^5}}$ 12. $log_5 \sqrt[3]{\frac{100x}{y^2 z^4}}$

In Exercises 13 – 24, express as single logarithms:

13. $2\log_2 x + 3\log_2 y$ 14. $\log_3 x - 2\log_3 y$

15. $3\log x - \log y$ 16. $log_5 x + \frac{2}{3} log_5 y$

17. $\frac{1}{3} log x^2 - 2 log y$ 18. $3\text{In } x + 2\text{In } y - \text{In } z$

19. $2\log_3 x - (3\log_3 y + \log_3 z)$ 20. $\frac{1}{2} log_5 x - \frac{2}{3} log_5 y - 3 log_5 z$

21. $\frac{3}{2}(2 log x - 6 log y + 4 log z)$ 22. $\frac{1}{4}(2\log_2 x + 3\log_2 y - 4\log_2 z)$

23. $3 + 2\log x - \frac{1}{4} \log y$ 24. $\frac{2}{3} log x - 3 log y + 1$

In Exercises 25 – 40, simplify each of the following logarithms:

25. $\log 100{,}000$ 26. $\log_2 128$ 27. $\log_3 243$ 28. $\log_6 216$

29. $log_4 \frac{1}{16}$ 30. $\log_5 125$ 31. $\frac{4}{5} log_2 32$ 32. $log_{\frac{1}{3}} 27$

33. $log_8 \frac{1}{2}$ 34. $log_2 \sqrt[3]{2}$ 35. $log_{\frac{1}{4}} 64$ 36. $-2 log_3 \frac{1}{3}$

37. $\frac{log_2 27}{log_2 9}$ 38. $\frac{In\ 8}{In\ 4}$ 39. $\frac{log\, 64}{log\, 8}$ 40. $\frac{log_3 8}{log_3 16}$

In Exercises 41 – 56, solve each of the following equations:

41. $64^x = 32$ 42. $4^{x+1} = 8$

43. $243^x = 9$ 44. $2^x = \frac{1}{32}$

45. $3^{2x+1} = \frac{1}{81}$ 46. $8^{x-7} = 16^{-x}$

47. $2^{\sqrt{3x}} = 8^x$

48. $\left(\frac{1}{27}\right)^{-x} = 9^{2x+3}$

49. $5^x = 8$

50. $4^x = 20$

51. $3^{2x} = 4$

52. $7^{3x} = 50$

53. $2^{2x+1} = 15$

54. $5^{2x-3} = 15$

55. $4^x = 3^{x+1}$

56. $3^{2x-3} = 12$

13.4 Change of Base

Occasionally, we need to evaluate logarithms with bases which are not 10 or *e*. In such cases, the change of base formula will be useful. The formula would help you use a calculator that has only the common logarithm and the natural logarithm keys to evaluate logarithms to other bases.

Suppose that

$$a^y = x$$

where a, x and y are positive real numbers such that $a \neq 1$ and $x \neq 1$.

Taking logarithm to base b of both sides gives

$$\log_b a^y = \log_b x$$

$$y \log_b a = \log_b x$$

$$y = \frac{\log_b x}{\log_b a}$$

Using the definition of logarithm, we write $a^y = x$ as $y = \log_a x$.

So, we have $\log_a x = \frac{\log_b x}{\log_b a}$

Change-of –Base Theorem of Logarithms

If x is any positive number and if a and b are positive real numbers, $a \neq 1$, $b \neq 1$, then

$$log_a x = \frac{log_b x}{log_b a}$$

Example

Evaluate:

(a) $\log_2 25$ (b) $\log_5 16$

Solution

(a)
$$log_2 25 = \frac{log_{10}25}{log_{10}2}$$
$$= \frac{1.3979}{0.3010}$$
$$= 4.644$$

You could obtain the same answer by using the natural logarithm in the change-of- base formula.

$$log_2 25 = \frac{In\ 25}{In\ 2}$$
$$= \frac{3.21888}{0.69315}$$
$$= 4.644$$

(b)
$$log_5 16 = \frac{log\,16}{log\,5}$$
$$= \frac{1.2041}{0.6990}$$
$$= 1.723$$

Another way to evaluate this is to use the natural logarithm in the change-of-base formula.

$$log_5 16 = \frac{In\ 16}{In\ 5}$$
$$= \frac{2.7723}{1.6094}$$
$$= 1.723$$

Exercise 13.4

Evaluate each of the following logarithms:

1. $\log_2 26$ 2. $\log_4 9$ 3. $\log_5 10$ 4. $\log_3 8$ 5. $\log_7 12$

6. In 15 7. $\log_8 32$ 8. $\log_9 8$ 9. $\log_6 10$ 10. $\log_5 16$

13.5 Logarithmic Equations

A logarithmic equation is an equation that contains a logarithmic expression. Examples of logarithmic equations include $\log_3 2x = 5$ and $\log x - \log 5 = \log 3x^2$.

Solving Logarithmic Equations

To solve a logarithmic equation, we use the fact that exponential functions and logarithmic functions are inverses of each other, and any property of logarithms that may be useful in simplifying the logarithmic equation.

It is important to check all possible solutions in the original equation in order to avoid answers that would result in taking logarithm of zero or a negative number. Recall that the logarithmic function is defined for positive real numbers.

Example 1

Solve $\log_3 x + \log_3 2 = 4$.

Solution

$\log_3 x + \log_3 2 = 4$ Original equation

Write the logarithmic expressions on the left as single logarithm.

$\log_3 2x = 4$

$2x = 3^4$ Rewrite as an exponential equation

$2x = 81$

$x = 40\frac{1}{2}$

The solution is $x = 40\frac{1}{2}$. Check this solution in the original equation for confirmation.

Example 2

Solve $\log_2 \frac{1}{2} x + \log_2 (x - 4) = 4$.

Solution

$\log_2 \frac{1}{2} x + \log_2 (x - 4) = 4$ Original equation

Write the logarithmic expressions on the left as single logarithm.

$\log_2 \frac{1}{2} x(x - 4) = 4$

$\frac{1}{2} x(x - 4) = 2^4$ Rewrite as an exponential equation

$x^2 - 4x - 32 = 0$ Simplify

$(x - 8)(x + 4) = 0$ Factor

The possible solutions are $x = -4$ and $x = 8$.

You need to check each solution in the original equation.

We will only check the solution $x = -4$.

Now substituting -4 for x in the original equation we get

$\log_2(-2) + \log_2(-8) \neq 4$

Because the logarithm of a negative number is undefined the only solution is $x = 8$.

Try checking to see whether $x = 8$ is a solution.

Example 3

Solve $\log(2x + 6) - \log(x + 3) = \log x$

Solution

$$\log(2x + 6) - \log(x + 3) = \log x \qquad \text{Original equation}$$

Write the logarithmic expressions on the left as a single logarithm.

$$\log\frac{2x+6}{x+3} = \log x$$

Hence
$$\frac{2x+6}{x+3} = x$$

$$x^2 + x - 6 = 0 \qquad \text{Simplify}$$

$$(x + 3)(x - 2) = 0 \qquad \text{Factor}$$

The possible solutions are $x = -3$ and $x = 2$.

The only solution for the equation is $x = 2$.

Check the two solutions in the original equation.

Exercise 13.5

In Exercises 1 – 18, solve each of the following:

1. $\log_3 x = 4$
2. $\log_x 25 = 2$
3. $\log_5 25 = x$
4. $x = \log_4 32$
5. $\log_x \frac{1}{8} = -3$
6. $\log_{32} x = \frac{3}{5}$
7. $\log_x \frac{9}{4} = -2$
8. $\log_5 x = 3$
9. $\log_2(x + 1) = 3$
10. $\log_5(2x - 1) = 2$
11. $\log_2 x + \log_2 8 = 6$
12. $\log x + \log 5 = 2$
13. $\log_x 3 = 2 - \log_x 27$
14. $\log_x 5 + 3 = \log x\, 320$
15. $\log_2 x + \log_x 2 = 2$
16. $\log_2 x = \log_4(x + 6)$
17. $\log_3 x - 2\log_x 3 = 1$
18. $\log_x 7 + 1 = \log_x 112 - 3$

In Exercises 19 – 30, solve each of the following equations:

19. $\log(2x + 1) - \log(x + 2) = \log x$

20. $\log_2 x + \log_2 (x + 2) = 3$

21. $\log_3 x + \log_3(2x + 3) = 2$

22. $\log_2 (x + 2) + \log_2 (x - 5) = 3$

23. $\log_3 (x + 1) - \log_3 (x - 2) = 2$

24. $\log (x + 2) - \log (2x - 1) = 1$

25. $\log (x + 5) - \log (x - 2) = \log 5$

26. $\log_3 (x + 12) - \log_3 (x - 3) = \log_3 6$

27. $\log_2 x + \log_2 (x + 2) - \log_2 3 = 4$

28. $\log_2 (x^2 - 1) - \log_2 (x - 2) = 3$

29. $2\log_3 x - \log_3 2 = \log_3 (5x - 12)$

30. $\log (x + 9) = 1 + \log (2x - 1)$

Review Exercises

In Exercises 1 – 6, solve the equation:

1. $2^{x-4} = 8^2$
2. $3^{3x} = 9$
3. $4^{3x-1} = 64$
4. $5^{2-x} = 25$
5. $4^{x+3} = 32^x$
6. $9^{x-2} = 234^{x+1}$

In Exercises 7 – 12, use the properties of logarithms to expand:

7. $\log\left(\frac{x^2y}{z^3}\right)$
8. $log\,\frac{x^5y^3}{z^2}$
9. $log_2\,\frac{9x^2y^3}{z^4}$
10. $\frac{1}{2}In\left(\frac{x^3y^5}{z^3}\right)$
11. $log_5\sqrt{\frac{x^5}{y^3z^7}}$
12. $log\sqrt[3]{\frac{x^4y^2}{z^4}}$

In Exercises 13 – 20, use the properties of logarithms to write each expression as a single logarithm.

13. $3\log x + 5\log y$

14. $2\log x - 3\log y$

15. $5\log x + 3\log y - 2\log z$

16. $\frac{1}{5}(3\log x - 4\log y)$

17. $\frac{1}{3}(2\log x + 4\log y) - 3\log z$

18. $2\text{In}\, x + \frac{1}{4}\text{In}\, y$

19. $\frac{1}{2}\text{In}\, x - 3\text{In}\, y + 2\text{In}\, z$

20. $3\log_2 x - 5 - 2\log_2 y$

In Exercises 21 – 29, evaluate the following logarithms.

21. $\log 1000$
22. $\log_2 64$
23. $\log_4 \frac{1}{27}$
24. $\log_3 \sqrt{9}$
25. $\log 0.001$
26. $\log_8 2$
27. $\log_a(\sqrt[3]{a})^4$
28. $\log_7 49$
29. $\log_9 \sqrt[3]{81}$

In Exercises 30 – 33, simplify each of the following.

30. $\frac{\log 128}{\log 32}$ 31. $\frac{\log_2 81}{\log_2 27}$ 32. $\frac{\ln 8}{\ln 16}$ 33. $\frac{\log_4 125}{\log_4 25}$

In Exercises 34 – 39, solve each exponential equation.

34. $3^x = 91$ 35. $5^x = 8$ 36. $2^{x+1} = 6$

37. $7^{3-x} = 15$ 38. $9^{3x} = 126$ 39. $8^{2x} = 35$

In Exercises 40 – 45, solve each logarithmic equation.

40. $\log_6 (x-5) + \log_6 x = 2$ 41. $2\log_4 (x+5) = 3$ 42. $\log x + \log (x-3) = 1$

43. $\log_3 x - \log_3 4 = 2$ 44. $\log_5 (x+3) - \log_5 x = 1$ 45. $2\log_4 x - \log_4 (x-1) = 1$

In Exercises 46 – 50, without using a calculator find the value of each expression.

46. $log_9 27\sqrt{3}$ 47. $log_2 25 \times log_5 8$

48. $log75 + 3log2 - log6$ 49. $\frac{log_4 64 + log_3 9 - log_2 16}{log_9 81}$

50. $\frac{2}{3}log64 - log27 - 3log\left(\frac{2}{3}\right) + log50$

51. Given $\log_3 5 = 1.465$, evaluate without using a calculator $\log_3 25 + \log_9 15$.

52. Find the values of x for which $\log_5 x - 2\log_x 5 = 1$.

Given that $\log 2 = 0.3010$ and $\log 3 = 0.4771$, find the value of:

53. $\log 15$ 54. $\log 1.8$

Solve the following system of equations

55. $2\log_2 x - 3\log_2 y = 7$

$\log_2 x - 2\log_2 y = 4$

56. $\log_3 x + \log_3 y = 3$

$\log_y x = 2$

57. Given that a and b are both positive and unequal, find a in terms of b if $\log_a b^2 + \log_b a = 3$

58. Given $\log_a x = 3$, find the value of:

$(a)\ \log_a x^2$ $(b)\ \log_a \frac{1}{\sqrt{x}}$

14 Variation

14.1 Direct Variation

The table below shows the total revenue (in dollars) obtained from selling a number of concerts tickets.

Number of tickets sold	100	200	300	400	500
Total revenue (in \$)	800	1600	2400	3200	4000

If you divide the total revenue by the number of tickets sold in each case, you get \$8. For instance,

$$\frac{1600}{200} = 8 \quad \text{and} \quad \frac{2400}{300} = 8$$

When x tickets are sold, the total revenue is \$ $8x$. If y represents the total revenue in dollars, then the equation relating the total revenue to the total number of tickets sold is

$$y = 8x$$

If the value of one variable y is found by multiplying the value of a second variable x by a constant k then the relationship is called a direct variation. The equation relating y and x is

$$y = k\,x$$

The number k is called the constant of proportionality or the constant of variation.

If $y = k\,x$ for some constant k, we say that " y varies directly as x" or " y is directly proportional to x".

Finding the Constant of Variation

To solve the equation $y = k\,x$, use the given values of x and y to find the value of the constant k, as shown in Example 1.

Example 1

y varies directly as the square of x. If y is 80 when x is 4, find the constant of variation k.

Solution

$y = kx^2$	General equation
$80 = k \times 4^2$	Substitute 80 for y and 4 for x
$80 = 16k$	Simplify
$5 = k$	Divide each sides by 16

Solving Direct Variation

You may solve direct variation problems using the method illustrated in Examples 2 and 3.

Example 2

y varies directly as x. If $y = 2$ when $x = 4$, find y when $x = 10$.

Solution

$y = kx$ General equation

$2 = 4k$ Substitute 2 for y and 4 for x.

$\frac{1}{2} = k$ Divide both sides by 4

So the equation relating y and x is

$y = \frac{1}{2}x$ Substitute ½ for k

When $x = 10$, y is

$y = \frac{1}{2}(10)$ Substitute 10 for x

$y = 5$ Simplify

Example 3

The volume of juice in a can V cm^3, varies directly as its height, H cm, The volume is 150 cm^3 when the height is 6 cm. Calculate the volume when the height is 9 cm.

Solution

Because the volume of juice in the can varies directly as its height,

$$V = kH$$

To find the value of the constant k, use the fact that V = 150 cm^3 when H = 6 cm.

$150 = k(6)$ Substitute 150 for V and 6 for H

$k = 25$ Divide each side by 6

So, the equation relating the volume and Height is

$$V = 25H$$

When H = 9, then the volume is

$$V = 25(9) = 225 \text{ cm}^3$$

Exercise 14.1

1. y varies directly as x. If $y = 40$ when $x = 8$, find the equation of variation

2. p varies directly as the square of r. If $p = 100$ when $r = 5$, find the equation of variation.

3. p varies directly as the cube root of r. If $p = 135$ when $r = 27$, find the equation of variation.

4. y varies directly as x. If $y = 12$ when $x = 4$, find y when $x = 2$.

5. y varies directly as the square of x. If $y = 32$ when $x = 4$, find y when $x = 3$.

6. y varies directly as the square root of x. If $y = 30$ when $x = 25$, find y when $x = 16$.

7. y varies directly as the square root of x. If $y = 40$ when $x = 25$, find x when $y = 24$.

8. q varies directly as the cube root of p. If $q = 6$ when $p = 27$, find p when $q = 10$.

9. y varies directly as the cube of x. If $y = 40$ when $x = 2$, find x when $y = 135$.

10. The weight of a steel bar, W grams, varies directly as its length, L centimeters. A 15-centimeter steel bar weighs 180 grams. Find the weight of a steel bar 26 centimeters long

11. The total revenue R is directly proportional to the number of units sold x. When 50 units are sold, the revenue is \$600. Find the revenue when 126 units are sold.

12. The cost, \$C of producing a booklet varies directly as the square root of the number N of pages it contains. A booklet with 25 pages cost \$ 2.40 to produce. Find the cost of producing a booklet with 36 pages.

13. The volume of juice in a can V cm^3, varies directly as its height, H cm. The volume is 240 cm^3 when the height is 8 cm. Calculate the volume when the height is 15 cm.

14. The distance a body falls from rest varies directly as the square of the time it falls, ignoring air resistance. If a ball falls 245 meters in 5 seconds, how far will the ball fall in 8 seconds?

15. The distance a spring stretches is proportional to the force on the spring. A force of 30 pounds stretches a spring 7.5 inches. How far will a force of 50 pounds stretch the spring?

14.2 Inverse Variation

Another type of variation is called inverse variation. If y varies directly as $1/x$, we say that " y varies inversely as x " or " y is inversely proportional to x."

Solving Inverse Variation

The method of solution is the same as the method you used to solve direct variation.

Example 1

y varies inversely as the cube of x. If $y = 24$ when $x = 2$, find x when $y = 3$.

Solution

Because y varies inversely as cube of x, we obtain

$$y = \frac{k}{x^3}$$ General equation

By substituting $x = 2$ and $y = 24$, we obtain

$$24 = \frac{k}{2^3}$$ Substitute 24 for y and 2 for x.

$$k = 192$$ Multiply each side by 8

So, $$y = \frac{192}{x^3}$$

To find x substitute 3 for y in the equation and obtain

$$3 = \frac{192}{x^3}$$ Substitute 3 for y.

$$x^3 = 64$$

$$x = 4$$ Take cube root of each side

Example 2

y varies inversely as the square root of x. If $y = 5$ when $x = 81$, find x when $y = 3$.

Solution

$$y = \frac{k}{\sqrt{x}}$$ General equation

By substituting 81 for x and 5 for y, we obtain

$$5 = \frac{k}{\sqrt{81}}$$

$$45 = k$$ Multiply each side by 9

So, $$y = \frac{45}{\sqrt{x}}$$

To find x, substitute 3 for y to obtain

$$3 = \frac{45}{\sqrt{x}}$$

$$\sqrt{x} = 15$$

$x = 225$ Square each side

Exercise 14.2

1. y varies inversely as x, and $y = 15$ when $x = 2$. Find y when $x = 6$.

2. p varies inversely as q, and $p = 8$ when $q = 3$. Find q when $p = 6$.

3. y varies inversely as x, and $y = 6$ when $x = 2$. Find x when $y = 3$.

4. p varies inversely as the square of q, and $p = 5$ when $q = 3$. Find q when $p = 125$.

5. y varies inversely as the square of x, and $y = 5$ when $x = 2$. Find x when $y = 45$.

6. y varies inversely as the square root of x, and $y = 20$ when $x = 25$. Find y when $x = 16$.

7. y varies as inversely as the square root of x, and $y = 6$ when $x = 25$. Find x when $y = 15$.

8. p varies inversely as the square root of q, and $p = 9$ when $q = 25$. Find q when $p = 27$.

9. p varies inversely as the cube of r, and $p = 4$ when $r = 3$. Find r when $p = 32$.

10. p varies inversely as the cube root of r, and $p = 9$ when $r = 8$. Find p when $r = 27$.

11. y varies inversely as the square of x, and $y = 40$ when $x = \frac{1}{2}$. Find x when $y = 250$.

12. p varies inversely as the square root of r, and $p = 4/5$ when $r = 25$. Find r when $p = 6$.

14.3 Joint Variation

If y varies directly as the product of x and z, we say that " y varies jointly as x and z "or " y is jointly proportional to x and z."

Example

z varies jointly as x and the square root of y. If z = 60, $x = 3$ and $y = 25$, find y when z = 80 and $x = 5$.

Solution

$$z = kx\sqrt{y}$$ General equation

$$60 = k(3)(\sqrt{25})$$ Substitute 3 for x and 25 for y

$$60 = 15k$$

$$4 = k$$ Divide each side by 15

So, $$z = 4x\sqrt{y}$$

By substituting z = 80 and $x = 5$, we obtain

$$80 = 4(5)(\sqrt{y})$$

$4 = \sqrt{y}$ Divide each side by 20

$16 = y$ Square each side

Exercise 14.3

1. y varies jointly as x and z, and $y = 36$ when $x = 3$ and $z = 4$. Find y when $x = 5$ and $z = 2$.

2. z varies jointly as x and y, and $z = 60$ when $x = 3$ and $y = 4$. Find z when $x = 6$ and $y = 3$.

3. y varies jointly as x and z, and $y = 45$ when $x = 3$ and $z = 5$. Find z when $x = 2$ and $y = 60$.

4. y varies jointly as x and the square of z, and $y = 100$ when $x = 2$ and $z = 5$. Find z when $y = 36$ and $x = 2$.

5. p varies jointly as q and the square root of r, and $p = 96$ when $q = 4$ and $r = 36$. Find r when $p = 40$ and $q = 2$.

14.4 Combined Variation

Some variation questions combine both direct variation and inverse variation.

Example

z varies directly as x and inversely as the square root of y. If $z = 12$ when $x = 4$ and $y = 25$, find y when $z = 30$ and $x = 6$.

Solution

$z = \frac{kx}{\sqrt{y}}$ General equation

$12 = \frac{k(4)}{\sqrt{25}}$ Substitute 12 for z, 4 for x and 25 for y

$15 = k$

So, $z = \frac{15x}{\sqrt{y}}$ Substitute 15 for k

$30 = \frac{15(6)}{\sqrt{y}}$ Substitute 30 for z and 6 for x

$\sqrt{y} = 3$

$y = 9$ Square each side

Exercise 14.4

1. y varies directly as x and inversely as z, and $y = 12$ when $x = 6$ and z = 2. Find y when $x = 8$ and z = 4.

2. z varies directly as x and inversely as the square of y, and z = 9 when $x = 3$ and $y = 2$. Find z when $x = 6$ and $y = 3$.

3. p varies directly as q and inversely as the square root of r, and $p = 6$ when $q = 3$ and $r = 25$. Find p when $q = 6$ and $r = 16$.

4. y varies directly as x and inversely as the square of z, and $y = 16$ when $x = 8$ and z = 2. Find z when $x = 4$ and $y = 2$.

5. p varies directly as q and inversely as the square root of r, and $p = 4$ when $q = 2$ and $r = 9$. Find r when $p = 12$ and $q = 8$.

6. p varies directly as q and inversely as the square root of r, and $p = 6$ when $q = 8$ and $r = 16$. Find r when $p = 15$ and $q = 35$.

7. z varies directly as x and inversely as y, and z = 36 when $x = 3$ and $y = 6$. Find x when z = 9 and $y = 32$.

8. r varies directly as the square of s and inversely as t, and $r = 18$ when $s = 3$ and $t = 2$. Find s when $r = 9$ and $t = 16$.

9. y varies directly as z and inversely as the square of x, and $y = 10$ when z = 5 and $x = 2$. Find x when $y = 9$ and z = 18.

10. w varies jointly as x and y, and inversely as z, and w = 12 when $x = 2$, $y = 3$ and z = 4. Find w when $x = 9$, $y = 5$ and z = 6.

Review Exercises

1. p varies inversely as the square root of r. If $p = 8$ when $r = 36$, find the equation of variation.

2. w varies jointly as x^2 and y. If w = 72 when $x = 3$ and $y = 2$, find the equation of variation.

3. p varies directly as q and inversely as the square root of r. If $p = 6$ when $q = 3$ and $r = 25$, find the equation of variation.

4. w varies jointly as x and y, and inversely as z. If w = 50 when $x = 2$, $y = 5$ and z = 3, find the equation of variation.

5. r varies directly as s. If $r = 12$ when $s = 4$, find r when $s = 11$.

6. z varies directly as the square of y. If z = 32 when $y = 4$, find z when $y = 5$.

7. x varies inversely as y. If $x = ½$ when $y = 10$, find x when $y = 15$.

8. p varies inversely as the square root of r. If $p = 12$ when $r = 25$, find r when $p = 15$.

9. z varies jointly as w and y. If z = 12 when w = 9 and $y = 4$, find z when w = 50 and $y = 6$.

10. z varies jointly as x and the square of y. If z = 135 when $x = 5$ and $y = 3$, find y when z = 192 and $x = 4$.

11. z varies directly as x and inversely as y. If z = 8 when $x = 4$ and $y = 5$, find z when $x = 6$ and $y = 15$.

12. w varies jointly as x and y and inversely as the square of z. If w = 24 when $x = 4$, $y = 6$ and z = 3, find z when w = 15, $x = 27$ and $y = 5$.

15 Sequences and Series

Sequences

A set of numbers each of which can be obtained from the preceding one by a rule is called a sequence. For instance, the sequence 5, 8, 11, 14, 17, 20 is obtained by adding 3 to a preceding number, whereas the sequence 4, 8, 16, 32, 64, 128 is obtained by multiplying a preceding number by 2. Each number in the list is called a term of the sequence.

A sequence can have a finite number of terms or an infinite number of terms. For instance, the set of positive even numbers less than 14, i.e. 2, 4, 6, 8, 10, 12, is a finite sequence. However, the set of positive even numbers, 2, 4, 6, 8, 10, 12, . . ., is an infinite sequence. The three dots indicate that the sequence never ends.

Because each term of a sequence is related to its position in the sequence, a sequence can be considered as a function that assigns a real number to each positive integer. For example, the sequence $a(n) = 3n - 2$ whose domain is the set of natural numbers, $n = 1, 2, 3, 4, \ldots$ can be written by listing it terms

$1, 4, 7, 10, \ldots, 3n - 2, \ldots$

For this sequence, we have a rule for the n th term, i.e. $3n - 2$, so it is easy to find any term of the sequence. Instead of writing $a(n)$ for the general term of a sequence, it is customary to write. a_n.

15.1 Arithmetic Sequences

A sequence is called an arithmetic sequence if each term after the first is obtained by adding or subtracting the same number. The difference of any two consecutive terms is a constant d called the common difference. The common difference of any two consecutive terms is $d = a_n - a_{n-1}$.

Example 1

Find the common difference and the next three terms of the sequence.

(a) 8, 11, 14, 17, ... (b) 15, 13, 11, 9, ...

Solution

(a) The common difference is $11 - 8 = 3$.

The next three terms are 20, 23 and 26.

(b) The common difference is $13 - 15 = -2$.

The next three terms are 7, 5 and 3.

The n th Term of an Arithmetic Sequence

Consecutive terms of an arithmetic sequence differ by the same number. If a is the first term of an arithmetic sequence and d is the common difference, then the second term is given by $a + d$, and the third term by $(a + d) + d = a + 2d$ and so on. That is

$a_1 = a$

$a_2 = a + d$

$a_3 = a + 2d$

$a_4 = a + 3d$

You can see from this pattern that the n th term is given by

$a_n = a + (n - 1)d.$

Example 2

Find the tenth term of the sequence 9, 13, 17, 21, ...

Solution

Begin by finding the common difference, d.

The common difference is $d = 13 - 9 = 4$.

Using $a = 9$, $d = 4$ and $n = 10$ in $a_n = a + (n - 1)d$ gives

$$\begin{aligned} a_{10} &= 9 + (10 - 1)(4) \\ &= 9 + 36 \\ &= 45 \end{aligned}$$

You can obtained the same result as follows.

$$a_{10} = a_4 + 6d = 21 + 6(4) = 45.$$

Example 3

Find the number of terms in the sequence

8, 17, 26, . . ., 143.

Solution

Begin by finding the common difference, d.

The common difference is $d = 17 - 8 = 9$.

Using $a = 8$, $d = 9$ and $a_n = 143$ in $a_n = a + (n - 1)d$ gives

$$143 = 8 + (n - 1)(9)$$

$$15 = n - 1$$

$$16 = n$$

The number of term is 16.

Example 4

The thirteenth term of an arithmetic sequence is 81 and the common difference is 6, find the fifteenth term.

Solution

Since $a_{13} = 81$ and $d = 6$, $a_{15} = 81 + 2d$

$$a_{15} = 81 + 2(6) = 93$$

Sum of an Arithmetic Sequence

If a is the first term of an arithmetic sequence, d the common difference and a_n the n th term, then the sum, S_n of the first n terms is given by

$$S_n = a + (a + d) + (a + 2d) + (a + 3d) + \ldots + a_n \qquad \text{(i)}$$

Writing S_n backward, we have

$$S_n = a_n + (a_n - \text{d}) + (a_n - 2d) + (a_n - 3d) + \ldots + a \qquad \text{(ii)}$$

Adding (i) and (ii) gives

$$2S_n = (a + a_n) + (a + a_n) + (a + a_n) + \ldots + (a + a_n)$$

Because, we are adding n terms we have

$$2S_n = n(a + a_n)$$

Dividing each side by 2, we get

$$S_n = \frac{n}{2}(a + a_n) \qquad (1)$$

Substituting $a + (n - 1)d$ for a_n in (1) gives

$$S_n = \frac{n}{2}[a + a + (n - 1)d]$$

$$= \frac{n}{2}[2a + (n - 1)d] \qquad (2)$$

Example 5

Find the sum of the arithmetic sequence 4 + 9 + 14 + … + 59.

Solution

Begin by finding the number of terms, n, of the sequence.

Using $a = 4$, $d = 5$ and $a_n = 59$ in $a_n = a + (n - 1)d$, we get

$$59 = 4 + (n - 1)(5)$$

$$11 = n - 1$$

$$12 = n$$

Now substitute $a = 4$, $n = 12$ and $a_n = 59$ into

$$S_n = \frac{n}{2}(a + a_n)$$

$$S_{12} = \frac{12}{2}(4 + 59)$$

$$= 6(63)$$

$$= 378$$

Example 6

Find the sum of the first twelve terms of the sequence 7 + 10 + 13 + …

Solution

Substitute $a = 7$, $d = 3$ and $n = 12$ into

$$S_n = \frac{n}{2}[2a + (n - 1)d]$$

$$S_{12} = \frac{12}{2}[2(7) + (12 - 1)(3)]$$

$$= (6)(14 + 33)$$

$$= 6(47)$$

$$= 282$$

Application of Arithmetic Sequence

Example 7

A theatre has 200 seats in the first row, 250 seats in the second row, 300 seats in the third row and so on in the same increasing pattern. If the theatre has 20 rows of seats, how many seats are in the theatre?

Solution

The number of seats in the first row, second row and the third row form an arithmetic sequence with $a = 200$ and $d = 50$. Since the number of seats increases in the same pattern you can find the number of seats in the theatre by using the formula $S_n = \frac{n}{2}[2a + n-1d$. Substituting $a = 200$, $d = 50$ and $n = 20$ in the formula gives

$$S_{20} = \frac{20}{2}[2(200) + (20 - 1)(50)]$$

$$= (10)(400 + 950)$$

$$= (10)(1350)$$

$$= 13{,}500$$

Example 8

A bookshop sells $10,000 worth of books during its first year of business. If the shop set a target of increasing annual sales by $2,500 each year for 9 years, how much will the shop sell during the first 10 years of business?

Solution

The total sales during the first 10 years can be found using the formula for the sum of an arithmetic sequence.

$$\text{Total sales} = \frac{n}{2}[2a + (n - 1)d]$$

$$= \frac{10}{2}[2(10{,}000) + 9(2{,}500)]$$

$$= 5(20{,}000 + 22{,}500)$$

$$= 5(42{,}500)$$

$$= 212{,}500$$

The total sales is $212,500.

Exercise 15.1

In Exercises 1 – 6, find the common difference of each linear sequence.

1. 4, 7, 10, 13 . . .
2. – 4, 0, 4, 8 . . .
3. 50, 44, 38, 32 . . .
4. 10, -2, -14, -26 . . .
5. 4, 9/2, 5, 11/2, 6, . . .
6. ½ , 5/4, 2, 11/4, …

In Exercises 7 – 12, determine whether the sequence is arithmetic. If so find the common difference.

7. 3, 5, 7, 9 . . . 8. 1, 3, 4, 8 . . . 9. 8, 6, 4, 2, 0 . . .

10. 45, 30, 26, 13 . . . 11. 5, 9, 13, 17 . . . 12. 2, 7/2, 5, 13/2, . . .

In Exercises 13 – 18, write the next two terms of each arithmetic sequence.

13. 7, 10, 13 . . . 14. 1, 6, 11 . . . 15. 6, 4, 2 . . .

16. 3/2, 4, 13/2, ... 17. – 5/4,– ½ , ¼ ,... 18. 4, 15/4, 7/2, ...

In Exercises 19 – 24, find the indicated term for each arithmetic sequence.

19. 2, 10, 18, 26 . . . 15 th 20. 9, 6, 3, 0 . . . 24 th

21. 1, 2 ½, 4, 5 ½ ... 18 th 22. 60, 58, 56, 54 . . . 30 th

23. 8, 7 ½ , 7, 6 ½ ... 21 st 24. – 6, – 4, – 2, 0, ... 27 th

In Exercises 25 – 30, find the number of terms in each arithmetic sequence.

25. 3, 5, 7 . . . 47 26. 48, 45, 42 . . .12

27. 3, – 8, –19 . . . –129 28. 1, 3½ , 6 ... 101

29. 10, 9, 8 . . . –20 30. 8, 13, 18, . . . , 103

31. The second and fifth terms of an arithmetic sequence are 7 and 19. Find the first term and the common difference.

32. The first term of an arithmetic sequence is 26 and the fourth term is 11. Find the common difference.

33. The sixth term of an arithmetic sequence is twice the third term and the first term is 5. Find the common difference and the tenth term.

34. The third term of an arithmetic sequence is 8 and the sixth term is 17. Find the first three terms of the sequence.

35. The ninth term of an arithmetic sequence is three times the second term and the fifth term is 13. Find the first three terms of the sequence.

36. A boy saved 20 ¢ of his pocket money on the first day of June, 25 ¢ on the second day, 30 ¢ on the third day and so on. Find the amount he saved on the last day of June.

37. A job is advertised at a starting salary of $25,000 with an annual increment of $1,200. Find the salary an employee will receive in the eighth year.

38. A man's starting salary was \$2,500 increasing to \$6,700 in 15 years. How much increase did he receive each year?

39. A man accepts a job in a factory that pays a starting salary of \$5,428 with annual increments of \$450. If he retires on a salary of \$10,378, how many years will he have worked at the factory?

40. An auditorium has 20 seats in the first row. Each successive row has two more seats than the previous row. How many seats are in the twelfth row?

In Exercises 41 – 46, find the sum of each arithmetic sequence.

41. $2 + 4 + 6 + \ldots + 100$

42. $3 + 8 + 13 + \ldots + 78$

43. $1 + 3\frac{1}{2} + 6 + \ldots + 101$

44. $-9 - 6 - 3 + \ldots + 51$

45. $76 + 72 + 68 + \ldots - 48$

46. $2 + 2\frac{1}{2} + 2\frac{1}{4} + \ldots + 12$

In Exercises 47 – 52, find the sum of the arithmetic sequence as far as the term indicated.

47. $5 + 11 + \ldots$ 12 th

48. $17 + 15 + \ldots$ 20 th

49. $7 + 8 + \ldots$ 10 th

50. $19 + 12 + \ldots$ 16 th

51. $1 + 2\frac{1}{2} + \ldots$ 12 th

52. $20 + 15 + \ldots$ 18 th

53. Find the sum of the terms between 1 and 100 divisible by 3.

54. Find the sum of the numbers from 1 to 100 inclusive which are not divisible by 4.

55. The sum of an arithmetic sequence is 144, the first term is 3 and the final term is 15. Find the common difference.

56. The sum of the first and last terms of an arithmetic sequence is 42. The sum of all the terms is 420. The second term is 4. Find the common difference.

57. The fifth term of an arithmetic sequence is 24 and the sum of the first five terms is 80. Find the first term, the common difference and the sum of the first fifteen terms of the sequence.

58. Find how many terms of the arithmetic sequence $3 + 5 + 7 + 9 + \ldots$ have sum of 624.

59. The sum of the first n terms of an arithmetic sequence is 360 and the sum of the first $(n - 1)$ th terms is 308. If the first term of the sequence is 8, find the n th term and the number of terms of the sequence.

60. A boy who works part-time earn 25 ¢ on the first day of the month, 50 ¢ on the second day, 75 ¢ on the third day and so on. Determine the total amount that he will earn during a 30-day mouth.

61. A man works in a company that pays \$8,750 the first year. If his salary is increased each year by \$450, find the total salary he will receive in the first 12 years.

62. A man accepted a job that offered him a starting salary of \$9200 and an annual increment of \$350. Find his salary in the sixth year and the total amount he will have received in the first 15 years.

63. A man starts work on an initial salary of \$2,500 per annum with the promise that his salary will increase yearly by \$250. In how many years will he received in total \$46,500?

64. A free-falling object will fall 12 meters during the first second, 44 more meters during the second second, 76 more meters during the third second and so on. What is the total distance the object will fall in 10 seconds if this pattern continues?

65. Milk tins are stacked in a pile. The top row has 15 milk tins and the bottom row has 21 milk tins. If the pile has 7 rows how many milk tins are in the pile?

66. A boy stacks logs so that there are 20 logs in the bottom layer and each layer contains one log less than the layer below it. How many logs are in the pile?

67. An auditorium has 18 seats in the first row. Each successive row has two more seats than the previous row. How many seats are in the first 15 rows?

68. Each swing of a pendulum is 25 centimeters shorter than the preceding swing. The first swing is 8 meters. Determine the distance traveled by the pendulum during the first 12 swings.

69. An oil company bores a hole 120 meters deep. Estimate the cost of boring if the cost is \$60 for drilling the first meter with an increase in cost of \$5 per meter for each succeeding meter.

15.2 Geometric Sequences

A sequence is called a geometric sequence if each term after the first is obtained by multiplying the preceding term by the same number. The ratio of any two consecutive terms is a constant r called the common ratio.

The common ratio of any two consecutive terms is

$$r = \frac{a_{n+1}}{a_n}$$

Example 1

Find the common ratio and the next three terms of each of the following geometric sequences.

(a) 2, 6, 18, ... (b) 24, 12, 6, ...

Solution

(a) The common ratio is 6/2 or 18/6 which gives 3. The next three terms are:

18(3) = 54, 54(3) = 162 and 162(3) = 486, i.e. 54, 162 and 486.

(b) The common ratio is 12/24 or 6/12 which gives ½. The next three terms are:

6(½) = 3, 3(½) = 3/2 and (3/2)(½) = ¾, i.e. 3, 3/2 and ¾ .

The nth Term of a Geometric Sequences

If a and r are the first term and common ratio respectively of a Geometric sequence then

$a_1 = a$

$a_2 = ar$

$a_3 = ar^2$

$a_4 = ar^3$

You can see from this pattern that the nth term is given by $a_n = ar^{n-1}$.

Example 2

Find the indicated term for each geometric sequence.

(a) 6, 12, 24, ... 7 th term (b) 27, 9, 3, ... 6 th term

Solution

(a) 6, 12, 24, ...

Begin by finding the common ratio.

The common ratio is $r = 12/6 = 24/12 = 2$.

Using $a = 6$, $r = 2$ and $n = 7$ in $a_n = ar^{n-1}$, we get

$$a_7 = 6(2^6) = 6(64) = 384$$

(b) 27, 9, 3, ...

The common ratio is $r = 9/27 = 3/9 = 1/3$.

Using $a = 27$, $r = 1/3$ and $n = 6$ in $a_n = ar^{n-1}$, we get

$$a_6 = 27(\tfrac{1}{3})^5$$

$$= \frac{27}{243}$$

$$= \frac{1}{9}$$

Example 3

Find a formula for the n th term of the geometric sequence whose common ratio is 2 and whose first term is 16. What is the sixth term of the sequence?

Solution

Using $a = 16$ and $r = 2$ in $a_n = ar^{n-1}$, we get

$$a_n = 16(2^{n-1})$$

$$= 2^{n+3}$$

The sixth term of the sequence is $a_6 = 2^{6+3} = 2^9 = 512$.

Example 4

The fourth term of a geometric sequence whose common ratio is ½ is 8, what is the sixth term of the sequence.

Solution

The sixth term of the sequence is $a_6 = a_4 \cdot r \cdot r = 8r^2$.

Substituting ½ for r gives $a_6 = 8\,(\tfrac{1}{2})^2 = 8(\tfrac{1}{4}) = 2$.

The Sum of a Geometric Sequence

If a and r are the first term and the common ratio of a Geometric sequence then the sum S_n of the first n terms is

$$S_n = a + ar + ar^2 + ar^3 + \ldots + ar^{n-1} \qquad (1)$$

Multiplying Equation (1) by r, where $n \neq 1$, gives

$$rS_n = ar + ar^2 + ar^3 + \ldots + ar^n \qquad (2)$$

Now, subtracting corresponding sides of Equation (2) from Equation (1) gives

$$S_n - rS_n = a - ar^n$$

Factoring we obtain

$$(1 - r)S_n = a(1 - r^n)$$

and dividing each side by $(1 - r)$ gives

$$S_n = \frac{a(1-r^n)}{1-r}$$

Multiplying both the numerator and the denominator by – 1, gives

$S_n = \frac{a(r^n - 1)}{r - 1}$

This is more convenient if r is greater than 1.

Example 5

Find the sum of the first six terms of the geometric sequence 16, 8, 4, ...

Solution

Using $a = 16$, $r = ½$ and $n = 6$ in $S_n = \frac{a(1-r^n)}{1-r}$, we get

$$S_n = \frac{a(1-r^n)}{1-r}$$

$$S_6 = \frac{16\left[1-\left(\frac{1}{2}\right)^6\right]}{1-\frac{1}{2}}$$

$$= 32\left(1 - \frac{1}{64}\right)$$

$$= 32\left(\frac{63}{64}\right)$$

$$= 31.5$$

Example 6

Find the sum of the first seven terms of the geometric sequence 2, 6, 18, ...

Solution

Here $a = 2$, $r = 3$ and $n = 7$, so

$$S_n = \frac{a(r^n - 1)}{r - 1}$$

$$S_7 = \frac{2(3^7 - 1)}{3 - 1}$$

$$= 2187 - 1$$

$$= 2186$$

Application of Geometric Sequences

Example 7

Mary deposits $32,000 at the end of each year for 6 years in an account paying 5% interest compounded annually. How much will she have on deposit after 6 years?

Solution

Here $a = 32{,}000$, $n = 6$ and $r = 1.05$. Using the formula for the sum of the first n terms of a geometric sequence gives

$$S_n = \frac{a(r^n - 1)}{1 - r}$$

$$S_6 = \frac{32{,}000[(1.05)^6 - 1]}{1.05 - 1}$$

$$= 640{,}000(0.340096)$$

$$= 217{,}661.44$$

The deposit after 6 years is \$217,661.44.

Exercise 15.2

In Exercises 1 – 6, find the common ratio of each geometric sequence.

1. 7, 14, 28, 56 . . .
2. 2, 6, 18, 54 . . .
3. – 5, – 0.5, – 0.05, – 0.005 . . .
4. ⅓, – 1/9, 1/27, – 1/81, …
5. 75, 15, 3, 3/5, …
6. 12, – 4, 4/3, – 4/9, …

In Exercises 7 – 12, determine whether the sequence is geometric sequence.

7. 64, 32, 16, 8 . . .
8. 10, 15, 20, 25 . . .
9. 5, 10, 20, 40 . . .
10. 54, −18, 6, – 2 . . .
11. 1, 8, 27, 64, 125 . . .
12. 1, – ⅔, 4/9, – 9/27, …

In Exercises 13 – 17, find the next two terms of each geometric sequence.

13. 4, 8, 16 . . .
14. 6, 3, 3/2, …
15. 5, -10, 20 . . .
16. 1, – ½ , ¼ …
17. 1, 0.2, 0.04 . ..

In Exercises 18 – 23, find the indicated term of the geometric sequence.

18. 2, 6, 18 . . . 6 th
19. 8, 4, 2 . . . 7 th
20. 6, 2, ⅔, … 5 th
21. 4, – 8, 16 . . . 9 th
22. 2, 1, ½ … 5 th
23. 12, – 6, 3 ... 6 th

In Exercises 24 – 29, find the number of terms of the geometric sequence

24. 2, 4, 8... 256
25. $9, 3, 1 \ldots \frac{1}{243}$
26. $32, 16, 8 \ldots \frac{1}{8}$
27. $\frac{1}{27}, \frac{1}{9}, \frac{1}{3}, \ldots 81$
28. $\frac{1}{32}, \frac{1}{16}, \frac{1}{8} \ldots 16$
29. $\frac{3}{4}, \frac{3}{2}, 3 \ldots, 24$

30. The sixth term of a geometric sequence is 192 and the third term is 24. Find the first term and the common ratio.

31. The third term of a geometric sequence is 1 and the fifth is 9. Find two possible values of the common ratio and hence find two possible values for the first term.

32. The second, fourth and eight terms of an arithmetic sequence form consecutive terms of a geometric sequence and the first term is 8. Find the common ratio of the exponential sequence.

33. The second, third and fourth terms of a geometric sequence are $(n + 6)$, n and $(n - 4)$. Find n and hence find the first term of the sequence.

34. A man bought a new machine for \$76,800. Each year the machine loses 25 % of its value, find its value at the end of the fifth year.

35. A village of 5,000 people is growing at the rate of 2 % per year. Estimate the population of the village at the end of 20 years.

36. A deposit of \$500 is made in an account that earns 7 % interest compounded yearly. How much will be in the account after 20 years.

37. At the end of each year the value of a machine with an initial cost of \$32,000 is three-fourths what it was at the beginning of the year. Find the value of the machine 3 years after it was purchased.

In Exercises 38 – 43, find the sum of the geometric sequence as far as the terms indicated.

38. $2 + 4 + 8 + \ldots$ 8 terms

39. $9 + 3 + 1 + \ldots$ 6 terms

40. $5 + 1 + 1/5 + \ldots$ 5 terms

41. $8 - 4 + 2 - \ldots$ 6 terms

42. $2 + 1 + ½ + \ldots$ 10 terms

43. $3 - 6 + 12 - \ldots$ 8 terms

The following are the n term of a geometric sequence. Find the common ratio and write down the first four terms. Also find the sum of the first n terms.

44. $-5\left(\frac{1}{5}\right)^n$

45. $\frac{3^{n-1}}{9}$

46. $\frac{2^n}{4}$

47. $\frac{3^{n-1}}{2^n}$

48. Find the sixth term and the sum of the first 8 terms of the sequence 3, 6, 12...

49. The sum of the first 3 terms of a geometric sequence is 21. If the first term is 3, find the possible values of the common ratio and hence find the possible values for the sum of the first 5 terms.

50. The fifth and second terms of a geometric sequence are 81 and 24. What is the first term and the sum of the first five terms?

51. How many terms of the sequence 1, 3, 9, 27 ... must be taken for the sum to exceed 1,000?

52. How many terms of the sequence 1, ½ , ¼ … must be taken for the sum to exceed 1.995?

53. At the beginning of each year a man invests $100 in an account paying 5 % interest compounded annually. How much will he have on deposit at the end of the tenth year?

54. A man invests $1,000 at 5 % interest compounded annually in an account. What will be the total amount in the account and the interest at the end of the tenth year?

55. A man accepts a job that pays a salary of $3,000 the first year. During the next 19 years he received a 5 % raise each year. What would his total salary be over a 20-year period?

56. A substance losses half its mass each hour. If there are initially 300 grams of the substance, find the number of hours after which only 37.5 grams of the substance remain.

57. In February 2000 the population of a village was about 2,810 people. If the population grows at a rate of 2.2 % per year, estimate the population after 10 years.

15.3 Series

A summation of all the terms of an infinite sequence is called a series. The sum of the first n terms of a series is called the n th partial sum of the sequence.

Infinite Series

If the terms of an infinite sequence a_1, a_2, a_3 …, a_n, … are added together we get the series $a_1 + a_2 + a_3 + \ldots + a_n + \ldots$. The n th partial sum of this sequence is

$$S_n = a_1 + a_2 + a_3 + \ldots + a_n$$

written in summation notation as

$$\sum_{i=1}^{n} a_i$$

where i is the index of summation, n is the upper limit of summation, and 1 is the lower limit of summation.

Any letter can be used as the index of summation. It is customary to use the letters i, j, k and n. Note that the summation notation is an instruction to add terms of a sequence. It also helps you generate the terms of the sequence before you find the actual sum.

Example 1

Find the partial sum:

(a) $\sum_{i=1}^{5} 3i$ (b) $\sum_{k=1}^{6} \frac{1}{3}(-3)^k$

Solution

(a)

$$\sum_{i=1}^{5} 3i = 3+6+9+12+15=45$$

Because the series is arithmetic, you can obtain the same result by using the formula for the sum of the first n terms of an arithmetic sequence. Here $a = 3$, $a_n = 15$ and $n = 5$, so using the formula $S_n = \frac{n}{2}(a + a_n)$, gives

$$\sum_{i=1}^{5} 3i = \tfrac{5}{2}(3+15) = 45$$

(b)

$$\sum_{k=1}^{6} \tfrac{1}{3}(-3)^k = -1+3-9+27-81+243=162$$

Because the series is geometric, an alternative method uses the formula for the sum of the first n terms of the geometric sequence. Here, $a = -1$ and $r = -3$, so we have

$$\sum \tfrac{1}{3}(-3)^k = \frac{-1[(-3)^6-1]}{-3-1} = 162$$

Convergent and Divergent Series

A series of terms whose sum approaches a finite value as the number of terms is increased indefinitely is called a convergent series, and the finite value is called its sum to infinity. If the sum of a series does not tend to a limit, the series does not converge. The series is then said to be divergent.

Infinite Geometric Series

The sum of the first n terms of a geometric sequence $a + ar + ar^2 + \ldots + ar^n$ is

$$S_n = \frac{a(1-r^n)}{1-r}$$

If r lies between -1 and $+1$, then as n become larger and larger r^n gets closer and closer to 0. That is,

$$\lim_{n\to\infty} r^n = 0$$

So, if $|r| < 1$, the sum of an infinite geometric series $a_1 + a_2 + a_3 + \ldots + a_n + \ldots$ is

$$S = \frac{a}{1-r}.$$

Example 2

Determine if the following series converge. Give the sum of the series

$$2 + \frac{2}{3} + \frac{2}{9} + \frac{2}{27} + \cdots$$

Solution

Here $a = 2$ and $r = 1/3$. Since r is in the interval $(-1, 1)$,the series converges.

The sum of the sequence is

$$\frac{a}{1-r} = \frac{2}{1-1/3}$$

$$= 3$$

Example 3

Express the repeating decimal 0.151515... as a fraction

Solution

The repeating decimal 0.151515... is the sum of an infinite number of terms:

$$0.151515\ldots = 0.15 + 0.0015 + 0.000015 + \ldots$$

$$= \frac{15}{100} + \frac{15}{10000} + \frac{15}{1000000} + \cdots$$

We need to find the sum to infinity of this series.

$$S_\infty = \frac{\frac{15}{100}}{1 - \frac{1}{100}}$$

$$= \frac{15}{99}$$

$$= \frac{5}{33}$$

Exercise 15.3

In Exercises 1 – 8, find the partial sum.

1. $\sum_{k=1}^{6} 2k$

2. $\sum_{k=1}^{5} 4k$

3. $\sum_{i=1}^{6} (3i+2)$

4. $\sum_{i=1}^{8} (4i-1)$

5. $\sum_{n=1}^{7} \frac{1}{2}(-2)^n$

6. $\sum_{k=1}^{3} 27(\frac{1}{3})^k$

7. $\sum_{j=1}^{8} \frac{3}{4}(2)^j$

8. $\sum_{k=1}^{7} 6(2)^k$

In Exercises 9 – 20, calculate the sum to infinity of each geometric sequence.

9. $9 + 6 + 4 + \ldots$

10. $12 + 6 + 3 + \ldots$

11. $20 - 10 + 5 - \ldots$

12. $1 + \frac{1}{3} + \frac{1}{9} + \cdots$

13. $1 - \frac{1}{2} + \frac{1}{4} - \ldots$

14. $0.3 + 0.03 + 0.003 + \ldots$

15. $1 + 0.9 + 0.81 + \ldots$

16. $\frac{5}{10} + \frac{5}{100} + \frac{5}{1{,}000} + \cdots$

17. $12 + 4 + \frac{4}{3} + \cdots$

18. $\frac{5}{4} + \frac{5}{8} + \frac{5}{16} + \cdots$

19. $\frac{1}{3} - \frac{2}{9} + \frac{4}{27} - \cdots$

20. $4 + 2.4 + 1.44 + \ldots$

In Exercises 21 – 26, express each recurring decimal as a fraction.

21. $0.\dot{6}$

22. $0.6\dot{3}$

23. $0.\dot{2}\dot{7}$

24. $0.1\dot{8}\dot{1}$

25. $5.\dot{2}$

26. $3.\dot{7}\dot{2}$

27. If the sum to infinity of a geometric sequence is two times the first term, what is the common ratio?

28. The sum to infinity of a geometric sequence is 27 and the second term is 6. Find the two possible values of the common ratio and the corresponding first term.

29. On each swing a certain pendulum travels 90 % as far as on its previous swing. If the first swing is 10 meters determine the total distance traveled by the pendulum by the time it comes to rest.

15.4 Recursive Sequences

A sequence can be defined by assigning a value to the first term, or the first few terms, and specifying the n th term by an equation involving one or more of the terms preceding it. This type of sequence is called a recursive sequence, and the equation is called a recursion formula.

Example 1

Compute the first six terms of the recursive sequence $a_n = 2a_{n-1} + 1$, $a_0 = 1$ for all integers $n \geq 1$.

Solution

The first term is given as $a_0 = 1$. To get the second term, use $n = 1$in the formula to get $a_1 = 2a_0 + 1 = 2(1) + 1 = 3$. To get the third term, we use $n = 2$ in the formula to get $a_2 = 2a_1 + 1 = 2(3) + 1 = 7$.

Notice that to get the third term we need to use the value of the second term. The following terms are obtained in a similar way.

$$a_3 = 2a_2 + 1 = 2(7) + 1 = 15$$

$$a_4 = 2a_3 + 1 = 2(15) + 1 = 31$$

$$a_5 = 2a_4 + 1 = 2(31) + 1 = 63$$

The first six terms of this sequence are 1, 3, 7, 15, 31 and 63.

Example 2

Given $a_1 = 1$ and $a_{n+1} = 2 + a_n$, find the first five terms of recursive sequence and an expression for a_n in terms of n. Use your expression to find a_{20}.

Solution

We need to find a_2, a_3, a_4 and a_5. So,

$$a_2 = 2 + a_1 = 2 + 1 = 3$$

$$a_3 = 2 + a_2 = 2 + 3 = 5$$

$$a_4 = 2 + a_3 = 2 + 5 = 7$$

and $$a_5 = 2 + a_4 = 2 + 7 = 9$$

So, the first five terms of this sequence are 1, 3, 5, 7 and 9.

Notice that this is a linear sequence.

Substitute $a = 1$ and $d = 2$ in $a_n = a + (n - 1)d$ to get

$$a_n = 1 + (n - 1)(2) = 2n - 1$$

So, $$a_{20} = 2(20) - 1 = 39.$$

Exercise 15.4

In Exercises 1 – 12, find the first five terms of the sequence defined recursively for all integers $n \geq 1$.

1. $a_n = a_{n-1} + n$, $a_0 = 1$
2. $a_{n+1} = 2 + a_n$, $a_1 = 1$
3. $a_n = na_{n-1} - 1$, $a_0 = 1$
4. $a_{n+1} = n + a_n$, $a_1 = -2$
5. $a_n = a_{n-1} + 2^n$, $a_0 = -1$
6. $a_{n+1} = n - a_n$, $a_1 = 1$
7. $a_n = a_{n-1} + n^2$, $a_0 = 1$
8. $a_{n+1} = n + 2a_n$, $a_1 = -3$
9. $a_{n+1} = n - a_n$, $a_1 = 1$
10. $a_{n+2} = a_n + a_{n+1}$, $a_1 = 1$, $a_2 = 1$
11. $a_{n+1} = n + 3a_n$, $a_1 = 1$
12. $a_{n+2} = a_{n+1} + na_n$, $a_1 = -1$, $a_2 = 1$

In Exercises 13 – 20, find the first five terms of the recursive sequence and an expression for the n th term in terms of n.

13. $a_{n+1} = a_n + 3$, $a_1 = 2$
14. $a_{n+1} = a_n + 6$, $a_1 = 8$
15. $a_{n+1} = a_n - 2$, $a_1 = 3$
16. $a_{n+1} = a_n - 5$, $a_1 = 21$
17. $a_n = a_{n-1} + 5$, $a_0 = 3$
18. $a_n = a_{n-1} + 5$, $a_1 = -12$
19. $a_n = 2a_{n-1}$, $a_0 = 1$
20. $a_n = -2a_{n-1}$, $a_0 = 4$

21. Given that $a_n = 4n + 1$ for $n \geq 1$. Find a_1, a_2 and a_3. Find also, the sum of the first 20 terms of the sequence.

22. A sequence is defined by the recurrence relation $a_n = ka_{n-1} + m$. If $a_1 = 3$, $a_2 = 5$ and $a_3 = 7$, find

a. k and m

b. a_0 and a_4

Review Exercises

Determine whether the sequence is arithmetic. For each sequence that is arithmetic find the common difference.

1. $-3, 1, 5, 9, 13 \ldots$
2. $7, 5, 11, 13, 15 \ldots$
3. $5, 2, -1, -4, -7, \ldots$
4. $2.3, 3, 3.7, 4.4 \ldots$
5. $32, 16, 8, 4 \ldots$
6. $-8, 2, 12, 22, 32 \ldots$
7. $-20, -16, -12, -8 \ldots$
8. $-6, -4, 0, 1, 2 \ldots$

Determine whether the sequence is geometric. For each sequence that is geometric find the common ratio.

9. $-4, 8, -16, 32, -64 \ldots$

10. 18, 6, 3, ⅔, 2/9, …

11. 48, 24, 12, 6, 3 . . .

12. $-\frac{7}{8}, \frac{7}{12}, -\frac{7}{18}, \frac{7}{27} \ldots$

13. 6, 8, 10, 12, 14 . . .

14. ⅓,1, 3, 9, 27, …

15. $10, 4, 1, \frac{1}{5}, \frac{2}{25}, \ldots$

16. $15, 9, \frac{27}{5}, \frac{81}{25}, \frac{243}{125} \ldots$

Write down the terms indicated in each of the following sequence:

17. 7, 9, 11... 20 th

18. 3, 6, 12... 8 th

19. 4, −2, 1... 6 th

20. 5, 12, 19... 10 th

21. 9, 6, 4... 5 th

22. $12, 4, \frac{4}{3} \ldots$ 6 th

23. − 5, 2, 9 .... 12 th

24. $15, 3, \frac{3}{5} \ldots$ 5 th

Find the number of terms of each of the following sequence:

25. 2, 6, 10 ... 38

26. $6, 3, \frac{3}{2} \cdots \frac{3}{256}$

27. 5, 8, 11... 50

28. 3, 8, 13 ... 73

29. 2, 6, 18 ... 486

30. 10, 8, 6 ... −12

31. 5, 20, 80 ... 1,280

32. $4, 2, 1 \ldots \frac{1}{128}$

Find the sum of each of the following sequence as far as the term indicated:

33. 1, −3, 9.... 10 th

34. 5, 12, 19 ... 12 th

35. 9, 6, 4.... 12 th

36. 2, 8, 14, 20... 25 th

37. 3, − 6, 12... 12 th

38. 4, 12, 36... 8 th

39. −50, − 38, − 26... 50 th

40. 3.2, 3.8, 4.4... 12 th

Find the partial sum.

41. $\sum_{k=1}^{6} 3k$

42. $\sum_{k=1}^{8} 8(2)^k$

43. $\sum_{k=1}^{10} (6k-10)$

44. $\sum_{n=1}^{7} \frac{3}{2}(4)^n$

45. $\sum_{j=1}^{6} (2j+5)$

46. $\sum_{j=1}^{6} 128(\frac{1}{2})^j$

47. $\sum_{j=1}^{7} 4(3)^j$

48. $\sum_{n=1}^{8} (10-3n)$

Calculate the sum to infinity of each of the following geometric sequence:

49. $9 + 3 + 1 + \ldots$

50. $24 + 12 + 6 + \ldots$

51. $0.5 + 0.05 + 0.005 + \ldots$

52. $1 + \frac{2}{3} + \frac{4}{9} + \cdots$

Express the following recurring decimals as fractions:

53. $0.\dot{5}$

54. $0.\dot{7}$

55. $0.9\dot{3}$

56. $0.51\dot{6}$

57. $0.\dot{8}\dot{1}$

58. $0.\dot{4}8\dot{6}$

Find the first five terms of the following sequence defined recursively for all integers $n \geq 1$.

59. $a_{n+1} = n + a_n,\ a_1 = -2$

60. $a_n = a_{n-1} + 2^n,\ a_0 = -1$

61. $a_{n+1} = n + 3a_n,\ a_1 = -2$

62. $a_n = a_{n-1} + n^2,\ a_0 = 1$

63. The first term of an arithmetic sequence is 4, and the last term is 31. If the sum of the sequence is 175, find the number of terms in the sequence and the common difference.

64. Find the sum of the first n terms of the arithmetic sequence $3 + 5 + 7 + \ldots$. Find the value of n for which the sum of the first $2n$ terms will exceed the sum of the first n terms by 161.

65. The sum of n terms of a series is $2n^2$ for $n = 1, 2, 3, \ldots$. Show that the series is an arithmetic sequence and find its n th term.

66. The sum to n terms of a certain arithmetic sequence is $3n(n + 2)$. What is:

a. its n th term?

b. its common difference?

67. Find how many terms of the sequence $5 + 9 + 13 + 17 + \ldots$ have a sum of 495.

68. A geometric sequence is given by 27, 9, 3, ... Find:

a. the n th term

b. the sum of the first n terms

c. the sum for large values of n of the sequence

69. The third and the fifth terms of a geometric sequence of positive terms are 2 and ½ respectively. Find:

a. the common ratio

b. the first term

c. the sum of the first n terms of the sequence

70. Find the least number of terms of the geometric sequence $3 + 5 + 8\frac{1}{3} + \cdots$ that will give a sum greater than 100.

71. The n th term a_n of a sequence is given by $a_n = 2 \times 3^{n-1}$.

a. Write down the first four terms of the sequence.

b. Calculate the least value of n for which $a_n > 900$.

72. The fifth, ninth and sixteenth terms of arithmetic sequence are consecutive terms of a geometric sequence and the sum of the first and fourth terms is 17. Find the common ratio.

73. The numbers $a_1, a_2, a_3 \ldots$ satisfy the recurrence relation

$$a_{n+1} = a_n + (\tfrac{3}{4})^n$$

for all possible integers $n \geq 1$. If $a_1 = 1$, find:

a. a_2, a_3 and a_4

b. an expression for a_n in terms of n

c. the value of a_n for large values of n

74 a. Given that $a_0 = 5$ and $a_n = a_{n-1} + 4$ for $n > 1$ find:

i. a_1, a_2 and a_3

ii. a formula for a_n in terms of n

b. Use your formula for a_n to deduce a formula for S_n, the sum of the first n terms of the sequence $a_1, a_2, \ldots, a_n$. Hence find S_{32}.

75. The n th term a_n of a sequence is given by

$$a_n = 5 + 7 + 9 + 11 + \ldots + (2n + 3)$$

Express a_n in the form $a_n = pn(qn + r)$, where p, q and r are constants. Hence find the least value of n for which $a_n \geq 96$.

76. Each row in a small theatre has one seat less than the seats in the preceding row. There are 20 seats in the front row and 20 rows of seats on the floor of the theatre. One night the theatre took in $15000 for the sale of all seats, how much was charged per ticket?

77. A man is given a starting salary of $12,750 and is told he will receive $900 raise each year. How much will he receive in the 10 th year? How much will he receive in the first 10 years?

78. A job is advertised at a starting salary of $12,000 with an annual increase of 5 %. If a man is employed now, what will be his salary in the 10 th year? Find the salary he will receive in the first 10 years.

79. A boy stacks logs so that there are 20 logs in the bottom layer and each layer contains one log less than the layer below it. If he stopped stacking the logs after completing the layer containing eight logs, how many logs are in the pile?

80. Each swing of a pendulum is 25 centimeters shorter than the preceding swing. The first swing is 8 meters.

a. Find the length of the twelfth swing.

b. Determine the distance traveled by the pendulum during the first 12 swings.

81. The number of a certain type of bacteria doubles every hour. If there are initially 1,000 bacteria after how many hours will the number of bacteria reached 128,000?

82. A company buys a machine for $75,000. During the next 4 years the machine depreciates at the rate of 15 % per year. Find the value of the machine at the end of the fourth year.

83. A man invests \$10,000 in a savings account paying 6 % interest per year, compounded annually. Find the amount in his account at the end of the 8 th year.

84. A man invests \$500 at the end of each year in an account paying 9 % interest per year, compounded annually. How much will he have in the account after his tenth investment?

85. The temperature of water in an ice cube tray placed in a freezer is 70° F. The temperature of the water at any time is 20 % less than it was one hour earlier. Find the temperature of the water six hours after it is placed in the freezer.

16 The Binomial Theorem

16.1 Pascal's Triangle

You learned in Section 3.2 to expand $(x + y)^2$, using the distributive property to get $x^2 + 2xy + y^2$. Multiplying this result by $(x + y)$ will give the expansion of $(x + y)^3$, as shown below.

$$\begin{aligned}(x + y)^3 &= (x^2 + 2xy + y^2)(x + y)\\ &= x^3 + 2x^2y + xy^2 + x^2y + 2xy^2 + y^3\\ &= x^3 + 3x^2y + 3xy^2 + y^3\end{aligned}$$

Multiplying binomials raised to higher powers using the distributive property can be a little cumbersome. In this Chapter, we will describe techniques for raising a binomial to a power quickly and easily.

Let us look at the expansions of $(x + y)^n$ for some values of n.

$(x + y)^0 = 1$

$(x + y)^1 = x + y$

$(x + y)^2 = x^2 + 2xy + y^2$

$(x + y)^3 = x^3 + 3x^2y + 3xy^2 + y^3$

$(x + y)^4 = x^4 + 4x^3y + 6x^2y^2 + 4xy^3 + y^4$

A triangular display of the numerical coefficient as shown below is called Pascal's Triangle.

	n = 0
1	n = 0
1 1	n = 1
1 2 1	n = 2
1 3 3 1	n = 3
1 4 6 4 1	n = 4

The first and last numbers in each row of the Pascal's Triangle are 1. As you can see, every other number in each row is formed by adding the two numbers immediately above the number. The values in the next row of the Pascal's Triangle are:

$$1 \quad (1 + 4) = 5 \quad (4 + 6) = 10 \quad (6 + 4) = 10 \quad (4 + 1) = 5 \quad 1$$

Looking at the preceding expansions carefully you might notice that

1. The expansion of $(x + y)^n$, where n is any positive integer, is a finite series with $(n + 1)$ terms.

2. The sum of the powers of each term is n.

3. The power of x decrease by 1 in successive terms, whereas the powers of y increase by 1.

4. The term in y^r is the $(r + 1)$ th term.

5. The first term is x^n and the last term is y^n.

Using Pascal's Triangle

Example 1

Expand $(2x - y)^4$

Solution

Obtain the coefficients 1, 4, 6, 4, 1 from the fifth row of the Pascal's Triangle.

The expansion can be worked out as follows.

$$(2x - y)^4 = (2x)^4 + 4(2x)^3(-y) + 6(2x)^2(-y)^2 + 4(2x)(-y)^3 + (-y)^4$$
$$= 16x^4 - 32x^3y + 24x^2y^2 - 8xy^3 + y^4$$

Example 2

Expand $(3x + 2y)^3$.

Solution

The coefficients from the Pascal's Triangle are: 1, 3, 3, 1.

$$(3x + 2y)^3 = (3x)^3 + 3(3x)^2(2y) + 3(3x)(2y)^2 + (2y)^3$$
$$= 27x^3 + 54x^2y + 36xy^2 + 8y^3$$

Example 3

Expand $(3 + x)^4$ and use the first three terms to find $(2.98)^4$

Solution

$$(3 + x)^4 = (3)^4 + 4(3^3)(x) + 6(3^2)(x^2) + 4(3)(x^3) + x^4$$
$$= 81 + 108x + 54x^2 + 12x^3 + x^4$$

If x is made small, successive terms may be so small that they do not affect the answer to the required degree of accuracy.

We write $(2.98)^4$ as $(3 - 0.02)^4 = [3 + (-0.02)]^4$.

Replacing x with -0.02 in the expansion gives

$$
\begin{aligned}
(2.98)^4 &= 81 + 108(-0.02) + 54(-0.02)^2 \\
&= 81 - 2.16 + 0.0216 \\
&= 78.8616
\end{aligned}
$$

Note: When finding an approximation, take the first few terms of the expansion that will give you at least one place more than the required degree of accuracy.

Exercise 16.1

Expand:

1. $(1 + 2x)^6$
2. $(x - 2y)^4$
3. $(1 - x)^5$
4. $(2x + 3y)^4$
5. $(4x - 1)^3$
6. $(3x + y)^4$
7. $(1 - 3x)^4$
8. $(2a + x)^6$
9. $(2a - 3x)^4$
10. Expand $(1 - x)^5$ and use the expansion to evaluate $(0.98)^5$, correct to three significant figures.
11. Using the expansion of $(1 + x)^4$ find the value of $(1.01)^4$ correct to five significant figures.
12. Expand $(1 - 2x)^5$ and then use the expansion to evaluate $(0.98)^5$ correct to three significant figures.

16.2 The Binomial Theorem

The binomial theorem can be used to expand binomial expression of the form $(x + y)^n$, where n is a positive integer. The binomial expansion for a positive integer is given as

$$(x + y)^n = x^n + {}^nC_1x^{n-1}y + {}^nC_2x^{n-2}y^2 + \cdots + {}^nC_rx^{n-r}y^r + \ldots + y^n$$

where ${}^nC_r = \frac{n!}{(n-r)!r!}$.

${}^nC_r$ is called the binomial coefficient of the expansion.

Other notations commonly used for ${}^nC_r$ are ${}_nC_r$, $C(n, r)$ and $\binom{n}{r}$.

The symbol $n!$, read as n factorial, is defined as

$n! = n(n-1)(n-2) \ldots 3 \cdot 2 \cdot 1.$

For example,

$3! = 3 \cdot 2 \cdot 1 = 6$

$6! = 6 \cdot 5 \cdot 4 \cdot 3 \cdot 2 \cdot 1 = 720$

Note that $0! = 1$ and $1! = 1$.

Example 1

Find the following binomial coefficients

(a) ${}^8C_3$ (b) ${}^{10}C_4$

Solution

(a) $${}^8C_3 = \frac{8!}{5!3!} = \frac{8 \cdot 7 \cdot 6 \cdot 5!}{5! \cdot 3 \cdot 2 \cdot 1} = \frac{8 \cdot 7 \cdot 6}{3 \cdot 2 \cdot 1} = 8 \cdot 7 = 56$$

(b) $${}^{10}C_4 = \frac{10!}{6!4!} = \frac{10 \cdot 9 \cdot 8 \cdot 7 \cdot 6!}{6! \cdot 4 \cdot 3 \cdot 2 \cdot 1} = \frac{10 \cdot 9 \cdot 8 \cdot 7}{4 \cdot 3 \cdot 2 \cdot 1} = 10 \cdot 3 \cdot 7 = 210$$

Example 2

Expand $(x-2y)^4$.

Solution

$$\begin{aligned}(x-2y)^4 &= x^4 + {}^4C_1(x^3)(-2y) + {}^4C_2(x^2)(-2y)^2 + {}^4C_3(x)(-2y)^3 + (-2y)^4 \\ &= x^4 - 8x^3y + 24x^2y^2 - 32xy^3 + 16y^4\end{aligned}$$

Example 3

Write down the first five terms of the expansion $(1+x)^7$.

Solution

$$\begin{aligned}(1+x)^7 &= 1 + {}^7C_1x + {}^7C_2x^2 + {}^7C_3x^3 + {}^7C_4x^4 \\ &= 1 + 7x + 21x^2 + 35x^3 + 35x^4\end{aligned}$$

Example 4

Write down the term in x^6 in the expansion of $(x-2y)^9$.

Solution

The complete expansion is not necessary since we need only the term in x^6. We would use the fact that the $(r + 1)$ th term is given by ${}^nC_r x^{n-r} y^r$. The term in x^6 is the fourth term.

Here $r + 1 = 4$, so $r = 3$. Because $n = 9$ the term in x^6 is

$${}^9C_3(x^6)(-2y)^3 = -672x^6y^3$$

Example 5

Find the coefficient of the fifth term of the expansion of $(2x + 3y)^7$

Solution

In this case $n = 7$ and $r + 1 = 5$, so $r = 4$. The 5 th term is

$${}^7C_4(2x)^3(3y)^4 = 22{,}680x^3y^4$$

The coefficient is 22,680

Exercise 16.2

Evaluate the following binomial coefficients:

1. ${}^9C_6$ 2. ${}^7C_0$ 3. ${}^{10}C_1$ 4. ${}^6C_4$

5. ${}^7C_3$ 6. ${}^{10}C_5$ 7. ${}^{12}C_9$ 8. ${}^{12}C_7$

Write down the term indicated in the binomial expansions of the following:

9. $(1 + 4x)^6$, 3 rd term

10. $(3x - 2)^7$, 4 th term

11. $\left(1 - \frac{x}{2}\right)^{10}$, 6 th term

12. $(2 + x)^9$, 5 th term

13. $(2x + 3y)^8$, 5 th term

14. $(3 - 2x)^6$, 4 th term

15. $(2x - y)^{12}$, term containing x^4.

16. $\left(2x - \frac{1}{x}\right)^9$, term containing x^3.

17. $(2x - 3y)^{10}$, term containing y^8.

18. $\left(3x^2 - \frac{1}{2x}\right)^7$, term containing x^2.

Write down the first four terms in the binomial expansion of:

19. $(1 - x)^{10}$

20. $(1 + 2x)^9$

21. $(2 - x)^{10}$

22. $(1 + \frac{1}{2}x)^6$

23. $(3 - \frac{2}{3}x)^7$

24. $(2 + \frac{1}{2}x)^7$

Use the binomial theorem to find the values of:

25. $(1.01)^{12}$, correct to 3 d.p.

26. $(0.98)^{10}$, correct to 3 s.f.

27. $(2.01)^{9}$, correct to 6 s.f.

28. $(1.99)^{12}$, correct to 3 d.p.

29. Find the term independent of x in the expansion of $(x^2 - 2/x)^6$

30. Find the term in x in the expansion of $(x - 2/x)^5$.

31. Find the middle term in the expansion of $(x - 1/y)^8$.

32. Write down the first four terms of the binomial expansion of $(1 - 2x)^{10}$ and use it to find the value of $(0.98)^{10}$, correct to four decimal places.

33. Write down the first four terms of the binomial expansion of $(1 + x/5)^6$ and use it to find the value of $(1.01)^6$, correct to four significant figures.

Find each of the following, up to the term indicated:

34. $(1 + x - x^2)^7$, x^2 term

35. $(1 - 2x + 3x^2)^6$, x^2 term

36. $(1 - x)^7(1 + x)$, x^3 term

37. $(1 + 2x)^5(1 - x)$, x^3 term

16.3 Binomial Theorem for any power

If n is not a positive integer the binomial theorem is given by

$$(1 + x)^n = 1 + nx + \frac{n(n-1)}{2!}x^2 + \frac{n(n-1)(n-2)}{3!}x^3 + \cdots$$

In general this series continue indefinitely. However, when n is a positive integer the series will have $(n + 1)$ terms.

We can adapt this formula to expand binomial expressions of the form $(a + x)^n$.

Example 1

Expand $(1 + 2x)^{-1}$ as far as the term in x^3.

Solution

$$(1 + 2x)^{-1} = 1 + (-1)(2x) + \frac{(-1)(-1-1)}{2!}(2x)^2 + \frac{(-1)(-1-1)(-1-2)}{3!}(2x)^3$$

$$= 1 - 2x + 4x^2 - 8x^3$$

Example 2

Expand $(1 - 3x)^{1/3}$ up to the term in x^3. Use your result to find $\sqrt[3]{0.97}$ to two decimal places.

Solution

$$(1-3x)^{1/3} = 1+\frac{1}{3}(-3x)+\frac{\frac{1}{3}\left(\frac{1}{3}-1\right)}{2!}(-3x)^2+\frac{\frac{1}{3}\left(\frac{1}{3}-1\right)\left(\frac{1}{3}-2\right)}{3!}(-3x)^3$$

$$= 1 - x - x^2 - \frac{5}{3}x^3$$

Now $\sqrt[3]{0.97} = (1-0.03)^{\frac{1}{3}} = [1-3(0.01)]^{\frac{1}{3}}$.

Substitute $x = 0.01$ in the expansion.

$$\sqrt[3]{0.97} = 1 - 0.01 - (0.01)^2 - \frac{5}{3}(0.01)^3$$

$= 1 - 0.01 - 0.0001 - 0.00000167$

$= 0.98989833$

$= 0.99$

Example 3

Expand $(2 + x)^5$.

Solution

Rewrite $(2 + x)^5$ in the form

$$\left[2\left(1+\frac{x}{2}\right)\right]^5 = 32\left(1+\frac{x}{2}\right)^5$$

Using the $(1 + x)^n$ expansion with $n = 5$ and replacing x by $x/2$ we get

$$(2+x)^5 = 32\left[1+5\left(\frac{x}{2}\right)+\frac{5\cdot 4}{2!}\left(\frac{x}{2}\right)^2+\frac{5\cdot 4\cdot 3}{3!}\left(\frac{x}{2}\right)^3+\frac{5\cdot 4\cdot 3\cdot 2}{4!}\left(\frac{x}{2}\right)^4+\left(\frac{x}{2}\right)^5\right]$$

$= 32 + 80x + 80x^2 + 40x^3 + 10x^4 + x^5$

Exercise 16.3

Expand each of the following expressions in ascending powers of x as far as the term in x^3:

1. $(1-x)^{-2}$	2. $(1+3x)^{-\frac{1}{3}}$	3. $(1-x)^{\frac{1}{3}}$
4. $(1+2x)^{\frac{1}{2}}$	5. $\left(1-\frac{x}{2}\right)^{-2}$	6. $(1+x)^{-\frac{2}{3}}$
7. $(1+3x)^{-1}$	8. $(1+x)^{-\frac{1}{2}}$	9. $\left(1+\frac{x}{2}\right)^{-2}$
10. $(1+2x)^{-\frac{1}{2}}$	11. $(2-x)^{\frac{1}{2}}$	12. $(3+x)^{-1}$

Use the binomial theorem to evaluate:

13. $\sqrt{1.01}$ correct to five decimal places

14. $(1.03)^{-2}$ correct to four decimal places

15. $\sqrt{0.998}$ correct to three decimal places

16. $\sqrt[3]{0.97}$ correct to four decimal places

17. Find the first four terms of the expansion of $(1 + x)^{1/3}$ in ascending powers of x. Use your expansion to find $\sqrt[3]{64.8}$ correct to the four significant figures.

18. Find the expansion for $\sqrt{1+x}$ up to the term in x^3. Hence find an approximation for $\sqrt{17}$.

Review Exercises

Evaluate the following binomial coefficients:

1. ${}^{13}C_0$ 2. ${}^{6}C_2$ 3. ${}^{8}C_4$ 4. ${}^{9}C_3$

5. ${}^{12}C_5$ 6. ${}^{10}C_6$ 7. ${}^{7}C_4$ 8. ${}^{15}C_{12}$

Expand the following:

9. $(x+2)^3$ 10. $(x+3)^5$ 11. $(x-y)^4$

12. $(2x-1)^5$ 13. $(2x+y)^6$ 14. $(1-2x)^6$

15. $(x+4)^3$ 16. $(2x-y)^5$ 17. $(3x+2y)^4$

Find in ascending powers of x the first 4 terms in the expansion of the following:

18. $(2+x)^9$ 19. $(3-x)^7$

20. $(1+2x)^{12}$ 21. $(1-3x)^8$

Find the specified terms in the expansion of the following:

22. $(x-y)^{10}$, 4 th term

23. $(3x+y)^{12}$, 10 th term

24. $(3x+y)^7$, 3 rd term

25. $(a-4b)^9$, 6 th term

26. $(x+2y)^8$, 6 th term

27. $(x-3y)^{12}$, 4 th term

Find the coefficient of the terms indicated in the expansion of:

28. $(x-1)^{10}$, term containing x^7

29. $(x+y)^{15}$, term containing y^{11}

30. $(2x-y)^{12}$, term containing y^9

31. $(x+y)^{10}$, term containing y^3

32. $(x^2+3)^4$, term containing x^4

33. $(x-3)^{12}$, term containing x^9

Find the term independent of x in the expansion of the following:

34. $\left(x^2-\frac{2}{x}\right)^6$

35. $\left(x^3+\frac{1}{2x}\right)^8$

36. $\left(2x-\frac{1}{x^4}\right)^{10}$

37. Write down the binomial expansion of $(x+y)^4$. Use your expansion to evaluate, correct to three decimal places $(1.99)^4$.

38. (a) Using the Binomial Theorem write down the expansion of $(1+x)^7$.

(b) Use your result in (a) to evaluate $(0.997)^7$ correct to five decimal places.

39. (a) Write down and simplify the first five terms of the binomial expansion of $(1-⅓x)^8$.

(b) Use your expansion to evaluate $(0.997)^8$, correct to five decimal places.

40. Expand $(1-2x)^6$ and then use the expansion to evaluate $(0.94)^6$ correct to three significant figures.

41. The coefficient of x^2 in the expansion of $(2+px)^6$ is 15. Find the value of p.

42. The coefficient of x^2 in the expansion of $(1+px)^5$ is 90. Find the value of p.

Find the coefficient of x^3 in each of the following expansions:

43. $(3-x)(2+¼x)^6$

44. $(1+2x)(2-x)^8$

45. $(2+x)^6(1-x)^2$

46. Write down the expansion of $(1 + x)^{1/3}$ as far as the term in x^3. Use your expansion to approximate $\sqrt[3]{1010}$.

47. Expand $(1 + 2x)^{½}$ in ascending powers of x as far as the fourth term. Substituting 0.04 for x in your expansion find the value of $\sqrt{3}$.

48. Expand $(1 - ½\, x)^{-1/2}$ in ascending powers of x as far as the fourth term. Substituting 1/5 for x in your expansion find the value of $\sqrt{10}$.

17 Circles and Parabolas

17.1 Circles

A circle is the set of all points (x, y) in a plane that are a given distance r from a fixed point, called the center of the circle. The distance r between the center of the circle and any point on the circle is called the radius of the circle. The perimeter of the circle is called the circumference of the circle.

Equation of a Circle

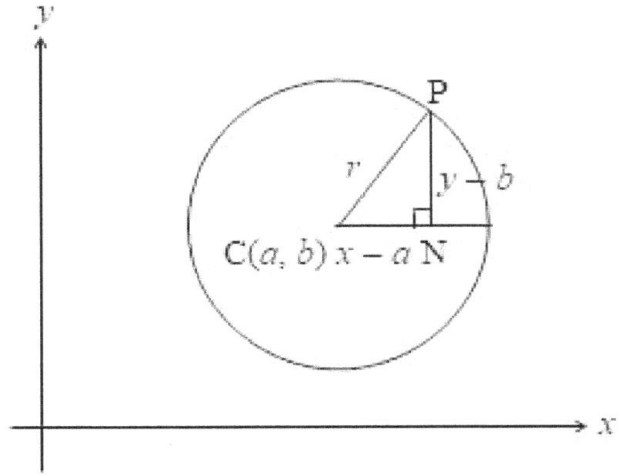

Figure 17.1

Consider a circle of radius r with center at the point C(a, b), as shown in Figure 17.1. If P(x, y) is a point on the circle, then by the distance formula

$$\sqrt{(x-a)^2+(y-b)^2}=r$$

By squaring each side of this equation, we obtain the standard form of the equation of a circle

$$(x-a)^2+(y-b)^2=r^2$$

From this result, we see that the standard form of the equation of a circle with center at the origin (0, 0) is

$$x^2+y^2=r^2$$

General Equation of a Circle

Expanding the left hand side of $(x-a)^2+(y-b)^2=r^2$ yield

$$x^2+y^2-2ax-2by+a^2+b^2-r^2=0$$

The following equation is the general equation of a circle

$$x^2 + y^2 + 2gx + 2fy + c = 0$$

where $a = -g$, $b = -f$ and $c = a^2 + b^2 - r^2$.

You can see that the center of the circle is $(-g, -f)$ and that the radius is $r = \sqrt{g^2 + f^2 - c}$.

Observe that the coefficients of x^2 and y^2 are equal and there is no term in xy.

Example 1

Find the equation of a circle whose center is (2, 3) and whose radius is 2.

Solution

$(x - a)^2 + (y - b)^2 = r^2$	Standard form
$(x - 2)^2 + (y - 3)^2 = 2^2$	Substitute for a, b and r.
$x^2 - 4x + 4 + y^2 - 6y + 9 = 4$	Expand.
$x^2 + y^2 - 4x - 6y + 9 = 0$	Equation of circle

Example 2

Write the equation $x^2 + y^2 + 6x - 4y - 3 = 0$ in standard form, and find the radius and the coordinates of the center of the circle.

Solution

$x^2 + y^2 + 6x - 4y - 3 = 0$	Original equation
$x^2 + 6x + y^2 - 4y = 3$	Rearrange the equation.
$x^2 + 6x + 9 + y^2 - 4y + 4 = 3 + 9 + 4$	Complete the square
$(x + 3)^2 + (y - 2)^2 = 16$	Factor
$(x + 3)^2 + (y - 2)^2 = 4^2$	Standard form

In this case, $a = -3$, $b = 2$ and $r = 4$.

Hence, the center of the circle is (−3, 2), and the radius is 4.

Example 3

The point (1, 5) lies on a circle whose center is (4, 3). Find the equation of the circle.

Solution

The radius r of the circle is the distance between (1, 5) and (4, 3).

$$r = \sqrt{(1-4)^2 + (5-3)^2}$$

$$= \sqrt{9+4}$$

$$= \sqrt{13}$$

Substitute $a = 4$, $b = 3$ and $r = \sqrt{13}$ in $(x - a)^2 + (y - b)^2 = r^2$.

This gives $(x - 4)^2 + (y - 3)^2 = 13$

$x^2 + y^2 - 8x - 6y + 12 = 0$ Equation of circle

Exercise 17.1

Which of the following equations represent circles?

1. $x^2 - y^2 + 2x + 3y + 5 = 0$
2. $x^2 + y^2 - 8x + 9y + 12 = 0$
3. $x^2 + y^2 + 6x + xy + 3 = 0$
4. $3x^2 + 3y^2 + 7x - 8y - 20 = 0$
5. $2x^2 + 3y^2 - 6x + 8y = 0$
6. $x^2 + y^2 - 6x = 0$
7. $x^2 + y^2 + 3xy - 6 = 0$
8. $2x^2 + 2y^2 - x + y = 0$

Find the equation of the circle with:

9. center (3, 2), radius 4
10. center (4, 3), radius 6
11. center (−2, 3), radius 5
12. center (2,− 5), radius 3
13. center (0, −5), radius 2
14. center (− 5, −3), radius 5
15. center (3, 0), radius 1
16. center (5, −2), radius 3/2
17. center (−3, 4), radius 5
18. center (3, −2), radius 7
19. center (5, 4), radius 3
20. center (−4, −3), radius 1/3

Find the center and radius of the circle given by the equation.

21. $x^2 + y^2 - 6x + 2y - 6 = 0$
22. $x^2 + y^2 - 4x - 2y + 1 = 0$
23. $x^2 + y^2 - 6x + 8y - 11 = 0$
24. $x^2 + y^2 + 6x - 4y - 3 = 0$
25. $x^2 + y^2 - 2x - 6y - 6 = 0$
26. $x^2 + y^2 - 2x + 6y - 15 = 0$
27. $x^2 + y^2 - 4x + 2y + 1 = 0$
28. $x^2 + y^2 + 8x + 4y - 5 = 0$
29. $x^2 + y^2 + 2x = 15$
30. $x^2 + y^2 + 2x + 6y + 6 = 0$
31. $x^2 + y^2 + 8y - 9 = 0$
32. $x^2 + y^2 - 14x + 8y + 56 = 0$

Find the coordinates of the center and the radius of each of the following circles.

33. $2x^2 + 2y^2 - 3x + 2y + 1 = 0$

34. $3x^2 + 3y^2 + 6x - 3y - 2 = 0$

35. $4x^2 + 4y^2 + 8x - 4y - 11 = 0$

36. $9x^2 + 9y^2 - 12x - 18y - 23 = 0$

37. $4x^2 + 4y^2 - 16x + 8y + 11 = 0$

38. $2x^2 + 2y^2 + 12x - 8y + 17 = 0$

39. $9x^2 + 9y^2 - 12x - 6y - 31 = 0$

40. $3x^2 + 3y^2 + 12x - 18y - 35 = 0$

41. The point (3, 4) lies on a circle whose center is (1, 2). Find the equation of the circle.

42. A circle whose center is (−2, 3) passes through the midpoint of the line joining the points (−3, 1) and (5, 3). Find the equation of the circle.

43. Find the equation of the circle whose diameter is the line joining (−3, 2) to (7, 4).

44. Find the equation of the circle whose center is (1,−4) and passes through the point where the line $3x + 4y + 6 = 0$ cuts the x-axis.

45. A circle whose center is in the second quadrant touches the y-axis at the point (0, 4) and also touches the x-axis at (−4, 0). Find the equation of the circle.

46. Find the equation of the circle whose center lies on the line $x + y = 4$ and which passes through the points (−1, 0) and (3, 0).

47. Find the equation of the circle whose center is (−3, 2) and touches the y-axis.

48. Find the equation of the circle which passes through (9, 5), (8, −2) and (2, 6).

17.2 Loci

A locus is defined as the path of a moving point satisfying a set of given conditions. The plural of locus is loci.

Note that:

1. Every point on the locus satisfies the given condition.

2. Every point which satisfies the given condition lies on the locus.

The locus of points can be described by an equation as illustrated by the example below.

Example 1

Find the locus of a point which moves so that its distance from the point (3, −1) is 2 units.

Solution

Let P(x, y) be any point on the locus. Then by the distance formula

$$\sqrt{(x-3)^2+(y+1)^2}=2$$

$$(x-3)^2+(y+1)^2=4 \qquad \text{Square each side}$$

$$x^2-6x+9+y^2+2y+1=4 \qquad \text{Expand.}$$

$$x^2+y^2-6x+2y+6=0 \qquad \text{Simplify.}$$

Therefore the locus of the point is

$$x^2+y^2-6x+2y+6=0.$$

Example 2

Find the locus of a point which moves so that the sum of the squares of its distances from (−3, 0) and (3, 0) is 38 units.

Solution

Let P(x, y) be any point on the locus. Then by the distance formula.

$$(x+3)^2+y^2+(x-3)^2+y^2=38$$

$$x^2+6x+9+y^2+x^2-6x+9+y^2=38$$

$$x^2+y^2=10$$

Exercise 17.2

1. Find the locus of a point which moves so that its distance from the point (2, 1) is 3 units.

2. Find the locus of a point which moves so that its distance from the point (−3, −2) is 5 units.

3. Find the locus of a point which is equidistant from the points (1, 2) and (−3, 1).

4. Find the locus of a point which is equidistant from the points (−1, 4) and (7, 3).

5. What is the locus of a point which moves so that its distance from the point (−3, 2) is equal to its distance from the point (1, 2).

6. Find the locus of a point which is equidistant from the origin and the line $x = -3$.

7. Find the locus of a point which is equidistant from the point (0, −1) and the line $y = 1$.

8. Find the locus of a point which moves so that its distance from the point A (−3, 0) is twice its distance from the origin.

9. A point P which moves so that its distance from A (1, 2) is twice its distance from B (5, −4). Find the locus of P.

10. Find the locus of a point which moves so that its distance from the point (0, 8) is twice its distance from the line $y = 2$.

11. Find the locus of a point which moves so that the sum of the squares of its distance from the points (−1, 1) and (3, 2) is 15 units.

12. A is at (3, −2) and B is at (5, 4). Find the locus of P which moves so that $PA^2 + PB^2 = 80$. Describe the locus of P.

13. A is at (3, −2) and B is at (2, 5). P moves so that PA is perpendicular to PB. Describe the locus of P.

14. A point P moves so that its distance from the point (3, 0) is half its distance from the line $x - 5 = 0$. Find the locus of P.

15. A is the point (2, 0) and B is the point (−2, 0). Find the locus of a point which moves so that the sum of the squares of its distance from A and B is equal to four times the squares of its distance from the origin.

16. A straight line of length 5 units moves with its ends A and B on the axes. Perpendiculars to the axes, erected at the points A and B, meets at P. Find the locus of P.

17.3 The Parabola

Recall from Section 10.3 that the graph of a quadratic function of the form $y = ax^2 + bx + c$ is a parabola that opens upward if a is positive and downward if a is negative. In this section, we give a definition of a parabola, and use this definition to derive the standard form of the equation of a parabola.

Definition of a Parabola

A parabola is the set of all points that are equidistant from a fixed line called the directrix, and a fixed point called the focus.

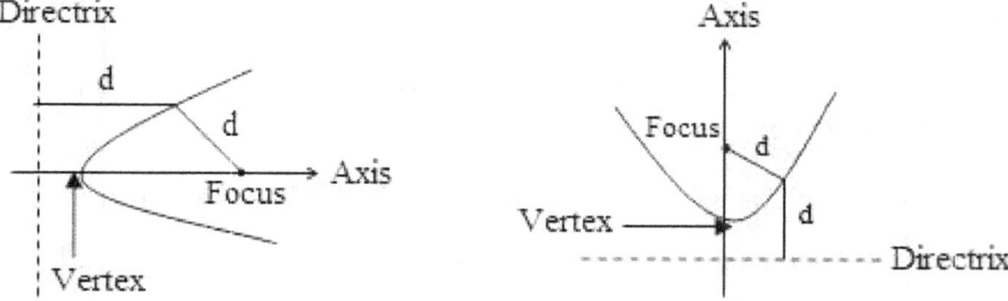

Figure 17.2

The line through the focus perpendicular to the directrix is called the axis of the parabola. A parabola is symmetric with respect to its axis. The midpoint between the focus and the directrix is called the vertex. Note that a parabola can open to the left or open downward.

Equation of a Parabola

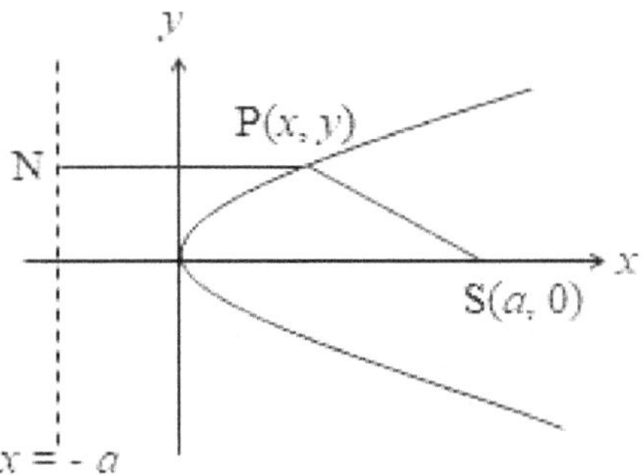

Figure 17.3

Figure 17.3 shows a parabola which opens to the right with its vertex at the origin and whose directrix is parallel to the y-axis. Let the focus of the parabola be $(a, 0)$ and the equation of the directrix be $x = -a$. By the definition of the parabola the distance PS and PN are equal. Thus $PS^2 = PN^2$.

So $$(x-a)^2 + y^2 = (x+a)^2$$

Simplifying this we get

$$y^2 = 4ax$$

The standard equation of a parabola with the vertex at the origin and focus at $(a, 0)$ is $y^2 = 4ax$. Similarly, we can show that the equation of a parabola that opens up with vertex at the origin and focus at $(0, a)$ is $x^2 = 4ay$.

Using the definition of a parabola we can also show that the standard equation of a parabola with a vertex at (p, q), focus $(p + a, q)$ and directrix $x = p - a$ is $(y - q)^2 = 4a(x - p)$, and the equation of a parabola with vertex at (p, q), focus $(p, q + a)$ and directrix $y = q - a$ is $(x - p)^2 = 4a(y - q)$.

Note that, if the focus of a parabola is above or to the right of the vertex, a is positive. If the focus is below or to the left of the vertex, a is negative.

Example 1

Identify the vertex and focus of the parabola $y^2 = 20x$.

Solution

Rewrite $y^2 = 20x$ as $y^2 = 4(5)x$.

Comparing this with the standard equation you can see that $a = 5$. Hence the focus is (5, 0) and the vertex is the origin.

Example 2

Identify the vertex and focus of the parabola $x^2 = -8y$.

Solution

Rewrite $x^2 = -8y$ as $x^2 = 4(-2)y$.

You can see that $a = -2$. The focus is (0, −2) and the vertex is at the origin

Example 3

Identify the vertex, focus and the directrix of the parabola $y^2 - 6y - 4x + 17 = 0$.

Solution

$$y^2 - 6y - 4x + 17 = 0 \qquad \text{Original equation}$$

$$y^2 - 6y = 4x - 17$$

$$y^2 - 6y + 9 = 4x - 17 + 9 \qquad \text{Complete the square}$$

$$(y - 3)^2 = 4x - 8 \qquad \text{Factor}$$

$$(y - 3)^2 = 4(x - 2) \qquad \text{Factor}$$

From $(y - q)^2 = 4a(x - p)$, you can see that the vertex is (2, 3) and $a = 1$.

So, the focus is (3, 3) and the directrix is $x = 1$.

Example 4

Find the equation of the parabola with its vertex at the origin and focus (2, 0).

Solution

Because the focus is (2, 0), we have $a = 2$.

Substitute $a = 2$ in $y^2 = 4ax$ to obtain

$y^2 = 4(2)x = 8x$.

Example 5

Find the equation of the parabola with its vertex at the origin and focus (0, 3).

Solution

Because the focus is (0, 3), we have $a = 3$.

Substitute $a = 3$ in $x^2 = 4ay$ to obtain

$x^2 = 4(3)y = 12y.$

Exercise 17.3

Find the focus of each of the following parabolas:

1. $y^2 = 4x$ 2. $y^2 = -6x$

3. $y^2 = 10x$ 4. $y^2 = -5x$

Find the focus of each of the following parabolas:

5. $x^2 = -4y$ 6. $x^2 = 6y$

7. $x^2 = -12y$ 8. $x^2 = 5y$

Write down the equation of the directrix of each of the following parabolas:

9. $y^2 = 8x$ 10. $y^2 = -4x$

11. $y^2 = 12x$ 12. $y^2 = -3x$

Write down the equation of the directrix of each of the following parabolas:

13. $x^2 = -8y$ 14. $x^2 = 4y$

15. $x^2 = -16y$ 16. $x^2 = 3y$

Write down the equation of the parabola with its vertex at the origin and the following focus:

17. (3, 0) 18. (−2, 0)

19. (7, 0) 20. (− 4, 0)

Write down the equation of the parabola with its vertex at the origin and the following focus:

21. (0, 1) 22. (0, 2)

3. (0, −3) 24. (0, −2)

Identify the vertex, focus and the directrix of the parabola.

25. $(y + 2)^2 + 8(x - 1) = 0$

26. $(y - 2) + (x + 3)^2 = 0$

27. $4y = x^2 - 2x + 5$

28. $x = \frac{1}{4}(y^2 + 2y + 33)$

29. $x^2 + 6x + 8y + 25 = 0$

30. $y^2 - 6y - 6x = 0$

Use the definition of a parabola to find the equation of a parabola with:

31. directrix $y = 0$ and focus (−2, −2)

32. directrix $y = 1$ and focus (3, 5)

33. directrix $x = -1$ and focus (-3, 4)

34. directrix $x = 2$ and focus (6, 3)

Write the standard form of the equation of the parabola with the given vertex and focus:

35. Vertex (4, 2); Focus (2, 2)

36. Vertex (−1, 3); Focus (−1, 1)

37. Vertex (0, 3); Focus (0, 5)

38. Vertex (3, −1); Focus (4, −1)

Review Exercises

Find the equation of the circle with:

1. center (4, 3); radius 10
2. center (−2, 5); radius 6
3. center (5, −3); radius 9
4. center (−5, −2); radius 5/2

Find the equation of the circle with:

5. center (−2, 1) which passes through the point (0, 1)

6. center (3, 2) which passes through the point (4, 6)

7. center (5, 3) which passes through the point (8, 7)

8. center (−3, −5) which passes through the point (−1, −3)

Find the center and radius of each of the following circles:

9. $x^2 + y^2 - 4x - 2y + 1 = 0$

10. $x^2 + y^2 + 6x - 4y - 3 = 0$

11. $x^2 + y^2 - 2x + 6y - 15 = 0$

12. $x^2 + y^2 - 14x + 8y + 56 = 0$

13. Find the radius and the coordinates of the center of the circle $9x^2 + 9y^2 + 12x - 6y - 4 = 0$.

14. Find the equation of the circle with center (4, 5) which passes through the y-intercept of the line $3x - 2y + 6 = 0$.

15. The equation of a circle is of the form $x^2 + y^2 - 4x + 6y + c = 0$, where c is a constant. If the circle has radius 3 find the value of c.

16. P moves so that its distance from the origin is twice its distance from (2, −3). Prove that P moves in a circle and find its center and radius.

17. A point P moves so that its distance from A (−1, 2) is three times its distance from B (0, −2).

18. A is at (−1, 3) and B is at (2, 1). P moves so that PA is perpendicular to PB. Prove that P moves in a circle and find its center and radius.

19. A is the point (4, 3) and B is the point (−2, −3). Find the locus of a point P which moves so that $PA^2 + PB^2 = 50$.

Write the standard form of the equation of the parabola with its vertex at the origin:

20. Focus (0, − ¾) 21. Focus (3, −80)

22. Focus (0, 1) 23. Focus (−4, 0)

Identify the focus and the directrix of each of the following parabolas:

24. $y = \frac{1}{2} x^2$ 25. $y^2 = 3x$

26. $x^2 + 10y = 0$

27. $x - y^2 = 0$

Identify the vertex, focus and the directrix of the parabola.

28. $(x - 1)^2 = -8(y + 2)$

29. $(y - 2)^2 = -(x + 3)$

30. $4x - y^2 - 2y - 33 = 0$

Write the standard form of the equation of the parabola with the given vertex and focus.

31. Vertex: $(-1, 2)$; Focus: $(-1, 0)$

32. Vertex: $(-2, 1)$; Focus: $(-5, 1)$

18 Matrices

18.1 Matrix Algebra

A matrix (plural matrices) is a rectangular array of numbers arranged in rows and columns and commonly enclosed in square brackets. An example is

$$\begin{bmatrix} 3 & -1 & 5 \\ -2 & 0 & 1 \end{bmatrix}$$

Each number in a matrix is called an element or entry of the matrix. The example above has two rows and three columns, written briefly as 2 × 3 (read "2 by 3 matrix"). The expression 2 × 3 is called the dimension (or size) of the matrix. Note that the number of rows is always given first. In general, the dimension of a matrix with m rows and n columns is $m \times n$.

Other examples are:

$$\begin{bmatrix} 5 & 6 \end{bmatrix}, \quad \begin{bmatrix} 4 \\ -3 \end{bmatrix} \quad \begin{bmatrix} 2 & -3 \\ -1 & 0 \\ 3 & 4 \end{bmatrix} \quad \text{and} \quad \begin{bmatrix} 4 & 5 & 6 & 0 \\ -1 & 0 & 1 & -2 \\ 2 & -3 & 1 & 3 \end{bmatrix}$$

A matrix with only one row is called a row matrix (or a row vector) and a matrix with only one column is called a column matrix (or a column vector).

Square Matrix

If a matrix has the same number of rows as it has columns, it is called a square matrix. For example,

$$\begin{bmatrix} -3 & 1 \\ 2 & 0 \end{bmatrix}$$

is a 2 × 2 square matrix and

$$\begin{bmatrix} 0 & 4 & -1 \\ -2 & 0 & 1 \\ 3 & 2 & 1 \end{bmatrix}$$

is a 3 × 3 square matrix.

Equal Matrices

Two matrices A and B are equal if they are of the same dimension and if corresponding elements are equal. Thus, if $A = \begin{bmatrix} a & b \\ c & d \end{bmatrix}$ and $B = \begin{bmatrix} e & f \\ g & h \end{bmatrix}$

then A = B if and only if $a = e$, $b = f$, $c = g$ and $d = h$.

Example 1

Given that $\begin{bmatrix} 3x + y & 5 \\ -2 & -2x + y \end{bmatrix} = \begin{bmatrix} -1 & 5 \\ x & 3z \end{bmatrix}$ find the values of x, y and z.

Solution

Because the two matrices are equal

$$3x + y = -1 \qquad (1)$$

and

$$x = -2 \qquad (2)$$

Substituting $x = -2$ into (1) gives

$$3(-2) + y = -1$$

$$y = 5$$

Also

$$-2x + y = 3z$$

$$-2(-2) + 5 = 3z$$

$$z = 3$$

Adding Matrices

If A and B are two matrices of the same dimension, their sum A + B is obtained by adding the corresponding elements of A and B.

Example 2

Given $A = \begin{bmatrix} -1 & 2 \\ 9 & 4 \end{bmatrix}$ and $B = \begin{bmatrix} 2 & 3 \\ 4 & 8 \end{bmatrix}$, find

(a) A + B (b) B + A

Solution

(a)

$$A + B = \begin{bmatrix} -1 & 2 \\ 9 & 4 \end{bmatrix} + \begin{bmatrix} 2 & 3 \\ 4 & 8 \end{bmatrix}$$

$$= \begin{bmatrix} -1 + 2 & 2 + 3 \\ 9 + 4 & 4 + 8 \end{bmatrix}$$

$$= \begin{bmatrix} 1 & 5 \\ 13 & 12 \end{bmatrix}$$

(b) $$B + A = \begin{bmatrix} 2 & 3 \\ 4 & 8 \end{bmatrix} + \begin{bmatrix} -1 & 2 \\ 9 & 4 \end{bmatrix}.$$

$$= \begin{bmatrix} 2-1 & 3+2 \\ 4+9 & 8+4 \end{bmatrix}$$

$$= \begin{bmatrix} 1 & 5 \\ 13 & 12 \end{bmatrix}$$

The commutative and associative properties apply to matrix addition. If the matrices A, B and C have the same dimension, then A + B = B + A and (A + B) + C = A + (B + C).

Negative Matrices

If A is any matrix, the negative matrix of A, denoted by – A, is obtained by replacing each element in A by its negative. For example, If

$$A = \begin{bmatrix} a & b \\ c & d \end{bmatrix}, \text{ then } -A = \begin{bmatrix} -a & -b \\ -c & -d \end{bmatrix}.$$

Example 3

Find the negative of the matrix $A = \begin{bmatrix} -3 & 1 \\ 0 & -2 \end{bmatrix}$

Solution

$$-A = -\begin{bmatrix} -3 & 1 \\ 0 & -2 \end{bmatrix} = \begin{bmatrix} 3 & -1 \\ 0 & 2 \end{bmatrix}$$

Subtraction of Matrices

If A and B are two matrices of the same dimension, the difference A – B is obtained by subtracting the elements in B from the corresponding elements in A.

Example 4

Given $A = \begin{bmatrix} -2 & 3 \\ -8 & 7 \end{bmatrix}$ and $B = \begin{bmatrix} -2 & 4 \\ -10 & 6 \end{bmatrix}$, find

(a) A – B (b) B – A

Solution

(a) $$A - B = \begin{bmatrix} -2 & 3 \\ -8 & 7 \end{bmatrix} - \begin{bmatrix} -2 & 4 \\ -10 & 6 \end{bmatrix}$$

$$= \begin{bmatrix} -2 + 2 & 3 - 4 \\ -8 + 10 & 7 - 6 \end{bmatrix}$$

$$= \begin{bmatrix} 0 & -1 \\ 2 & 1 \end{bmatrix}$$

(b) $$B - A = \begin{bmatrix} -2 & 4 \\ -10 & 6 \end{bmatrix} - \begin{bmatrix} -2 & 3 \\ -8 & 7 \end{bmatrix}$$

$$= \begin{bmatrix} -2 + 2 & 4 - 3 \\ -10 + 8 & 6 - 7 \end{bmatrix}$$

$$= \begin{bmatrix} 0 & 1 \\ -2 & -1 \end{bmatrix}$$

Observe that A – B ≠ B – A. In general, matrix subtraction is not commutative.

Zero Matrix (Null Matrix)

A matrix in which all entries are 0 is called a zero matrix. The symbol O is use to represent the zero matrix of any dimension. For example,

$$\begin{bmatrix} 0 & 0 \\ 0 & 0 \end{bmatrix}$$

is a 2 × 2 zero matrix.

Note that, for any matrix A, and the zero matrix O of the same dimension A + O = O + A = A and A – A = O.

Example 5

Given $A = \begin{bmatrix} -3 & 4 \\ 5 & -6 \end{bmatrix}$, find

(a) A + O (b) A – A

Solution

(a) $$A + O = \begin{bmatrix} -3 & 4 \\ 5 & -6 \end{bmatrix} + \begin{bmatrix} 0 & 0 \\ 0 & 0 \end{bmatrix}$$

$$= \begin{bmatrix} -3 + 0 & 4 + 0 \\ 5 + 0 & -6 + 0 \end{bmatrix}$$

$$= \begin{bmatrix} -3 & 4 \\ 5 & -6 \end{bmatrix}$$

(b)
$$A - A = \begin{bmatrix} -3 & 4 \\ 5 & -6 \end{bmatrix} - \begin{bmatrix} -3 & 4 \\ 5 & -6 \end{bmatrix}$$
$$= \begin{bmatrix} -3+3 & 4-4 \\ 5-5 & -6+6 \end{bmatrix}$$
$$= \begin{bmatrix} 0 & 0 \\ 0 & 0 \end{bmatrix}$$

Multiplying a Matrix by a Scalar

If $A = \begin{bmatrix} a & b \\ c & d \end{bmatrix}$, then *k*A, where *k* is a scalar is obtained by multiplying each element by *k*,

i.e. $kA = \begin{bmatrix} ka & kb \\ kc & kd \end{bmatrix}$

Example 6

Given $A = \begin{bmatrix} -1 & 0 \\ 2 & -3 \end{bmatrix}$ find – 2A.

Solution

$$-2A = -2\begin{bmatrix} -1 & 0 \\ 2 & -3 \end{bmatrix}$$
$$= \begin{bmatrix} -2\cdot -1 & -2\cdot 0 \\ -2\cdot 2 & -2\cdot -3 \end{bmatrix}$$
$$= \begin{bmatrix} 2 & 0 \\ -4 & 6 \end{bmatrix}$$

Exercise 18.1

Find the values of *x*, *y* and z.

1. $\begin{bmatrix} 2 & x+y \\ 2y & 6 \end{bmatrix} = \begin{bmatrix} x-y & 4 \\ z & 6 \end{bmatrix}$

2. $\begin{bmatrix} 3 & x-2y \\ 5 & -2 \end{bmatrix} = \begin{bmatrix} 3 & -4 \\ x+y & -2 \end{bmatrix}$

Work out:

3. $\begin{bmatrix} 5 & 3 \\ 1 & 2 \end{bmatrix} + \begin{bmatrix} 6 & 4 \\ 1 & 3 \end{bmatrix}$

4. $\begin{bmatrix} 0 & 2 \\ 4 & -1 \end{bmatrix} + \begin{bmatrix} 6 & -1 \\ 5 & 2 \end{bmatrix}$

5. $\begin{bmatrix} 4 & 2 \\ 3 & -1 \end{bmatrix} + \begin{bmatrix} 1 & 3 \\ -2 & 2 \end{bmatrix}$

6. $\begin{bmatrix} 3 & -1 & 0 \\ -2 & 0 & 1 \end{bmatrix} + \begin{bmatrix} -5 & 2 & -1 \\ 4 & 3 & -4 \end{bmatrix}$

7. $\begin{bmatrix} -4 & 2 & 3 \\ 5 & -3 & 1 \end{bmatrix} + \begin{bmatrix} 5 & -3 & -2 \\ -3 & 7 & -3 \end{bmatrix}$

9. $\begin{bmatrix} 0 & 5 \\ 3 & -4 \\ -2 & 1 \end{bmatrix} + \begin{bmatrix} -1 & -6 \\ 2 & 6 \\ 3 & -2 \end{bmatrix}$

9. $\begin{bmatrix} -7 & 8 & 6 \\ 5 & 4 & -3 \\ 2 & -5 & 1 \end{bmatrix} + \begin{bmatrix} 9 & -6 & -10 \\ -3 & 0 & 5 \\ -3 & 6 & -2 \end{bmatrix}$

10. Find the values of *a*, *b*, *c* and *d* so that

$$\begin{bmatrix} a & b \\ c & d \end{bmatrix} + \begin{bmatrix} -2 & 3 \\ 2 & 1 \end{bmatrix} = \begin{bmatrix} -1 & 2 \\ -3 & 4 \end{bmatrix}$$

11, Find the values of *w*, *x*, *y* and z so that

$$\begin{bmatrix} -1 & 5 \\ 1 & -2 \end{bmatrix} + \begin{bmatrix} w & x \\ y & z \end{bmatrix} = \begin{bmatrix} -2 & 3 \\ 6 & 1 \end{bmatrix}$$

12. Find the values of *x* and *y* so that

$$\begin{bmatrix} 2x & 5 \\ -1 & 3x \end{bmatrix} + \begin{bmatrix} 3y & -3 \\ -5 & -2y \end{bmatrix} = \begin{bmatrix} 7 & 2 \\ -6 & 4 \end{bmatrix}$$

13. Find the values of *x* and *y* so that

$$\begin{bmatrix} -3x & -3 \\ 2 & 4x \end{bmatrix} + \begin{bmatrix} 2y & 5 \\ -2 & -5y \end{bmatrix} = \begin{bmatrix} -1 & 2 \\ 0 & -1 \end{bmatrix}$$

Write down the negative matrix of each of the following matrices:

14. $\begin{bmatrix} -2 & 3 \\ 1 & -4 \end{bmatrix}$

15. $\begin{bmatrix} -3 & -2 \\ 5 & -1 \end{bmatrix}$

16. $\begin{bmatrix} 6 & 0 \\ 0 & -6 \end{bmatrix}$

17. $\begin{bmatrix} 3 & -2 & 0 \\ -1 & 5 & -2 \end{bmatrix}$

18. $\begin{bmatrix} -4 & 0 \\ 3 & 1 \\ -2 & -1 \end{bmatrix}$

19. $\begin{bmatrix} 2 & -5 & 3 \\ -3 & 0 & -2 \\ 4 & -3 & 2 \end{bmatrix}$

Work out:

20. $\begin{bmatrix} -1 & 1 \\ 4 & 2 \end{bmatrix} - \begin{bmatrix} -2 & 0 \\ 4 & 1 \end{bmatrix}$

21. $\begin{bmatrix} 4 & 2 \\ -1 & 3 \end{bmatrix} - \begin{bmatrix} 1 & 8 \\ 1 & 0 \end{bmatrix}$

22. $\begin{bmatrix} 3 & -6 \\ -2 & 1 \end{bmatrix} - \begin{bmatrix} -4 & -2 \\ 1 & -3 \end{bmatrix}$

23. $\begin{bmatrix} -2 & 4 \\ -8 & 5 \end{bmatrix} - \begin{bmatrix} -2 & 0 \\ -5 & 3 \end{bmatrix}$

24. $\begin{bmatrix} 0 & 3 & -1 \\ 1 & -2 & 0 \end{bmatrix} - \begin{bmatrix} -1 & -5 & 2 \\ -4 & 4 & 3 \end{bmatrix}$

25. $\begin{bmatrix} 5 & -3 & 1 \\ -4 & 2 & 1 \end{bmatrix} - \begin{bmatrix} -3 & -7 & -3 \\ -5 & 3 & 0 \end{bmatrix}$

26. $\begin{bmatrix} 5 & 0 \\ -4 & 3 \\ 1 & -2 \end{bmatrix} - \begin{bmatrix} -6 & -1 \\ 6 & 2 \\ -2 & 3 \end{bmatrix}$

27. $\begin{bmatrix} 2 & -5 & 1 \\ -7 & 8 & 6 \\ 2 & -5 & 1 \end{bmatrix} - \begin{bmatrix} -3 & -6 & 2 \\ -3 & 6 & 4 \\ -5 & 7 & -2 \end{bmatrix}$

Work out:

28. $3\begin{bmatrix} 1 & 2 \\ -2 & 1 \end{bmatrix}$

29. $-2\begin{bmatrix} -3 & 1 \\ 1 & -2 \end{bmatrix}$

30. $4\begin{bmatrix} -1 & 0 \\ 0 & 1 \end{bmatrix}$

31. $-\frac{1}{2}\begin{bmatrix} -2 & 6 \\ 4 & 0 \end{bmatrix}$

32. $\frac{2}{3}\begin{bmatrix} -6 & 0 \\ 12 & -9 \end{bmatrix}$

33. $-2\begin{bmatrix} 2 & -1 & -3 \\ 0 & 3 & -2 \end{bmatrix}$

34. $3\begin{bmatrix} 0 & 1 \\ -1 & 2 \\ -2 & 3 \end{bmatrix}$

35. $-\frac{1}{3}\begin{bmatrix} 3 & 0 & -6 \\ -3 & 9 & -3 \\ 0 & -3 & 6 \end{bmatrix}$

Given that $A = \begin{bmatrix} -3 & 2 \\ 2 & 0 \end{bmatrix}$, $B = \begin{bmatrix} -2 & 1 \\ 1 & 5 \end{bmatrix}$ and $C = \begin{bmatrix} 0 & -3 \\ 1 & 2 \end{bmatrix}$,find

36. 2A + B

37. 3C – 2B

38. 2C – 3A

39. 2(B – A) + C

40. (A + B) – 3C

18.2 Multiplying Matrices

You can find the product AB of two matrices A and B if the number of columns of A equals the number of rows of B. The product AB will have the same number of rows as A and the same number of column as B. For example, the dimension of the product of a 3 × 4 matrix and a 4 × 2 matrix will be a 3 × 2 matrix.

If A and B are two 2 × 2 square matrices the product AB is obtained as follows. Multiply the first element in the first row in A by the first element in the first column in B. Next multiply the second element in the first row in A by the second element in the first column in B. Then add the two results to give the first element in the first row and first column of the matrix AB. The element in the first row and second column of AB is obtained in the same manner using the first row of matrix A and the second column of matrix B. Similarly, the elements in the second row, first column and the second row, second column are obtained by multiplying the second row of matrix A by the first column and second column of matrix B respectively.

For example, if $A = \begin{bmatrix} a & b \\ c & d \end{bmatrix}$ and $B = \begin{bmatrix} e & f \\ g & h \end{bmatrix}$ then

$$AB = \begin{bmatrix} a & b \\ c & d \end{bmatrix}\begin{bmatrix} e & f \\ g & h \end{bmatrix}$$

$$= \begin{bmatrix} ae + bg & af + bh \\ ce + dg & cf + dh \end{bmatrix}$$

Example 1

Given $A = \begin{bmatrix} 2 & 0 \\ -3 & 1 \end{bmatrix}$ and $B = \begin{bmatrix} -1 & 3 \\ -2 & 4 \end{bmatrix}$ find

(a) AB (b) BA

Solution

(a)

$$AB = \begin{bmatrix} 2 & 0 \\ -3 & 1 \end{bmatrix}\begin{bmatrix} -1 & 3 \\ -2 & 4 \end{bmatrix}$$

$$= \begin{bmatrix} 2\cdot -1 + 0\cdot -2 & 2\cdot 3 + 0\cdot 4 \\ -3\cdot -1 + 1\cdot -2 & -3\cdot 3 + 1\cdot 4 \end{bmatrix}$$

$$= \begin{bmatrix} -2 + 0 & 6 + 0 \\ 3 - 2 & -9 + 4 \end{bmatrix}$$

$$= \begin{bmatrix} -2 & 6 \\ 1 & -5 \end{bmatrix}$$

(b)

$$BA = \begin{bmatrix} -1 & 3 \\ -2 & 4 \end{bmatrix}\begin{bmatrix} 2 & 0 \\ -3 & 1 \end{bmatrix}$$

$$= \begin{bmatrix} -11 & 3 \\ -16 & 4 \end{bmatrix}$$

The order of matrix multiplication is important. If we can find the product AB, the product BA may not be defined. From the above example, you would see that even if both AB and BA are defined, they may not be equal. Matrix multiplication is not commutative. It is possible to find two matrices, A and B, with non-zero elements such that AB = 0.

If A is a square matrix then AA is defined and is denoted by A^2. Similarly, AAA = A^3.

Example 2

Given $A = \begin{bmatrix} -2 & 3 \\ -4 & 6 \end{bmatrix}$ and $B = \begin{bmatrix} 3 & -6 \\ 2 & -4 \end{bmatrix}$, find AB.

Solution

$$AB = \begin{bmatrix} -2 & 3 \\ -4 & 6 \end{bmatrix}\begin{bmatrix} 3 & -6 \\ 2 & -4 \end{bmatrix} = \begin{bmatrix} 0 & 0 \\ 0 & 0 \end{bmatrix}$$

Identity Matrix (Unit Matrix)

A square matrix whose diagonal entries are 1s and all other entries are 0s is called an identity matrix (or unit matrix). The symbol I is use to represent a matrix of any dimension.

The 2 × 2 identity matrix is

$$\begin{bmatrix} 1 & 0 \\ 0 & 1 \end{bmatrix}$$

and the 3 × 3 identity matrix is

$$\begin{bmatrix} 1 & 0 & 0 \\ 0 & 1 & 0 \\ 0 & 0 & 1 \end{bmatrix}.$$

In general, if A is a square matrix of dimension $n \times n$, then IA = AI = A.

Example 3

Given $A = \begin{bmatrix} -2 & 3 \\ 1 & 4 \end{bmatrix}$, find

(a) AI (b) IA

Solution

(a)
$$AI = \begin{bmatrix} -2 & 3 \\ 1 & 4 \end{bmatrix}\begin{bmatrix} 1 & 0 \\ 0 & 1 \end{bmatrix}$$
$$= \begin{bmatrix} -2 & 3 \\ 1 & 4 \end{bmatrix}$$

(b)
$$IA = \begin{bmatrix} 1 & 0 \\ 0 & 1 \end{bmatrix}\begin{bmatrix} -2 & 3 \\ 1 & 4 \end{bmatrix}$$
$$= \begin{bmatrix} -2 & 3 \\ 1 & 4 \end{bmatrix}$$

Exercise 18.2

Evaluate:

1. $\begin{bmatrix} 1 & 2 \\ 0 & 3 \end{bmatrix}\begin{bmatrix} 3 & 2 \\ 4 & -1 \end{bmatrix}$

2. $\begin{bmatrix} 2 & -3 \\ -5 & 1 \end{bmatrix}\begin{bmatrix} 1 & 2 \\ 3 & 2 \end{bmatrix}$

3. $\begin{bmatrix} 1 & 3 \\ -2 & 5 \end{bmatrix}\begin{bmatrix} 0 & -2 \\ 2 & -4 \end{bmatrix}$

4. $\begin{bmatrix} 3 & 4 \\ 0 & 1 \end{bmatrix}\begin{bmatrix} 6 & 2 \\ 5 & 0 \end{bmatrix}$

5. $\begin{bmatrix} 1 & 2 \\ 2 & 4 \end{bmatrix}\begin{bmatrix} 2 \\ 3 \end{bmatrix}$

6. $\begin{bmatrix} 1 & 3 \\ 2 & 5 \end{bmatrix}\begin{bmatrix} 2 \\ -1 \end{bmatrix}$

7. $\begin{bmatrix} 1 & -3 \\ -2 & 1 \end{bmatrix}\begin{bmatrix} 1 \\ 5 \end{bmatrix}$

8. $\begin{bmatrix} 4 & -3 \\ -2 & 1 \end{bmatrix}\begin{bmatrix} -4 \\ -5 \end{bmatrix}$

9. $\begin{bmatrix} 3 & -2 & 1 \\ 1 & 0 & 2 \end{bmatrix}\begin{bmatrix} -1 & 0 \\ 2 & 1 \\ -3 & -2 \end{bmatrix}$

10. $\begin{bmatrix} 4 & -2 \\ 2 & 1 \\ -1 & 0 \end{bmatrix}\begin{bmatrix} 3 & -1 \\ 2 & 0 \end{bmatrix}$

11. $\begin{bmatrix} -2 & 1 & -1 \\ 3 & 4 & 0 \\ -1 & 2 & -2 \end{bmatrix}\begin{bmatrix} 0 & 2 \\ 1 & -2 \\ 0 & 1 \end{bmatrix}$

12. $\begin{bmatrix} 4 & 5 & -3 \\ 0 & 2 & 1 \\ -1 & 0 & 2 \end{bmatrix}\begin{bmatrix} 0 & 1 & 2 \\ 4 & -1 & 0 \\ 3 & 2 & 1 \end{bmatrix}$

Evaluate:

13. $\begin{bmatrix} 6 & -3 \\ -4 & 2 \end{bmatrix}\begin{bmatrix} 4 & 3 \\ 8 & 6 \end{bmatrix}$ 14. $\begin{bmatrix} 3 & 2 \\ -3 & -2 \end{bmatrix}\begin{bmatrix} 2 & -4 \\ -3 & 6 \end{bmatrix}$

Find A^2 and A^3 for each of the following matrices.

15. $\begin{bmatrix} 1 & -1 \\ 3 & 2 \end{bmatrix}$ 16. $\begin{bmatrix} -3 & -1 \\ 2 & 1 \end{bmatrix}$ 17. $\begin{bmatrix} 1 & 0 \\ 2 & -1 \end{bmatrix}$

18. Find the values of x, y and z so that

$$\begin{bmatrix} 2 & 3 \\ -3 & 1 \end{bmatrix}\begin{bmatrix} x & 2 \\ -1 & y \end{bmatrix} = \begin{bmatrix} 1 & z \\ -7 & z \end{bmatrix}$$

19. Find the values of x, y and z so that

$$\begin{bmatrix} y & 3 \\ x & 1 \end{bmatrix}\begin{bmatrix} 2 & 3 \\ 1 & -2 \end{bmatrix} = \begin{bmatrix} -1 & -12 \\ z & z \end{bmatrix}$$

20. Find the values a, b, c, and d so that

$$\begin{bmatrix} 2 & 3 \\ 3 & -2 \end{bmatrix}\begin{bmatrix} a & b \\ c & d \end{bmatrix} = \begin{bmatrix} -7 & 8 \\ -4 & -1 \end{bmatrix}$$

21. Find the values a, b, c and d so that

$$\begin{bmatrix} -3 & 2 \\ 1 & -2 \end{bmatrix}\begin{bmatrix} a & b \\ c & d \end{bmatrix} = \begin{bmatrix} 0 & 5 \\ -4 & 1 \end{bmatrix}$$

18.3 Inverse of a Matrix

Determinants

Associated with each square matrix is a real number, called the determinant. The determinant of a matrix A is denoted by det A or |A|, and is defined for 2 × 2 square matrix

$$A = \begin{bmatrix} a & b \\ c & d \end{bmatrix} \text{ by } \begin{vmatrix} a & b \\ c & d \end{vmatrix} = ad - bc$$

Example 1

Find the determinant of the matrix $\begin{bmatrix} 3 & -4 \\ -2 & 2 \end{bmatrix}$

Solution

$$\begin{vmatrix} 3 & -4 \\ -2 & 2 \end{vmatrix} = (3)(2) - (-4)(-2)$$

$$= 6 - 8$$

$$= -2$$

Inverse Matrix

If A and B are two square matrices of the same dimension, then the matrix B is the inverse matrix of A, if AB = BA = I, where I is the identity matrix. The inverse matrix of A is denoted by A^{-1}, read "A inverse''.

A square matrix may or may not have an inverse. A matrix which has an inverse is said to be non-singular (or invertible). A matrix that has no inverse is said to be singular. The determinant of a singular matrix is zero. If

$$A = \begin{bmatrix} a & b \\ c & d \end{bmatrix}$$

and $ad - bc \neq 0$, then A^{-1} is obtained by interchanging the elements on the leading diagonal, changing the signs of the other two elements and finally multiplying by $1/(ad - bc)$, i.e.

$$A^{-1} = \frac{1}{ad - bc}\begin{bmatrix} d & -b \\ -c & a \end{bmatrix}$$

Observe that $ad - bc$ is the determinant of the matrix A.

Example 2

Find the inverse of $A = \begin{bmatrix} 3 & 2 \\ 1 & 4 \end{bmatrix}$

Solution

Begin by finding the determinant of A

$$\begin{vmatrix} 3 & 2 \\ 1 & 4 \end{vmatrix} = (3)(4) - (2)(1) = 10$$

Hence $$A^{-1} = \frac{1}{10}\begin{bmatrix} 4 & -2 \\ -1 & 3 \end{bmatrix}$$

$$= \begin{bmatrix} 0.4 & -0.2 \\ -0.1 & 0.3 \end{bmatrix}$$

Solving a System of two Linear Equations Using Inverses

One method of solving system of two linear equations involves the use of an inverse matrix. Consider the following systems of equation

$$a_{11}x_1 + a_{12}x_2 = b_1$$

$$a_{21}x_1 + a_{22}x_2 = b_2$$

The system of equations can be written in matrix form as

$$\begin{bmatrix} a_{11} & a_{12} \\ a_{21} & a_{22} \end{bmatrix} \begin{bmatrix} x \\ y \end{bmatrix} = \begin{bmatrix} b_1 \\ b_2 \end{bmatrix}.$$

If we let $A = \begin{bmatrix} a_{11} & a_{12} \\ a_{21} & a_{22} \end{bmatrix}$ $X = \begin{bmatrix} x_1 \\ x_2 \end{bmatrix}$ and $B = \begin{bmatrix} b_1 \\ b_2 \end{bmatrix}$, then the system of equations can be written as

$$AX = B$$

If A has an inverse matrix A^{-1}, then multiplying each side of the equation by A^{-1}, we get

$$A^{-1}(AX) = A^{-1}B$$

$$(A^{-1}A)X = A^{-1}B$$

$$IX = A^{-1}B$$

$$X = A^{-1}B$$

In general, if A is a square matrix and the inverse matrix A^{-1} exists, then a system of equations $AX = B$ has a unique solution, given by

$$X = A^{-1}B.$$

Example 3

Solve the system of equations

$$2x + y = 7$$

$$3x + 2y = 12$$

Solution

Begin by writing the system of equation in matrix form

$$\begin{bmatrix} 2 & 1 \\ 3 & 2 \end{bmatrix} \begin{bmatrix} x \\ y \end{bmatrix} = \begin{bmatrix} 7 \\ 12 \end{bmatrix}$$

Here $\qquad A = \begin{bmatrix} 2 & 1 \\ 3 & 2 \end{bmatrix}$, $X = \begin{bmatrix} x \\ y \end{bmatrix}$ and $B = \begin{bmatrix} 7 \\ 12 \end{bmatrix}$

Next find the inverse matrix of A.

$$|A| = \begin{vmatrix} 2 & 1 \\ 3 & 2 \end{vmatrix} = 2(2) - 1(3) = 1$$

So, $\qquad A^{-1} = \begin{bmatrix} 2 & -1 \\ -3 & 2 \end{bmatrix}$

Finally, multiple each side of the matrix equation by inverse matrix to get

$$\begin{bmatrix} x \\ y \end{bmatrix} = \begin{bmatrix} 2 & -1 \\ -3 & 2 \end{bmatrix} \begin{bmatrix} 7 \\ 12 \end{bmatrix}$$

$$= \begin{bmatrix} 2 \\ 3 \end{bmatrix}$$

Thus, the solution to this system is (2, 3).

Exercise 18.3

Show that the given matrices are inverses of each other.

1. $\begin{bmatrix} 1 & 2 \\ 2 & 3 \end{bmatrix}, \begin{bmatrix} -3 & 2 \\ 2 & -1 \end{bmatrix}$

2. $\begin{bmatrix} 3 & 2 \\ 7 & 5 \end{bmatrix}, \begin{bmatrix} 5 & -2 \\ -7 & 3 \end{bmatrix}$

3. $\begin{bmatrix} 1 & -2 \\ -1 & 3 \end{bmatrix}, \begin{bmatrix} 3 & 2 \\ 1 & 1 \end{bmatrix}$

4. $\begin{bmatrix} 1 & -2 \\ -1 & 3 \end{bmatrix}, \begin{bmatrix} 3 & 2 \\ 1 & 1 \end{bmatrix}$

Evaluate each determinant.

5. $\begin{vmatrix} 4 & 3 \\ 3 & 2 \end{vmatrix}$

6. $\begin{vmatrix} 2 & 1 \\ 4 & 3 \end{vmatrix}$

7. $\begin{vmatrix} -2 & 3 \\ -2 & 2 \end{vmatrix}$

8. $\begin{vmatrix} -3 & -2 \\ 4 & 3 \end{vmatrix}$

9. $\begin{vmatrix} 6 & 3 \\ 2 & 1 \end{vmatrix}$

10. $\begin{vmatrix} 6 & 4 \\ 5 & 3 \end{vmatrix}$

If $A = \begin{bmatrix} -2 & 1 \\ 1 & 3 \end{bmatrix}$ and $B = \begin{bmatrix} -1 & 2 \\ 5 & 4 \end{bmatrix}$, find:

11. A|

12. |B|

13. |AB|

14. |BA|

15. |A||B|

Solve the equation.

16. $\begin{vmatrix} x & x \\ 3 & 4 \end{vmatrix} = -5$

17. $\begin{vmatrix} 1 & x \\ x & 3 \end{vmatrix} = 2$

18. $\begin{vmatrix} 4 & 5-x \\ 2-x & 1 \end{vmatrix} = 0$

Determine whether the given matrix has an inverse.

19. $\begin{bmatrix} 5 & 4 \\ 3 & 2 \end{bmatrix}$

20. $\begin{bmatrix} 7 & 3 \\ -2 & 1 \end{bmatrix}$

21. $\begin{bmatrix} 1 & 2 \\ 3 & 6 \end{bmatrix}$

22. $\begin{bmatrix} 8 & 4 \\ -4 & -2 \end{bmatrix}$

23. $\begin{bmatrix} -3 & 2 \\ -1 & 3 \end{bmatrix}$

24. $\begin{bmatrix} 3 & -2 \\ 6 & 4 \end{bmatrix}$

Find the inverse of each matrix.

25. $\begin{bmatrix} -2 & 3 \\ 2 & 0 \end{bmatrix}$

26. $\begin{bmatrix} 4 & 5 \\ 2 & 3 \end{bmatrix}$

27. $\begin{bmatrix} -2 & -1 \\ 4 & 3 \end{bmatrix}$

28. $\begin{bmatrix} 2 & 4 \\ 3 & 5 \end{bmatrix}$

29. $\begin{bmatrix} 5 & -2 \\ 0 & 2 \end{bmatrix}$

30. $\begin{bmatrix} 3 & 2 \\ -1 & 2 \end{bmatrix}$

Use matrices to solve the system of linear equations.

31. $5x + 4y = 3$
$3x + 2y = 1$

32. $x + 2y = 1$
$4x + y = 4$

33. $-x + 2y = 4$
$-2x + 3y = 5$

34. $7x + 3y = -8$
$5x + 2y = -6$

35. $3x + 2y = 5$
$-2x + 3y = -12$

36. $2x - 3y = -5$
$-x + 2y = 4$

18.4 3 × 3 Matrices

Minor of an element

The minor of an element in a 3 × 3 matrix is the determinant of the 2 × 2 matrix obtained by deleting the row and column in which the element occurs. We denote the minor of a_{ij} by Mij, where a_{ij} represents the entry in the i th row and j th column.

For the 3 × 3 determinant

$$\begin{vmatrix} a_{11} & a_{12} & a_{13} \\ a_{21} & a_{22} & a_{23} \\ a_{31} & a_{32} & a_{33} \end{vmatrix},$$

the minor of a_{23} is

$$\begin{vmatrix} a_{11} & a_{12} \\ a_{31} & a_{32} \end{vmatrix},$$

obtained by deleting the second row and the third column.

Example 1

If $A = \begin{bmatrix} 1 & 2 & 0 \\ 0 & 1 & 4 \\ 1 & 2 & 6 \end{bmatrix}$ find the minors M_{12}, M_{23}, M_{33} of A.

Solution

We have

$$M_{12} = \begin{vmatrix} 0 & 4 \\ 1 & 6 \end{vmatrix} = (0)(6) - (1)(4) = -4$$

$$M_{23} = \begin{vmatrix} 1 & 2 \\ 1 & 2 \end{vmatrix} = (1)(2) - (1)(2) = 0$$

$$M_{33} = \begin{vmatrix} 1 & 2 \\ 0 & 1 \end{vmatrix} = (1)(1) - (0)(2) = 1$$

Cofactor

The cofactor of an element is the minor of the element together with the sign obtained by raising − 1 to a power that is the sum of the numbers indicating the number of the row and column in which the element appears. For example, the sign of the minor of the element in the second row and third column is given by $(-1)^{2+3}$, i.e. − 1. We denote the cofactor of an element a_{ij} by A_{ij}, where a_{ij} represents the entry in the i th row and j th column.

For example, the cofactor of a_{ij} is $(-1)^{i+j}$ (minor of a_{ij}), i.e.

$A_{ij} = (-1)^{i+j} M_{ij}$.

Notice that, the cofactor of an element is a signed minor.

For the 3 × 3 determinant

$$\begin{vmatrix} a_{11} & a_{12} & a_{13} \\ a_{21} & a_{22} & a_{23} \\ a_{31} & a_{32} & a_{33} \end{vmatrix}$$

the cofactor of a_{23} is

$$(-1)^{2+3} \begin{vmatrix} a_{11} & a_{12} \\ a_{31} & a_{32} \end{vmatrix} = -\begin{vmatrix} a_{11} & a_{12} \\ a_{31} & a_{32} \end{vmatrix}$$

and the cofactor of

$$a_{22} = (-1)^{2+2} \begin{vmatrix} a_{11} & a_{13} \\ a_{31} & a_{33} \end{vmatrix} = +\begin{vmatrix} a_{11} & a_{13} \\ a_{31} & a_{33} \end{vmatrix}$$

Example 2

Given $A = \begin{bmatrix} 1 & -1 & 2 \\ 2 & 1 & 0 \\ -3 & 3 & 4 \end{bmatrix}$, find A_{12} and A_{33}.

Solution

$$A_{12} = (-1)^{1+2}\begin{vmatrix} 2 & 0 \\ -3 & 4 \end{vmatrix}$$

$$= -\,[(2)(4) - (-3)(0)]$$

$$= -8$$

$$A_{33} = (-1)^{3+3}\begin{vmatrix} 1 & -1 \\ 2 & 1 \end{vmatrix} = (1)(1) - (2)(-1) = 3$$

The associated signs for the minors follow the alternating pattern shown in the diagram below.

$$\begin{vmatrix} + & - & + \\ - & + & - \\ + & - & + \end{vmatrix}$$

You could use the diagram or the exponent method whichever is easier for you to determine the sign in front of the minor.

Determinant of 3 × 3 Matrices

The value of the determinant of 3 × 3 matrix is the sum of three products obtained by multiplying each element of any one row (or each element of any one column) by its cofactor. Using the first row, we have

$$\begin{vmatrix} a_{11} & a_{12} & a_{13} \\ a_{21} & a_{22} & a_{23} \\ a_{31} & a_{32} & a_{33} \end{vmatrix} = a_{11}\begin{vmatrix} a_{22} & a_{23} \\ a_{32} & a_{33} \end{vmatrix} - a_{12}\begin{vmatrix} a_{21} & a_{23} \\ a_{31} & a_{33} \end{vmatrix} + a_{13}\begin{vmatrix} a_{21} & a_{22} \\ a_{31} & a_{32} \end{vmatrix}$$

Writing the determinant of a 3 × 3 matrix in terms of the minors is called expansion by minors. The value of a determinant does not depend on the choice of the row or column used in the expansion. You can reduce the amount of work needed to compute the value of the determinant if you expand across a row or column that has element equal to 0. A zero element will always yield a zero term.

Example 3

Given $A = \begin{bmatrix} 1 & 0 & 1 \\ 2 & 1 & 2 \\ 0 & 4 & 6 \end{bmatrix}$, find the determinant of A.

Solution

$$\begin{vmatrix} 1 & 0 & 1 \\ 2 & 1 & 2 \\ 0 & 4 & 6 \end{vmatrix} = 1\begin{vmatrix} 1 & 2 \\ 4 & 6 \end{vmatrix} - 0\begin{vmatrix} 2 & 2 \\ 0 & 6 \end{vmatrix} + 1\begin{vmatrix} 2 & 1 \\ 0 & 4 \end{vmatrix}$$

$$= (6 - 8) - 0 + (8 - 0)$$

$$= -2 + 8$$

$$= 6$$

The Inverse of 3× 3 Matrices

Let

$$M = \begin{bmatrix} a_{11} & a_{12} & a_{13} \\ a_{21} & a_{22} & a_{23} \\ a_{31} & a_{32} & a_{33} \end{bmatrix}$$

To calculate the inverse of a matrix, we proceed as follows. First, we find the cofactors of each element, and then interchange rows and columns of the matrix of cofactors to obtain the matrix

$$M^* = \begin{bmatrix} A_{11} & A_{21} & A_{31} \\ A_{12} & A_{22} & A_{32} \\ A_{13} & A_{23} & A_{33} \end{bmatrix}$$

called the adjoint of M.

Then divide the adjoint by the value of the determinant to give the inverse matrix

$$M^{-1} = \frac{1}{|M|} M^*.$$

Example 4

Calculate the inverse of the matrix $A = \begin{bmatrix} 1 & 0 & 1 \\ 2 & 1 & 2 \\ 0 & 4 & 6 \end{bmatrix}$

Solution

Begin by calculating the cofactors of A.

$$A_{11} = \begin{vmatrix} 1 & 2 \\ 4 & 6 \end{vmatrix} = -2$$

$$A_{21} = -\begin{vmatrix} 0 & 1 \\ 4 & 6 \end{vmatrix} = 4$$

$$A_{31} = \begin{vmatrix} 0 & 1 \\ 1 & 2 \end{vmatrix} = -1$$

$$A_{12} = -\begin{vmatrix} 2 & 2 \\ 0 & 6 \end{vmatrix} = -12$$

$$A_{22} = \begin{vmatrix} 1 & 1 \\ 0 & 6 \end{vmatrix} = 6$$

$$A_{32} = -\begin{vmatrix} 1 & 1 \\ 2 & 2 \end{vmatrix} = 0$$

$$A_{13} = \begin{vmatrix} 2 & 1 \\ 0 & 4 \end{vmatrix} = 8$$

$$A_{23} = -\begin{vmatrix} 1 & 0 \\ 0 & 4 \end{vmatrix} = -4$$

$$A_{33} = \begin{vmatrix} 1 & 0 \\ 2 & 1 \end{vmatrix} = 1$$

From Example 15, we know that the determinant of A is 6.

So,
$$A^{-1} = \frac{1}{6}\begin{bmatrix} -2 & 4 & -1 \\ -12 & 6 & 0 \\ 8 & -4 & 1 \end{bmatrix}$$

$$= \begin{bmatrix} -\frac{1}{3} & \frac{2}{3} & -\frac{1}{6} \\ -2 & 1 & 0 \\ \frac{4}{3} & -\frac{2}{3} & \frac{1}{6} \end{bmatrix}$$

Finding Inverse of a Matrix using Elementary Row Operations

Another way of finding the inverse of a matrix is to use elementary row operations.

Elementary Row Operations

1. Interchange two rows

2. Multiply or divide each element in a row by a nonzero constant

3. Add to or subtract from a row a multiple of another row.

Example 5

Calculate the inverse of the matrix $A = \begin{bmatrix} 1 & 0 & 1 \\ 2 & 1 & 2 \\ 0 & 4 & 6 \end{bmatrix}$

Solution

Write matrix A down with an identity matrix next to it. The vertical bar is used to separate the matrix and the identity matrix.

$$\left(\begin{array}{ccc|ccc} 1 & 0 & 1 & 1 & 0 & 0 \\ 2 & 1 & 2 & 0 & 1 & 0 \\ 0 & 4 & 6 & 0 & 0 & 1 \end{array}\right)$$

This is called the augmented matrix.

Now, using elementary row operations, write the matrix on the left side of the bar in reduced row-echelon form. If the identity matrix appears on the left side of the bar, then the matrix on the right side of the bar is the inverse matrix.

$$\left(\begin{array}{ccc|ccc} 1 & 0 & 1 & 1 & 0 & 0 \\ 2 & 1 & 2 & 0 & 1 & 0 \\ 0 & 4 & 6 & 0 & 0 & 1 \end{array}\right)$$

$$-2R_1 + R_2 \rightarrow \left(\begin{array}{ccc|ccc} 1 & 0 & 1 & 1 & 0 & 0 \\ 0 & 1 & 0 & -2 & 1 & 0 \\ 0 & 4 & 6 & 0 & 0 & 1 \end{array}\right)$$

$$-4R_2 + R_3 \rightarrow \left(\begin{array}{ccc|ccc} 1 & 0 & 1 & 1 & 0 & 0 \\ 0 & 1 & 0 & -2 & 1 & 0 \\ 0 & 0 & 6 & 8 & -4 & 1 \end{array}\right)$$

$$R_3 \div 6 \rightarrow \left(\begin{array}{ccc|ccc} 1 & 0 & 1 & 1 & 0 & 0 \\ 0 & 1 & 0 & -2 & 1 & 0 \\ 0 & 0 & 1 & \frac{4}{3} & -\frac{2}{3} & \frac{1}{6} \end{array}\right)$$

$$R_1 - R_3 \rightarrow \left(\begin{array}{ccc|ccc} 1 & 0 & 0 & -\frac{1}{3} & \frac{2}{3} & -\frac{1}{6} \\ 0 & 1 & 0 & -2 & 1 & 0 \\ 0 & 0 & 1 & \frac{4}{3} & -\frac{2}{3} & \frac{1}{6} \end{array}\right)$$

The matrix on the right-hand side of the vertical bar is the inverse matrix A^{-1}. So,

$$A^{-1} = \begin{pmatrix} -\frac{1}{3} & \frac{2}{3} & -\frac{1}{6} \\ -2 & 1 & 0 \\ \frac{4}{3} & -\frac{2}{3} & \frac{1}{6} \end{pmatrix}$$

Compare this answer with the one we obtain in Example 4.

Exercise 18.4

Find the cofactor of the given element.

1. $\begin{vmatrix} 2 & 3 & -2 \\ -1 & 5 & -4 \\ 3 & 1 & 0 \end{vmatrix}, -4$

2. $\begin{vmatrix} -1 & -4 & 7 \\ 0 & 1 & 0 \\ 2 & -3 & -2 \end{vmatrix}, 7$

3. $\begin{vmatrix} 5 & -3 & -3 \\ 7 & 5 & -7 \\ 0 & 6 & -1 \end{vmatrix}$, 6

4. $\begin{vmatrix} 3 & -1 & 2 \\ 1 & -1 & 2 \\ -2 & -3 & 10 \end{vmatrix}$, 3

Evaluate each determinant.

5. $\begin{vmatrix} -2 & 3 & 0 \\ 1 & -2 & 1 \\ 4 & 5 & 3 \end{vmatrix}$

6. $\begin{vmatrix} 0 & 0 & 2 \\ 1 & -1 & 4 \\ 3 & 4 & -2 \end{vmatrix}$

7. $\begin{vmatrix} -2 & 3 & -7 \\ 0 & -4 & -3 \\ 1 & 5 & 2 \end{vmatrix}$

8. $\begin{vmatrix} 2 & -1 & 4 \\ 4 & 0 & 2 \\ 3 & -2 & 7 \end{vmatrix}$

9. $\begin{vmatrix} 1 & 1 & 2 \\ 3 & 1 & 0 \\ -2 & 0 & 3 \end{vmatrix}$

10. $\begin{vmatrix} 2 & 1 & 3 \\ 1 & 4 & 4 \\ 1 & 0 & 2 \end{vmatrix}$

Find the inverse matrix, if it exists.

11. $\begin{bmatrix} 1 & 2 & 0 \\ -1 & 1 & 0 \\ 2 & 5 & 1 \end{bmatrix}$

12. $\begin{bmatrix} 1 & 1 & 2 \\ 3 & 1 & 0 \\ -2 & 0 & 3 \end{bmatrix}$

13. $\begin{bmatrix} 2 & 1 & 3 \\ 1 & 4 & 4 \\ 1 & 0 & 2 \end{bmatrix}$

14. $\begin{bmatrix} 1 & 4 & -2 \\ 3 & 6 & -6 \\ -2 & 1 & 4 \end{bmatrix}$

15. $\begin{bmatrix} -2 & 2 & 3 \\ 1 & -1 & 0 \\ 0 & 1 & 4 \end{bmatrix}$

16. $\begin{bmatrix} 2 & -1 & 0 \\ 4 & 2 & 1 \\ 4 & 2 & 1 \end{bmatrix}$

17. $\begin{bmatrix} 2 & 3 & -1 \\ 1 & 1 & 1 \\ 0 & 2 & -1 \end{bmatrix}$

18. $\begin{bmatrix} 1 & 4 & -2 \\ 3 & 6 & -6 \\ -2 & 1 & 4 \end{bmatrix}$

19. $\begin{bmatrix} 1 & 1 & -1 \\ 3 & -1 & 0 \\ 2 & -3 & 4 \end{bmatrix}$

20. $\begin{bmatrix} 2 & -1 & 0 \\ 4 & 2 & 1 \\ 4 & 2 & 1 \end{bmatrix}$

18.5 Application of 3 × 3 Determinants

Equation of a Straight Line

The equation of a straight line that passes through the points (x_1, y_1) and (x_2, y_2) can be found by using the expression

$$\begin{vmatrix} x & y & 1 \\ x_1 & y_1 & 1 \\ x_2 & y_2 & 1 \end{vmatrix} = 0$$

Example 1

Find the equation of the straight line that passes through the point (3, 2) and (4, 5).

Solution

$$\begin{vmatrix} x & y & 1 \\ 3 & 2 & 1 \\ 4 & 5 & 1 \end{vmatrix} = x\begin{vmatrix} 2 & 1 \\ 5 & 1 \end{vmatrix} - y\begin{vmatrix} 3 & 1 \\ 4 & 1 \end{vmatrix} + \begin{vmatrix} 3 & 2 \\ 4 & 5 \end{vmatrix}$$

$$= x(2 - 5) - y(3 - 4) + (15 - 8)$$

$$= -3x + y + 7$$

So, the equation of the straight line is

$$-3x + y + 7 = 0$$

$$y = 3x - 7$$

Area of Triangle

We can use the 3 × 3 determinant to find the area of a triangle with vertices (x_1, y_1), (x_2, y_2) and (x_3, y_3). The area is the absolute value of

$$\frac{1}{2}\begin{vmatrix} x_1 & y_1 & 1 \\ x_2 & y_2 & 1 \\ x_3 & y_3 & 1 \end{vmatrix}$$

Example 2

Find the area of triangle ABC with vertices A (2, −3), B (−4, 7) and B (8, 2)

Solution

Choose $(x_1, y_1) = (2, -3)$, $(x_2, y_2) = (-4, 7)$ and $(x_3, y_3) = (8, 2)$. So,

$$\frac{1}{2}\begin{vmatrix} 2 & -3 & 1 \\ -4 & 7 & 1 \\ 8 & 2 & 1 \end{vmatrix} = \frac{1}{2}\left[2\begin{vmatrix} 7 & 1 \\ 2 & 1 \end{vmatrix} + 3\begin{vmatrix} -4 & 1 \\ 8 & 1 \end{vmatrix} + \begin{vmatrix} -4 & 7 \\ 8 & 2 \end{vmatrix}\right]$$

$$= \tfrac{1}{2}\,[2(7 - 2) + 3(-4 - 8) + (-8 - 56)]$$

$$= \tfrac{1}{2}\,(10 - 36 - 64)$$

$$= \tfrac{1}{2}\,(-90)$$

$$= -45$$

Therefore the area of the triangle ABC is 45 square units.

Note that, three points (x_1, y_1), (x_2, y_2) and (x_3, y_3) are collinear if and only if

$$\begin{vmatrix} x_1 & y_1 & 1 \\ x_2 & y_2 & 1 \\ x_3 & y_3 & 1 \end{vmatrix} = 0$$

Exercise 18.5

Find the equation of the straight lines joining the following pair of points

1. (1, 6), (5, 9)
2. (−3, 4), (8, 1)
3. (3, 2), (7, −3)
4. (−1, −4), (4, −3)
5. (−2, −2), (1, 5)
6. (−3, −2), (−1, −5)

Find the area of the triangle with the given vertices.

7. (5, 3), (2, 1), (7, 12)
8. (−5, −3), (10, −7), (1, 1)
9. (4, 2), (8, −1), (−2, 5)
10. (0, 7), (5, 0), (−3, −4)
11. (−2, 1), (3, −1), (1, 6)
12. (1. 5), (−4, 2), (4, −4)

18.6 Using Matrices to solve Systems of Three Linear Equations

We state the Cramer's Rule for three equations and three variables without proof.

Consider the system of the linear equations

$$a_{11}x + a_{12}y + a_{13}z = b_1$$

$$a_{21}x + a_{22}y + a_{23}z = b_2$$

$$a_{31}x + a_{32}y + a_{33}z = b_3$$

If $D = \begin{vmatrix} a_{11} & a_{12} & a_{13} \\ a_{21} & a_{22} & a_{23} \\ a_{31} & a_{32} & a_{33} \end{vmatrix} \neq 0$, then the solution of the linear equation is given by

$$x = \frac{\begin{vmatrix} b_1 & a_{12} & a_{13} \\ b_2 & a_{22} & a_{23} \\ b_3 & a_{32} & a_{33} \end{vmatrix}}{D} = \frac{D_x}{D},$$

$$y = \frac{\begin{vmatrix} a_{11} & b_1 & a_{13} \\ a_{21} & b_2 & a_{23} \\ a_{31} & b_3 & a_{33} \end{vmatrix}}{D} = \frac{D_y}{D}$$

and $$z = \frac{\begin{vmatrix} a_{11} & a_{12} & b_1 \\ a_{21} & a_{22} & b_2 \\ a_{31} & a_{32} & b_3 \end{vmatrix}}{D} = \frac{D_z}{D}$$

Example 1

Solve the system of linear equations

$$2x + y - z = 1$$

$$3x - y + 2z = 10$$

$$4x - 2y - 3z = 9$$

Solution

Begin by finding D, D_x, D_y and D_Z.

$$D = \begin{vmatrix} 2 & 1 & -1 \\ 3 & -1 & 2 \\ 4 & -2 & -3 \end{vmatrix}$$

$$= 2(3 + 4) - 1(-9 - 8) - 1(-6 + 4)$$

$$= 33$$

$$D_x = \begin{vmatrix} 1 & 1 & -1 \\ 10 & -1 & 2 \\ 9 & -2 & -3 \end{vmatrix}$$

$$= (3 + 4) - (-30 - 18) - (-20 + 9)$$

$$= 66$$

$$D_y = \begin{vmatrix} 2 & 1 & -1 \\ 3 & 10 & 2 \\ 4 & 9 & -3 \end{vmatrix}$$

$$= 2(-30 - 18) - (-9 - 8) - (27 - 40)$$

$$= -66$$

$$D_z = \begin{vmatrix} 2 & 1 & 1 \\ 3 & -1 & 10 \\ 4 & -2 & 9 \end{vmatrix}$$

$$= 2(-9 + 20) - (27 - 40) + (-6 + 4)$$

$$= 33$$

So, $$x = \frac{D_x}{D} = \frac{66}{33} = 2,$$

$$y = \frac{D_y}{D} = \frac{-66}{33} = -2$$

and $$z = \frac{D_z}{D} = \frac{33}{33} = 1$$

The solution is (2, −2, 1).

Solving System of Equations by Gauss-Jordan Elimination Method

Consider the following system of three linear equations.

$$a_{11}x + a_{12}y + a_{13}z = b_1$$

$$a_{21}x + a_{22}y + a_{23}z = b_2$$

$$a_{31}x + a_{32}y + a_{33}z = b_3$$

Associated with this system of equations is the following augmented matrix of the system.

$$\left(\begin{array}{ccc|c} a_{11} & a_{12} & a_{13} & b_1 \\ a_{21} & a_{22} & a_{23} & b_2 \\ a_{31} & a_{32} & a_{33} & b_3 \end{array}\right)$$

We use elementary row operations to write the matrix on the left side of vertical bar in reduced row-echelon form. The solution of the system is then read from the final matrix.

A matrix in reduced row-echelon form has the following properties.

1. All rows consisting entirely of zeros occur at the bottom of the matrix'

2. For each row that does not consist entirely of zeros, the first nonzero entry is 1 (called a leading 1).

3. Each column that contains a leading 1 has zeros in all other entries.

4. The leading 1 in any row is to the left of any leading 1's in the rows below it.

The following examples best show how to solve systems of linear equations using the Gauss-Jordan elimination method.

Example 2

Solve the system of linear equations by using the Gauss-Jordan elimination method.

$$2x + y - z = 1$$

$$3x - y + 2z = 10$$

$$4x - 2y - 3z = 9$$

Solution

The augmented matrix of the system is the following

$$\left(\begin{array}{ccc|c} 2 & 1 & -1 & 1 \\ 3 & -1 & 2 & 10 \\ 4 & -2 & -3 & 9 \end{array}\right)$$

We will now perform row operations until we obtain a matrix in reduced row-echelon form.

$$\begin{array}{r} R_2 - R_1 \rightarrow \\ R_3 - R_2 \rightarrow \\ \\ \end{array} \left(\begin{array}{ccc|c} 1 & -2 & 3 & 9 \\ 1 & -1 & -5 & -1 \\ 4 & -2 & -3 & 9 \end{array}\right)$$

$$\begin{array}{r} \\ R_1 - R_2 \rightarrow \\ -4R_1 + R_3 \rightarrow \end{array} \left(\begin{array}{ccc|c} 1 & -2 & 3 & 9 \\ 0 & -1 & 8 & 10 \\ 0 & 6 & -15 & -27 \end{array}\right)$$

$$\begin{array}{r} -2R_2 + R_1 \rightarrow \\ -R_2 \rightarrow \\ 6R_2 + R_3 \rightarrow \end{array} \left(\begin{array}{ccc|c} 1 & 0 & -13 & -11 \\ 0 & 1 & -8 & -10 \\ 0 & 0 & 33 & 33 \end{array}\right)$$

$$\begin{array}{r} \\ \\ R_3 \div 33 \rightarrow \end{array} \left(\begin{array}{ccc|c} 1 & 0 & -13 & -11 \\ 0 & 1 & -8 & -10 \\ 0 & 0 & 1 & 1 \end{array}\right)$$

$$\begin{array}{r} 13R_3 + R_1 \rightarrow \\ 8R_3 + R_2 \rightarrow \\ \\ \end{array} \left(\begin{array}{ccc|c} 1 & 0 & 0 & 2 \\ 0 & 1 & 0 & -2 \\ 0 & 0 & 1 & 1 \end{array}\right)$$

From this final matrix, you can see that the solution is (2, –2, 1).

Example 3

Solve the system of linear equations by using the Gauss-Jordan elimination method.

$$3x + y - 2z = 2$$

$$6x + 3y - 5z = 1$$

$$-3x - y + 2z = 1$$

Solution

The augmented matrix of the system is the following

$$\left(\begin{array}{ccc|c} 3 & 1 & -2 & 2 \\ 6 & 3 & -5 & 1 \\ -3 & -1 & 2 & 1 \end{array}\right)$$

Now perform row operations on this augmented matrix to produce a reduced form.

$$-2R_1 + R_2 \rightarrow \left(\begin{array}{ccc|c} 3 & 1 & -2 & 2 \\ 0 & 1 & -1 & -3 \\ -3 & -1 & 2 & 1 \end{array}\right)$$

$$R_3 + R_1 \rightarrow \left(\begin{array}{ccc|c} 3 & 1 & -2 & 2 \\ 0 & 1 & -1 & -3 \\ 0 & 0 & 0 & 3 \end{array}\right)$$

Because the last row has all zeros on the left side of the vertical line and a nonzero number on the right side we have a contradiction, so we can stop the process. In this case, the system is inconsistent and has no solution.

Exercise 18.6

Use matrices to solve the system of linear equations.

1. $x + 2y - z = 2$

   $2x - 3y + z = -1$

   $4x + y + 2z = 12$

2. $4x + 3y - z = 12$

   $3x - y = 5$

   $x + 2y + 2z = 2$

3. $2x + 3y - 5z = 0$

   $3x + 2y - 4z = -2$

   $4x + y - z = 2$

4. $2x - 3y + 4z = 3$

   $4x - 5y - z = -2$

   $x + y + z = 3$

5. $4x - y + z = -5$

   $2x + 2y + 3z = 10$

   $5x - 2y + 6z = 1$

6. $4x - 2y + 3z = -2$

   $2x + 2y + 5z = 16$

   $8x - 5y - 2z = 4$

Review Exercises

Use the following matrices to compute the given expressions

$A = \begin{bmatrix} 1 & 2 \\ 0 & 4 \end{bmatrix}$ $B = \begin{bmatrix} 3 & -1 \\ 4 & 2 \end{bmatrix}$ $C = \begin{bmatrix} 2 & 3 \\ 4 & -2 \end{bmatrix}$

1. A + B
2. B + C
3. 2A – 3C
4. A + C
5. A – 2C
6. 3C – 4B
7. (A + B) – 2C
8. 4C + (A – B)
9. 2A – 5(C + B)

10. 2(A – B) – C 11. 3A + (C – B) 12. 2(B – A) – C

Use the following matrices to compute the given expression

$A = \begin{bmatrix} 3 & 2 \\ -1 & 4 \end{bmatrix}$ $B = \begin{bmatrix} 2 & 4 \\ 3 & -1 \end{bmatrix}$ $C = \begin{bmatrix} 2 & 0 \\ 4 & -2 \end{bmatrix}$ $D = \begin{bmatrix} 2 \\ 3 \end{bmatrix}$

13. AB 14.CD 15. BC

16. AD – BD 17. (A + I)C 18. (B – C)D

19. B(A + C) 20. A(B – C) 21. 2BD – AD

Given that the matrix $A = \begin{bmatrix} -3 & 1 \\ 0 & 2 \end{bmatrix}$ find:

22. A^2 23. A^3

Given that $A = \begin{bmatrix} 4 & -2 \\ 3 & -1 \end{bmatrix}$ and $B = \begin{bmatrix} 1 & 2 \\ -2 & 1 \end{bmatrix}$, find:

24. A^2 25. B^2 26. A^2B 27. B^2A

28. Find the values of *a*, *b*, *c* and *d* if

$$\begin{bmatrix} 1 & 3 \\ 1 & 4 \end{bmatrix}\begin{bmatrix} a & b \\ c & d \end{bmatrix} = \begin{bmatrix} 6 & -5 \\ 7 & -7 \end{bmatrix}.$$

29. Find the values of *x* and *y* if

$$\begin{bmatrix} -1 & -1 \\ 4 & 2 \end{bmatrix}\begin{bmatrix} x \\ y \end{bmatrix} + \begin{bmatrix} -1 \\ 2 \end{bmatrix} = \begin{bmatrix} -5 \\ 0 \end{bmatrix}.$$

30. If $A = \begin{bmatrix} 6 & 2 \\ 4 & y \end{bmatrix}$ and $B = \begin{bmatrix} 4 & 2 \\ x & 3 \end{bmatrix}$, find the values of *x* and *y*, given that A and B are commutative under matrix multiplication.

Find the value of each determinant:

31. $\begin{vmatrix} 3 & 1 \\ 4 & 2 \end{vmatrix}$ 32. $\begin{vmatrix} 6 & 1 \\ 5 & 2 \end{vmatrix}$ 33. $\begin{vmatrix} 8 & -3 \\ 4 & 2 \end{vmatrix}$

34. $\begin{vmatrix} 3 & 2 \\ -1 & 4 \end{vmatrix}$ 35. $\begin{vmatrix} 4 & -1 \\ 6 & -1 \end{vmatrix}$ 36. $\begin{vmatrix} 2 & 0 \\ 4 & -3 \end{vmatrix}$

37. $\begin{vmatrix} 1 & 4 & -2 \\ 3 & 2 & 0 \\ -1 & 4 & -3 \end{vmatrix}$ 38. $\begin{vmatrix} 4 & -1 & 2 \\ 6 & -1 & 0 \\ 1 & -3 & 4 \end{vmatrix}$ 39. $\begin{vmatrix} 1 & 4 & -2 \\ 3 & 6 & -6 \\ -2 & 1 & 4 \end{vmatrix}$

Solve the following equations:

40. $\begin{vmatrix} x & x \\ 4 & 3 \end{vmatrix} = 5$ 41. $\begin{vmatrix} x & 1 \\ 3 & x \end{vmatrix} = -2$

42. $\begin{vmatrix} 5-x & 4 \\ 1 & 2-x \end{vmatrix} = 0$

43. $\begin{vmatrix} 2 & x \\ -1 & x^2-3x-6 \end{vmatrix} = 0$

44. $\begin{vmatrix} x & -1 & 2 \\ 1 & x & 3 \\ 0 & 1 & 2 \end{vmatrix} = -9x$

45. $\begin{vmatrix} x & 2 & 3 \\ 1 & x & 0 \\ 6 & 1 & -2 \end{vmatrix} = 19-28x$

Find the inverse of each of the following matrices:

46. $\begin{bmatrix} 2 & 1 \\ 1 & 1 \end{bmatrix}$

47. $\begin{bmatrix} 3 & -1 \\ -2 & 1 \end{bmatrix}$

48. $\begin{bmatrix} 6 & 5 \\ 2 & 2 \end{bmatrix}$

49. $\begin{bmatrix} -4 & 1 \\ 6 & -2 \end{bmatrix}$

50. $\begin{bmatrix} -1 & -2 \\ 3 & 4 \end{bmatrix}$

51. $\begin{bmatrix} 1 & 2 \\ 2 & 3 \end{bmatrix}$

52. $\begin{bmatrix} 1 & 1 & 2 \\ 2 & 1 & 0 \\ 1 & 2 & 2 \end{bmatrix}$

53. $\begin{bmatrix} -1 & 1 & 0 \\ 1 & 0 & 2 \\ 3 & 1 & 0 \end{bmatrix}$

54. $\begin{bmatrix} 2 & -2 & 1 \\ 3 & 1 & -1 \\ 1 & -3 & 2 \end{bmatrix}$

Solve the following system of linear equations

55. $5x + 2y = 7$
$3x - y = 13$

56. $5x - y = 13$
$2x + 3y = 12$

57. $x + 3y = 5$
$2x - 3y = -8$

58. $x + 2y - z = -3$
$2x - 4y + z = -7$
$-2x + 2y - 3z = 4$

59. $x + 6y + 2z = 9$
$3x - 2y + 3z = -1$
$5x - 5y + 2z = 7$

60. $x - y + 2z = -4$
$3x + y - 4z = -6$
$2x + 3y - 4z = 4$

19 Trigonometry

19.1 Angles and the Radian Measure

An angle is formed by rotating a ray about a point called the vertex. The initial position of the ray is called the initial side of the angle, and the final location of the ray is called the terminal side of the angle, as shown in Figure 19.1(a).

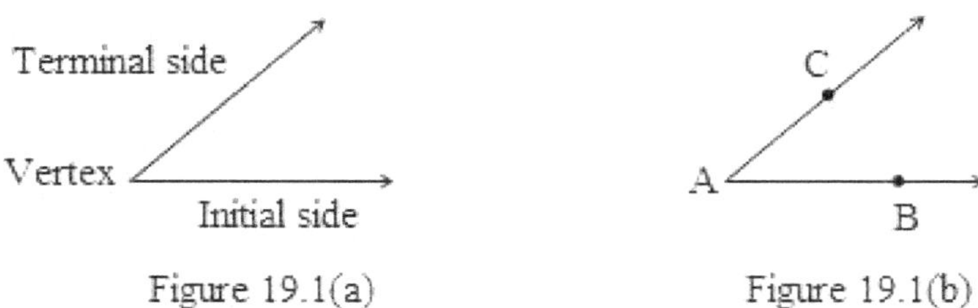

Figure 19.1(a) Figure 19.1(b)

An angle can be named by its vertex or by using three letters with the vertex letter in the middle. The angle shown in Figure 19.1(b) could be named $\angle A$, $\angle BAC$ or $\angle CAB$.

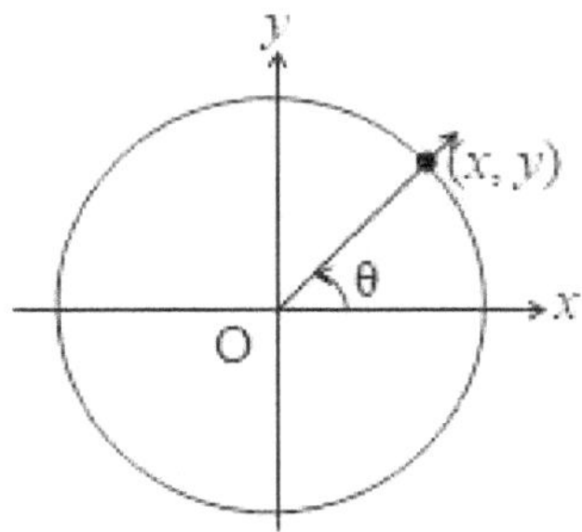

Figure 19.2

An angle whose vertex is the origin of a coordinate system and whose initial side is along the positive x-axis, as shown in Figure 19.2, is said to be in standard position. The angle, θ measured counterclockwise from the positive x-axis to the point (x, y) is positive, whereas the angle measured clockwise from the positive x-axis is negative. The sizes of angles are often indicated in degrees, denoted by the symbol °. In degree measure, a complete rotation of a ray results in an angle measure of 360°. An angle having a degree measure between 0° and 90° is called an acute angle. An angle measure of 90° is a right angle. An angle having measure more than 180° is an obtuse angle.

An angle in standard position is said to be in the quadrant of its terminal side. Angles between 0° and 90° lies in Quadrant I, angles between 90° and 180° lies in Quadrant II, angles between 180° and 270° lies in Quadrant III and angles between 270° and 360° lies in Quadrant IV.

The Radian Measure

An alternative unit for measuring angles is the radian measure.

Consider the circle of radius r, shown in Figure 19.3.

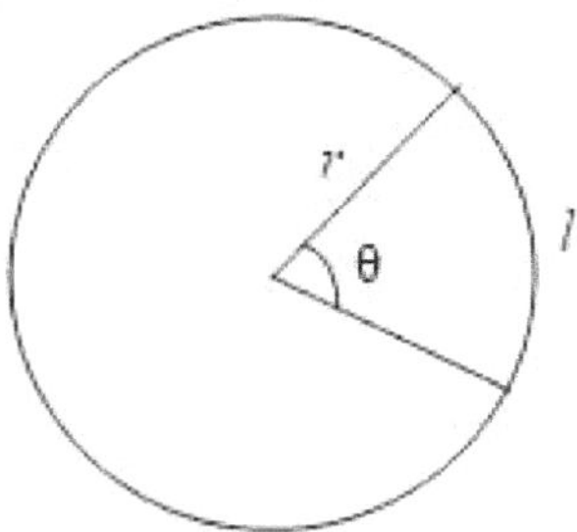

Figure 19.3

The angle θ cuts an arc of length l. The length of the arc of a circle is proportional to the radius of the circle. For any angle θ and radius r, the ratio of the length of the arc l to the radius of the circle is the same. So, we define radian measure on a circle as

$$\frac{Length\ of\ arc}{Radius} = \frac{l}{r}$$

If the length of the arc is r, then $\theta = r/r = 1$ radian. Thus, one radian is the measure of an angle θ that cuts an arc on the circle equal in length to the radius of the circle.

Because the circumference of a circle is $2\pi r$ units the radian measure of an angle of one full revolution is $2\pi r/r = 2\pi$.

It follows that 2π rad. = 360° so π rad. = 180°

Since π radian = 180° we have

$$1 \text{ radian} = \frac{180}{\pi} \approx 57.29578° \quad (5 \text{ d. p})$$

$$\text{Also, } 1° = \frac{\pi}{180} \approx 0.01745 \quad (5 \text{ d. p})$$

We can easily convert angles in degrees to radians or vice versa. To do this we use the equation π radian = 180° and the following conversion rules. The word radian is often omitted, so when no units of angle measure are specified, radian measure is implied.

Converting from degree measures to radians

To convert degrees to radians, multiply degrees by π/180, since 1° = π/180 radians.

Example 1

Convert the following angles to radian:

(a) 135^0 (b) 315^0

Solution

(a) $135° = 135 \times \frac{\pi}{180} = \frac{3}{4}\pi$

(b) $315° = 315 \times \frac{\pi}{180} = \frac{7}{4}\pi$

Often when we convert a degree measure to radian we give the answer in multiples of π.

Converting from radian measures to degrees

To convert radians to degrees, multiply radians by $180/\pi$, since 1 radian = $180°/\pi$.

Example 2

Convert the following angles to degrees:

(a) $\pi/4$ rad. (b) $3\pi/5$ rad.

Solution

(a) $\frac{1}{4}\pi = \frac{1}{4}\pi \times \frac{180}{\pi} = 45°$

(b) $\frac{3}{5}\pi = \frac{3}{5}\pi \times \frac{180}{\pi} = 108°$

Exercise 19.1

Convert the following degree measures to radians

1. 30^0 2. 60^0 3. 90^0

4. 120^0 5. 150^0 6. -210^0

7. 225^0 8. 270^0 9. 405^0

10.- 240^0 11. -330^0 12. 300^0

Convert the following radian measures to degrees

13. $\pi/5$ 14. $3\pi/4$ 15. $2\pi/3$

16. 3π 17. $-5\pi/4$ 18. $4\pi/5$

19. $-7\pi/5$ 20. $-8\pi/3$ 21. $4\pi/9$

22. $\pi/10$ 23. -4π 24. $5\pi/2$

19.2 Trigonometric Functions

The three basic trigonometric functions are sine, cosine and tangent, written briefly as sin, cos and tan respectively.

Consider the right triangle shown in Figure 19.4.

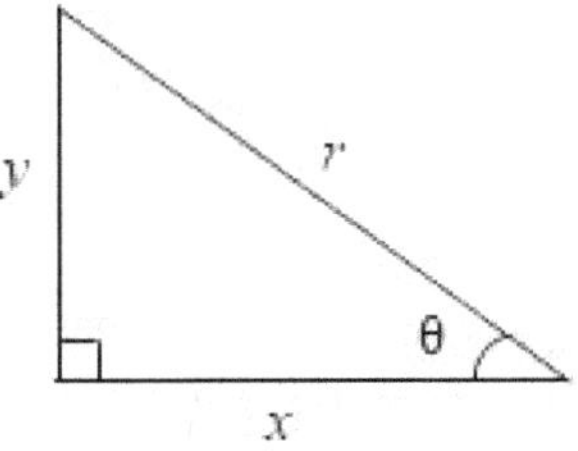

Figure 19.4

Let θ be an acute angle. Given the side opposite θ as y, the side adjacent as x and the hypotenuse r, we define sine, cosine and tangent of angle θ as:

$$\sin\theta = \frac{y}{r} \qquad \cos\theta = \frac{x}{r} \qquad \tan\theta = \frac{y}{x}$$

There are three other trigonometric functions: cotangent (cot), secant (sec) and cosecant (csc), defined by

$$\cot\theta = \frac{x}{y} \quad \text{or} \quad \cot\theta = \frac{1}{\tan\theta}$$

$$\sec\theta = \frac{r}{x} \quad \text{or} \quad \sec\theta = \frac{1}{\cos\theta}$$

$$\csc\theta = \frac{r}{y} \quad \text{or} \quad \csc\theta = \frac{1}{\sin\theta}$$

Notice that the cotangent, secant and cosecant are reciprocals of the tangent, cosine and sine functions respectively.

We have used the triangle to define the trigonometric functions. Another method for defining trigonometric functions is base on the unit circle.

Unit Circles

A unit circle is a circle with centre at the origin and radius 1 unit.

Consider the unit circle shown in Figure 19.5

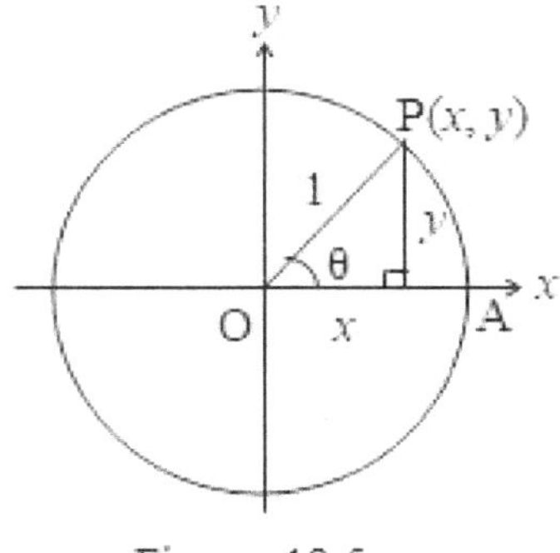

Figure 19.5

The point A with coordinates (1, 0) moves in a counterclockwise direction to the point P(x, y). The size of the angle formed by the line OP and the x-axis is θ. Because the length of OP is 1 unit, we define the following trigonometric ratios as

$$\sin\theta = \frac{y}{1} = y$$

$$\cos\theta = \frac{x}{1} = x$$

and
$$\tan\theta = \frac{y}{x} = \frac{\sin\theta}{\cos\theta}$$

In general, we can define cos θ and sin θ as the x- and y- coordinates respectively of a point on the unit circle.

Because the point A (1, 0) can be rotated several times around the circle, it follows that both sine and cosine can be defined for unlimited values of θ. You can see from the unit circle that the values of the x- and y- coordinates can only take on values from – 1 to 1, so for any angle θ we have $-1 \leq \sin\theta \leq 1$ and $-1 \leq \cos\theta \leq 1$.

Special Angles

You can find the values of the trigonometric functions for most angles by using a calculator with trigonometric keys. However, the exact values of trigonometric functions of some angles, called special angles, can be found without using a calculator.

Trigonometric functions of 0°, 90°, 180°, 270° and 360°

Consider the unit circle shown in Figure 19.6

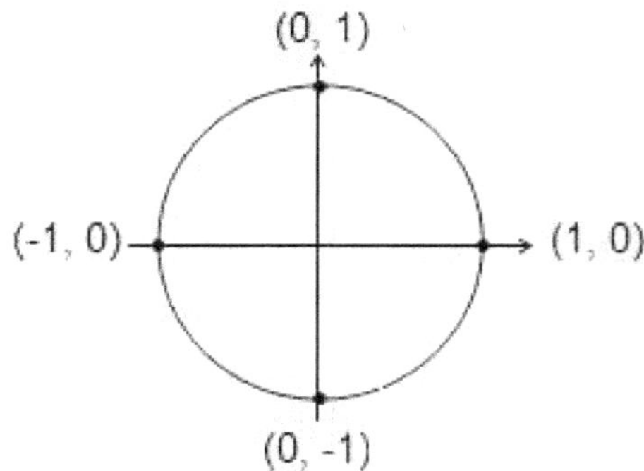

Figure 19.6

For each of the four points, we have the following.

$\sin 0° = 0,$ $\cos 0° = 1$ $\tan 0° = \frac{0}{1} = 0$

$\sin 90° = 1$ $\cos 90° = 0$ $\tan 90° = \frac{1}{0}$ undefined

$\sin 180° = 0$ $\cos 180° = -1$ $\tan 180° = \frac{0}{-1} = 0$

$\sin 270° = -1$ $\cos 270° = 0$ $\tan 270° = \frac{-1}{0}$ undefined

And $\sin 360° = 0$ $\cos 360° = 1$ $\tan 360° = \frac{0}{1} = 0$

Trigonometric Ratios for 45°

The triangle shown in Figure 19.7 is a right isosceles triangle. The equal sides are of length 1unit.

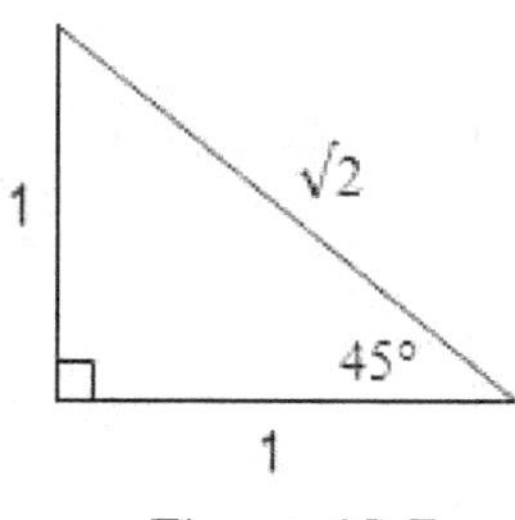

Figure 19.7

The Pythagorean Theorem gives the length of the hypotenuse as $\sqrt{1^2 + 1^2} = \sqrt{2}$.

You can obtain from the triangle in Figure 19.7 the following trigonometric function values:

$\sin 45° = \frac{1}{\sqrt{2}} = \frac{\sqrt{2}}{2}$

$\cos 45° = \frac{1}{\sqrt{2}} = \frac{\sqrt{2}}{2}$

$\tan 45° = 1$

Trigonometric Ratios for 30^0 and 60^0

The triangle shown in Figure 19.8 is an equilateral triangle of length 2 units. From plane geometry we know that any line drawn from a vertex perpendicular to a base bisects the base and the angle at the vertex.

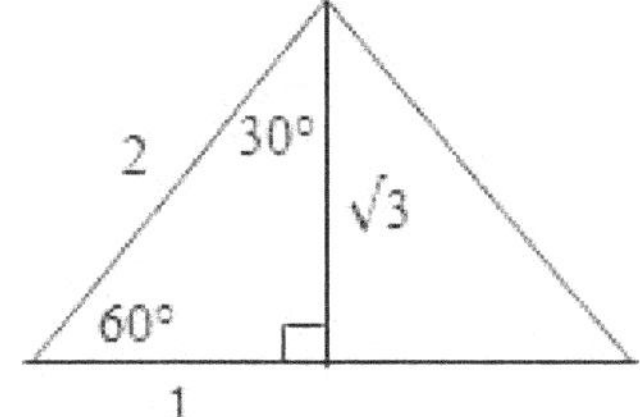

Figure 19.8

The Pythagorean Theorem gives the length of the altitude as $\sqrt{2^2 - 1^2} = \sqrt{3}$. You can obtain from the triangle in Figure 19.8 the following trigonometric function values:

$\sin 60° = \frac{\sqrt{3}}{2}$ $\qquad$ $\sin 30° = \frac{1}{2}$

$\cos 60° = \frac{1}{2}$ $\qquad$ $\cos 30° = \frac{\sqrt{3}}{2}$

$\tan 60° = \sqrt{3}$ $\qquad$ $\tan 30° = \frac{1}{\sqrt{3}} = \frac{\sqrt{3}}{3}$

In general, if θ is an acute angle, the following statements are true.

$\sin (90° - \theta) = \cos \theta$ $\qquad$ $\cos (90° - \theta) = \sin \theta$

$\tan (90° - \theta) = \cot \theta$ $\qquad$ $\cot (90° - \theta) = \tan \theta$

$\sec (90° - \theta) = \csc \theta$ $\qquad$ $\csc (90° - \theta) = \sec \theta$

Trigonometric Functions of any Angle

In degree measure 360° represents a complete rotation of a ray. A continuous rotation of a ray produces angles of measures larger than 360°.

The values of the trigonometric functions of an angle greater than 90° (or less than 0°) can be determined from the corresponding acute angle, called the reference angle, formed by the terminal side of the angle with the horizontal axis.

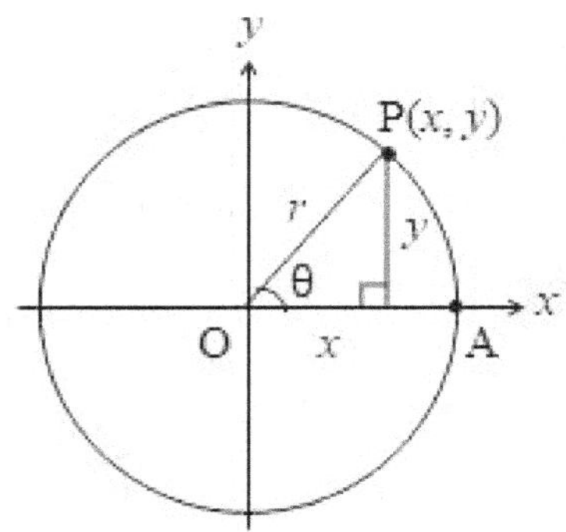

Figure 19.9

The values of the trigonometric functions of an angle are equal to the values of the same trigonometric functions of its reference angle, except that it may differ in sign, as the following discussion shows.

Figure 19.9 shows an angle θ in Quadrant I. In this quadrant, both *x*- and *y*- coordinates are both positive. Because the distance O to P can never be negative, *r* is always positive. It follows that

$$\sin\theta = \frac{y}{r} = + \qquad \cos\theta = \frac{x}{r} = + \qquad \tan\theta = \frac{y}{x} = +$$

In Quadrant II, *x* is negative but *y* is positive, so

$$\sin\theta = \frac{y}{r} = + \qquad \cos\theta = \frac{x}{r} = - \qquad \tan\theta = \frac{y}{x} = -$$

In Quadrant III, *x* is negative but *y* is also negative, so

$$\sin\theta = \frac{y}{r} = - \qquad \cos\theta = \frac{x}{r} = - \qquad \tan\theta = \frac{y}{x} = +$$

In Quadrant IV, *x* is positive and *y* is negative, so

$$\sin\theta = \frac{y}{r} = - \qquad \cos\theta = \frac{x}{r} = + \qquad \tan\theta = \frac{y}{x} = -$$

These results are summarized in Figure 19.10. The diagram shows the positive functions in each quadrant.

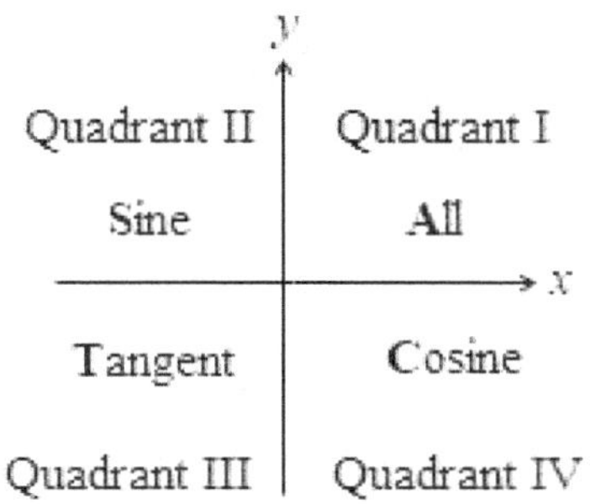

Figure 19.10

Starting from Quadrant IV and moving in the counterclockwise direction the bold-face letters spell the word CAST.

Example 1

Evaluate each trigonometric function.

(a) sin 150°　　(b) cos 225°

(c) tan 330°　　(d) cos 780°

Solution

(a) sin 150°

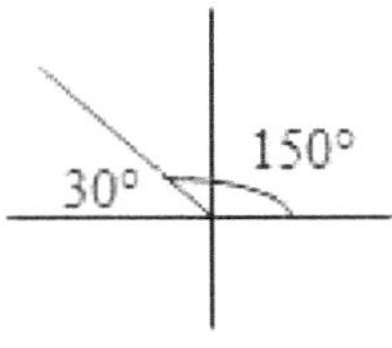

Figure 19.11

Because the terminal side of 150° lies in Quadrant II, as shown in Figure 19.11, the reference angle is 180° − 150° or 30°. The sine function is positive in Quadrant II, so

$$\sin 150° = \sin 30° = \tfrac{1}{2}$$

(b) cos 225°

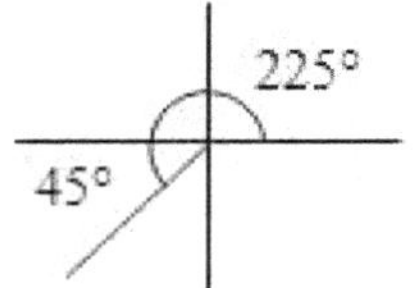

Figure 19.12

Because the terminal side of 225° lies in Quadrant III, as shown in Figure 19.12, the reference angle is 225° – 180° or 45°. The cosine function is negative in Quadrant III, so

$$\cos 225° = -\cos 45° = -\frac{\sqrt{2}}{2}$$

(c) tan 330°

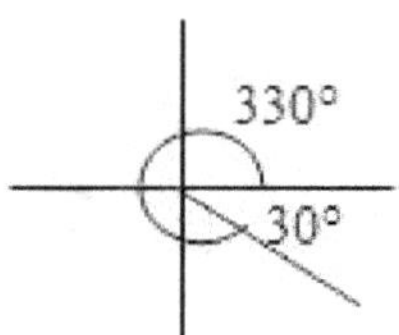

Figure 19.13

Because the terminal side of 330° lies in Quadrant IV, as shown in Figure 19.13, the reference angle is 360° – 330° or 30°. The tangent function is negative in Quadrant IV, so

$$\tan 330° = -\tan 30° = -\frac{\sqrt{3}}{3}$$

(d) cos 780°

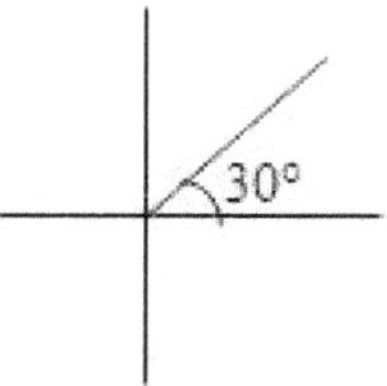

Figure 19.14

A complete rotation of a ray results in an angle of measure 360°. The angles 780° and 780° – 2(360°) = 60° have the same terminal side in Quadrant I, as shown in Figure 19.14. The cosine function is positive in this quadrant, so

$$\cos 780° = \cos 60° = ½$$

These results of Example 1 are summarized as follows.

1. If θ is the reference angle for a given angle in the second quadrant then:

$\sin(180 - \theta) = \sin\theta$

$\cos(180 - \theta) = -\cos\theta$

$\tan(180 - \theta) = -\tan\theta$

2. If θ is the reference angle for a given angle in the third quadrant then:

$\sin(180 + \theta) = -\sin\theta$

$\cos(180 + \theta) = -\cos\theta$

$\tan(180 + \theta) = \tan\theta$

3. If θ is the reference angle for a given angle in the fourth quadrant then:

$\sin(360 - \theta) = -\sin\theta$

$\cos(360 - \theta) = \cos\theta$

$\tan(360 - \theta) = -\tan\theta$

Trigonometric Functions of Negative Angles

Recall that angles measured clockwise from the positive x-axis are negative.

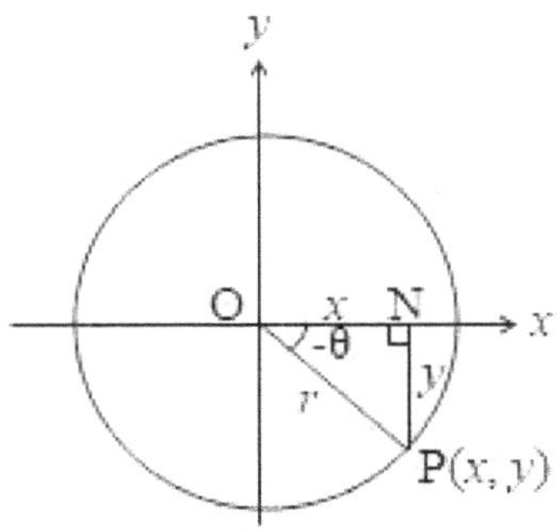

Figure 19.15

Using the definitions of the trigonometric functions and the triangle OPN, in Figure 19.15, you can see that

$$\sin(-\theta) = \frac{-y}{r} = -\sin\theta$$

$$\cos(-\theta) = \frac{x}{r} = \cos\theta$$

and
$$\tan(-\theta) = \frac{-y}{x} = -\tan\theta$$

Example 2

Evaluate each trigonometric function

(a) $\sin(-30°)$ (b) $\tan(-120°)$ (c) $\cos(-210°)$

Solution

(a) The angle -30^0 lies in Quadrant IV, and is coterminal with $330°$. The reference angle is $360° - 330° = 30°$, so

$$\sin(-30°) = -\sin 30° = -\tfrac{1}{2}$$

(b) The angle $-120°$ lies in the third quadrant, and is coterminal with $240°$. The reference angle is $240° - 180° = 60°$, so

$$\tan(-120°) = \tan 60° = \sqrt{3}$$

(c) The angle $-210°$ lies in the second quadrant, and is coterminal with $150°$. The reference angle is $180° - 150° = 30°$, so

$$\cos(-210°) = -\cos 30° = -\sqrt{3}/2$$

Exercise 19.2

Find the reference angle for the given angle.

1. 110^0 2. 200^0 3. 260^0

4. 320^0 5. 216^0 6. 480^0

7. -170^0 8. 563^0 9. 627^0

10. -401^0 11. -760^0 12. 1270^0

State the quadrant in which the angle lies.

13. -135^0 14. 265^0 15. 315^0

16. -215^0 17. 575^0 18. -330^0

19. 780^0 20. -420^0 21. 480^0

22. 600^0 23. -315 24. -240^0

Evaluate the following trigonometric functions without using a calculator.

25. $\cos 150°$ 26. $\tan 120°$ 27. $\sin 210°$

28. $\cot 225°$ 29. $\sin 135°$ 30. $\cos 270°$

31. sec (– 30°)　　32. csc (– 120°)　　33. cot (– 135°)

34. tan (– 60°)　　35. cos 405°　　36. sin (– 270°)

37. sec 600°　　38. tan 765°　　39. sin (– 540°)

Evaluate the following trigonometric functions without using a calculator:

40. cos (π/2)　　41. sin (π/4)　　42. tan (π/6)

43. csc (π/3)　　44. sec π　　45. cot (3π/4)

46. tan (2π/3)　　47. sin (3π/2)　　48. cos (2π/3)

49. sec (– 2π)　　50. cot (–4π/3)　　51. csc (– 5π/3)

52. cos (– 16π/3)　　53. sin (– 9π/4)　　54. tan (– 5π/3)

19.3 Relationships between Trigonometric Functions

If an angle θ corresponds to a point P(cos θ, sin θ) on a unit circle $x^2 + y^2 = 1$, we obtain the following identity

$$\cos^2 \theta + \sin^2 \theta = 1$$

Dividing this equation by $\cos^2\theta$ gives

$$1 + \tan^2 \theta = \sec^2 \theta \qquad (1)$$

Again dividing the equation by $\sin^2 \theta$ gives

$$1 + \cot^2 \theta = \csc^2 \theta \qquad (2)$$

Note that $\cos^2 \theta = (\cos \theta)^2$ and $\sin^2 \theta = (\sin \theta)^2$.

All the trigonometric identities discussed in this section and earlier sections are summarized for convenient reference.

$$\cot\theta = \frac{1}{\tan\ \theta}, \qquad \sec\theta = \frac{1}{\cos\ \theta},$$

$$csc\ \theta = \frac{1}{\sin\ \theta} \qquad \tan\theta = \frac{\sin\ \theta}{\cos\ \theta}$$

$$cot\theta = \frac{\cos\ \theta}{\sin\ \theta} \qquad cos^2\theta + sin^2\theta = 1$$

$$1 + tan^2\theta = sec^2\theta \qquad 1 + cot^2\theta = csc^2\theta$$

Using Trigonometric Identities

Example 1

Simplify:

(a) $sec^2\theta - tan^2\theta$ (b) $\frac{\cos\theta}{1-\sin\theta} + \frac{\cos\theta}{1+\sin\theta}$

Solution

(a) $\sec^2\theta - \tan^2\theta$

First, convert each term to sine and cosine.

$$sec^2\theta - tan^2\theta = \frac{1}{cos^2\theta} - \frac{sin^2\theta}{cos^2\theta}$$

$$= \frac{1-sin^2\theta}{cos^2\theta} \quad \text{Subtract fractions}$$

$$= \frac{cos^2\theta}{cos^2\theta} \quad \text{Use identity}$$

$$= 1 \quad \text{Simplify}$$

Note: From $\cos^2\theta + \sin^2\theta = 1$, we get $1 - \sin^2\theta = \cos^2\theta$.

(b) $\frac{\cos\theta}{1-\sin\theta} + \frac{\cos\theta}{1+\sin\theta}$

$$\frac{\cos\theta}{1-\sin\vartheta} + \frac{\cos\theta}{1+\sin\theta} = \frac{\cos\theta(1+\sin\theta)+\cos\theta(1-\sin\theta)}{(1-\sin\theta)(1+\sin\theta)} \quad \text{Add fractions.}$$

$$= \frac{2cos\theta}{1-sin^2\theta} \quad \text{Simplify}$$

$$= \frac{2}{\cos\theta} \quad \text{Use identity}$$

$$= 2\sec\theta \quad \text{Use identity}$$

Verifying Trigonometric Identities

The general method of procedure used in verifying a trigonometric identity is to choose one side of the equation and use the fundamental trigonometric identities to simplify it until you obtain the expression on the other side. On occasion, it is convenient to reduce both sides to identical form.

Example 2

Verify the identity

$$\tan\theta + \cot\theta = \frac{1}{\sin\theta\cos\theta}$$

Solution

We take the expression on the left hand side and try to obtain the expression on the right hand side. First, we rewrite the expression in terms of sine and cosine only.

$$\tan\theta + \cot\theta = \frac{\sin\theta}{\cos\theta} + \frac{\cos\theta}{\sin\theta}$$

$$= \frac{\sin^2\theta + \cos^2\theta}{\sin\theta\cos\theta} \qquad \text{Add fraction}$$

$$= \frac{1}{\sin\theta\cos\theta} \qquad \text{Use identity}$$

Hence $\tan\theta + \cot\theta = \frac{1}{\sin\theta\cos\theta}$

You can also start with the right hand side.

$$\frac{1}{\sin\theta\cos\theta} = \frac{\sin^2\theta + \cos^2\theta}{\sin\theta\cos\theta}$$

$$= \frac{\sin^2\theta}{\sin\theta\cos\theta} + \frac{\cos^2\theta}{\sin\theta\cos\theta}$$

$$= \frac{\sin\theta}{\cos\theta} + \frac{\cos\theta}{\sin\theta}$$

$$= \tan\theta + \cot\theta$$

Example 3

Verify the identity

$$\frac{\tan\theta}{\sec\theta - 1} - \frac{\tan\theta}{\sec\theta + 1} = 2\cot\theta$$

Solution

Taking the expression on the left hand side we have

$$\frac{\tan\theta}{\sec\theta - 1} - \frac{\tan\theta}{\sec\theta + 1} = \frac{\tan\theta(\sec\theta + 1) - \tan\theta(\sec\theta - 1)}{(\sec\theta - 1)(\sec\theta + 1)} \qquad \text{Subtract fractions}$$

$$= \frac{2\tan\theta}{\sec^2\theta - 1} \qquad \text{Simplify}$$

$$= \frac{2\tan\theta}{\tan^2\theta} \qquad \text{Use identity}$$

$$= \frac{2}{\tan\theta} \qquad \text{Simplify}$$

$$= 2\cot\theta \qquad \text{Use identity}$$

Hence $\frac{\tan\theta}{\sec\theta-1}-\frac{\tan\theta}{\sec\theta+1}=2\cot\theta$

Example 4

Verify the identity

$$\frac{\tan^2\theta}{1+\sec\theta}=\frac{1-\cos\theta}{\cos\theta}$$

Solution

In this example, we would work with each side separately. We start by taking the left hand side.

$$\frac{\tan^2\theta}{1+\sec\theta}=\frac{\sec^2\theta-1}{1+\sec\theta} \qquad \text{Use identity}$$

$$=\frac{(\sec\theta-1)(\sec\theta+1)}{1+\sec\theta} \qquad \text{Factor}$$

$$=\sec\theta-1 \qquad \text{Simplify}$$

Next, taking the right hand side we have

$$\frac{1-\cos\theta}{\cos\theta}=\frac{1}{\cos\theta}-1$$

$$=\sec\theta-1$$

Because both sides are equal to $\sec\theta-1$ it follows that $\frac{\tan^2\theta}{1+\sec\theta}=\frac{1-\cos\theta}{\cos\theta}$.

Exercise 19.3

Simplify the expression.

1. $\sin\theta\cot\theta$
2. $\sec\theta\cot\theta$
3. $(\sec\theta-1)(\sec\theta+1)$
4. $\frac{\sec^2\theta-1}{1-\cos^2\theta}$
5. $\frac{\sin^2\theta}{1-\cos\theta}$
6. $\frac{\cot^2\theta}{\csc\theta-1}$
7. $\frac{1-\cos^2\theta}{1-\sin^2\theta}$
8. $\frac{\cos\theta}{\sin\theta\cot^2\theta}$
9. $(1-\sin^2\theta)\sec^2\theta$
10. $\frac{\cos^2\theta}{1+\sin\theta}+\frac{\cos^2\theta}{1-\sin\theta}$
11. $\cot\theta+\frac{\sin\theta}{1+\cos\theta}$
12. $(\sin\theta+\cos\theta)^2-1$
13. $\frac{\sec^2\theta-1}{\sec^2\theta}$
14. $\frac{1}{1-\sin\theta}+\frac{1}{1+\sin\theta}$
15. $\frac{1+\sin\theta}{\cos\theta}+\frac{\cos\theta}{1+\sin\theta}$

Verify the identity.

16. $\sin\theta\cot\theta = \cos\theta$

17. $\cos\theta\, csc\,\theta = \cot\theta$

18. $(1 - cos^2\theta)csc^2\theta = 1$

19. $(1 - cos^2\theta)sec^2\theta = tan^2\theta$

20. $(1 + tan^2\theta)cos^2\theta = 1$

21 .$\sin^2\theta\,(1 + cot^2\theta) = 1$

22. $\frac{1}{sec^2\theta} + \frac{1}{csc^2\theta} = 1$

23. $\sec\theta - \tan\theta\sin\theta = \cos\theta$

24. $\frac{1}{1-\sin\theta} + \frac{1}{1+\sin\theta} = 2sec^2\theta$

25. $\frac{1-tan^2\theta}{1+tan^2\theta} = 1 - 2sin^2\theta$

26. $\frac{1}{1+sin^2\theta} + \frac{1}{1+csc^2\theta} = 1$

27. $\frac{2\tan\theta}{1+tan^2\theta} = 2\sin\theta\cos\theta$

28. $.(\tan\theta + \sec\theta)^2 = \frac{1+\sin\theta}{1-\sin\theta}$

29. $\frac{tan^2\theta}{1+tan^2\theta} \times \frac{1+cot^2\theta}{cot^2\theta} = sin^2\theta sec^2\theta$

30. $(\cos\theta + \sin\theta)^2 + (\cos\theta - \sin\theta)^2 = 2$

31. $\frac{\sec\theta - 1}{1-\cos\theta} = \sec\theta$

32. $(1 + \tan\theta)^2 + (1 - \tan\theta)^2 = 2sec^2\theta$

33. $\frac{\cos\theta\cot\theta}{1-\sin\theta} - 1 = \csc\theta$

34. $\frac{1+\sin\theta}{\cos\theta} + \frac{\cos\theta}{1+\sin\theta} = 2\sec\theta$

35. $\frac{1-cos^2\theta}{sec^2\theta-1} = 1 - sin^2\theta$

19.4 Compound Angles

Sum and Difference of Angles

The following identities for the sine and cosine functions involving sums or difference of angles are listed without proof.

$$\sin(A + B) = \sin A\cos B + \cos A\sin B$$

$$\sin(A - B) = \sin A\cos B - \cos A\sin B$$

$$\cos(A + B) = \cos A\cos B - \sin A\sin B$$

$$\cos(A - B) = \cos A\cos B + \sin A\sin B$$

Note that $\sin(A + B) \neq \sin A + \cos B$.

A numerical example will illustrate this: if A = 60°, B = 30° then

$\sin(A + B) = \sin(60 + 30)° = \sin 90° = 1$.

But $\sin A + \sin B = \sin 60° + \sin 30° = (\sqrt{3}/2) + ½ \neq 1$

You can use the sum and difference formulas for sine and cosine to derive a formula for tan (A + B).

$$\tan(A+B) = \frac{\sin(A+B)}{\cos(A+B)}$$

$$= \frac{\sin A\cos B+\cos A\sin B}{\cos A\cos B-\sin A\sin B}$$

Dividing both the numerator and denominator by cos A cos B we get

$$\tan(A+B) = \frac{\frac{\sin A}{\cos A}+\frac{\sin B}{\cos B}}{1-\frac{\sin A\sin B}{\cos A\cos B}}$$

$$= \frac{\tan A+\tan B}{1-\tan A\tan B}$$

Similarly, $\tan(A-B) = \frac{\tan A-\tan B}{1+\tan A\tan B}$

The six identities above can be used to find exact values of trigonometric functions involving sums or differences of special angles.

Example 1

Find, without using a calculator, the values of:

(a) cos (5π/12) (b) sin 15° (c) tan (− 75°)

Solution

(a) We write $\frac{5\pi}{12}$ as $\frac{\pi}{4}+\frac{\pi}{6}$, so

$$cos\frac{5\pi}{12} = \cos\left(\frac{\pi}{4}+\frac{\pi}{6}\right)$$

$$= cos\frac{\pi}{4}cos\frac{\pi}{6} - sin\frac{\pi}{4}sin\frac{\pi}{6}$$

$$= \frac{\sqrt{2}}{2}\cdot\frac{\sqrt{3}}{2} - \frac{\sqrt{2}}{2}\cdot\frac{1}{2}$$

$$= \frac{\sqrt{6}-\sqrt{2}}{4}$$

(b) We write 15° as 45° − 30°, so

$$\sin 15° = \sin(45° - 30°)$$

$$= \sin 45° \cos 30° - \cos 45° \sin 30°$$

$$= \frac{\sqrt{2}}{2}\cdot\frac{\sqrt{3}}{2} - \frac{\sqrt{2}}{2}\cdot\frac{1}{2}$$

$$= \frac{\sqrt{6} - \sqrt{2}}{4}$$

(c) We write – 75° as 45° – 120°, so

$$\tan(-75°) = \tan(45° - 120°)$$

$$= \frac{\tan 45° - \tan 120°}{1 + \tan 45° \tan 120°}$$

$$= \frac{1+\sqrt{3}}{1-\sqrt{3}}$$

$$= -\frac{1}{2}\left(1 + \sqrt{3}\right)^2$$

Example 2

Find the value of cos 75° cos 15° + sin 75° sin 15°, without using a calculator.

Solution

Using the formula for cos (A – B) backward gives

$$\cos 75° \cos 15° + \sin 75° \sin 15° = \cos(75 - 15)°$$

$$= \cos 60°$$

$$= ½$$

Example 3

If sin A = 3/5 and cos B = 5/13, where A is obtuse and B is acute, find without using a calculator the value of cos (A – B).

Solution

We draw a triangle with an angle A whose sine is 3/5 and a triangle with an angle B whose cosine is 5/13, as shown in Figure 19.16.

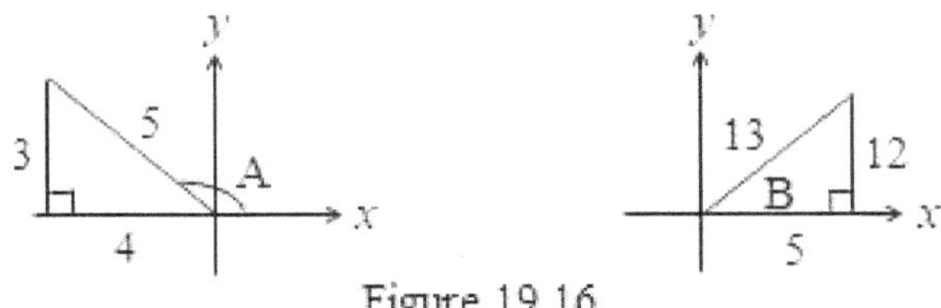

Figure 19.16

Because the cosine is negative in Quadrant II, cos A = – 4/5.

Now, $\cos(A-B) = \cos A \cos B + \sin A \sin B$

$$= -\frac{4}{5}\cdot\frac{5}{13}+\frac{3}{5}\cdot\frac{12}{13}$$

$$= -\frac{20}{65}+\frac{36}{65}$$

$$= \frac{16}{65}$$

Double Angles

If we put A = B in the formula for the sums of angles we obtain results for sin 2A, cos 2A and tan 2A.

$$\begin{aligned}\sin 2A &= \sin(A + A)\\ &= \sin A \cos A + \cos A \sin A\\ &= 2 \sin A \cos A\end{aligned}$$

$$\begin{aligned}\cos 2A &= \cos(A + A)\\ &= \cos A \cos A - \sin A \sin A\\ &= \cos^2 A - \sin^2 A \qquad (1)\end{aligned}$$

There are two other useful forms in which cos 2A may be expressed, one involving only cos A, and the other only sin A.

From $\sin^2 A + \cos^2 A = 1$ we obtain $\cos^2 A = 1 - \sin^2 A$.

Thus from (1) we get

$$\begin{aligned}\cos 2A &= (1 - \sin^2 A) - \sin^2 A\\ &= 1 - 2\sin^2 A \qquad (2)\end{aligned}$$

Again from (1) we get

$$\begin{aligned}\cos 2A &= \cos^2 A - (1 - \cos^2 A)\\ &= 2\cos^2 A - 1 \qquad (3)\end{aligned}$$

All three formulas for cos 2A are equivalent. Use the formula that works best with the information given.

$$\tan 2A = \tan (A + A)$$

$$= \frac{\tan A + \tan A}{1 - \tan A \tan A}$$

$$= \frac{2 \tan A}{1 - tan^2 A}$$

Example 4

Use the formula for sin 2A to find sin 60°.

Solution

$$\sin 60° = \sin (2 \times 30°)$$

$$= 2 \sin 30° \cos 30°$$

$$= 2 \times \frac{1}{2} \times \frac{\sqrt{3}}{2}$$

$$= \frac{\sqrt{3}}{2}$$

Example 5

If $\cos \theta = 3/5$, find $\cos 2\theta$.

Solution

$$\text{Now } \cos 2\theta = 2\cos^2 \theta - 1$$

$$= 2\left(\frac{3}{5}\right)^2 - 1$$

$$= \frac{18}{25} - 1$$

$$= -\frac{7}{25}$$

Deriving a Triple-Angle Formula

$\sin 3A = \sin (2A + A)$ Write 3A as 2A + A

$= \sin 2A \cos A + \cos 2A \sin A$ Use the sum of angles formula

$$= 2 \sin A \cos A \cdot \cos A + (1 - 2 \sin^2 A) \cdot \sin A$$

$$= 2 \sin A \cos^2 A + \sin A - 2\sin^3 A$$

$$= 2 \sin A (1 - 2 \sin^2 A) + \sin A - 2 \sin^3 A$$

$= 2 \sin A - 2 \sin^3 A + \sin A - 2 \sin^3 A$

$= 3 \sin A - 4\sin^3 A$

In a similar manner you can show that

$\cos A = 4 \cos^3 A - 3 \cos A$ and

$\tan 3A = \frac{3 \tan A - tan^3 A}{1 - 3tan^2 A}$

Exercise 19.4

Evaluate the trigonometric function without using a calculator.

1. $\cos 15°$ 2. $\sin 75°$ 3. $\tan 75°$ 4. $\tan 105°$ 5. $\sin 165°$

6. $\tan 15°$ 7. $\sin(-15°)$ 8. $\cos(-75°)$ 9. $\tan(-15°)$ 10. $\sin 195°$

Simplify the following trigonometric function.

11. $\sin(90° + x)$ 12. $\cos(360° + x)$

13. $\tan(180° - x)$ 14. $\cos(270° - x)$

Find the value of the expression without using a calculator.

15. $\sin 45° \cos 15° - \cos 45° \sin 15°$ 16. $\cos 80° \cos 20° + \sin 80° \sin 20°$

17. $\sin 38° \cos 7° + \cos 38° \sin 7°$ 18. $\cos 72° \cos 63° - \sin 72° \sin 63°$

19. $\frac{\tan 40° - \tan 10°}{1 + \tan 40° \tan 10°}$ 20. $\cos 25° \cos 5° - \sin 25° \sin 5°$

Simplify the expression.

21. $\frac{1}{2} \cos x + \frac{\sqrt{3}}{2} \sin x$ 22. $\frac{1}{\sqrt{2}} \sin x - \frac{1}{\sqrt{2}} \cos x$

23. $\frac{\sqrt{3} + \tan x}{1 - \sqrt{3} \tan x}$ 24. $\frac{1}{2} \cos 75° - \frac{\sqrt{3}}{2} \sin 75°$

If $\sin A = 4/5$ and $\sin B = 12/13$, where A is obtuse and B is acute, find without using a calculator the values of:

25. $\sin(A + B)$ 26. $\cos(A + B)$ 27. $\tan(A + B)$

If $\sin A = 3/5$ and $\cos B = 5/13$, where A is obtuse and B is acute, find without using a calculator, the values of:

28. $\sin(A + B)$ 29. $\cos(A - B)$ 30. $\tan(A + B)$

If cos A = 4/5, where 180° < A < 270° and tan B = 5/12, where 270° < B < 360°, find without using a calculator the values of:

31. sin (A + B) 32. tan (A – B) 33. cos (A + B)

Verify the identity.

34. sin (A + B) + sin (A – B) = 2 sin A cos B

35. cos (A + B) – cos (A – B) = – 2 sin A sin B

36. $\frac{\sin(A+B)}{\cos A \cos B} = \tan A + \tan B$

37. cos (A + B) cos (A – B) = $\cos^2$ A – $\sin^2$ B

38. sin (A + B) sin (A – B) = $\cos^2$ B – $\cos^2$ A

Evaluate the expression without using a calculator.

39. 2 sin 15° cos 15° 40. 2 $\cos^2$ 15° – 1

41. 1 – 2 $\sin^2$ 22 ½ ° 42. $\cos^2$ 67 ½ ° – $\sin^2$ 67 ½ °

43. 1 – 2 $\sin^2$ 75° 44. $\frac{2\tan 15°}{1-tan^2 15°}$

Find, without using a calculator, the values of $\sin 2\theta$ and $\cos 2\theta$ when:

45. $\cos\theta = \frac{12}{13}$ 46. $\sin\theta = \frac{8}{17}$ 47. $\tan\theta = \frac{12}{5}$

Verify the identity.

48. $\frac{\sin 2A}{1+\cos 2A} = \tan A$ 49. $\frac{\sin 2A}{1-\cos 2A} = \cot A$

50. $\frac{\cot A-\tan A}{\cot A+\tan A} = \cos 2A$ 51. tan A + cot A = 2 csc 2A

52. $\cos^4$ A – $\sin^4$ A = cos 2A 53. cot A – tan A = 2 cot 2A

54. $\frac{2-sec^2 A}{sec^2 A} = \cos 2A$ 55. $\frac{2\cot A}{1+cot^2 A} = \sin 2A$

Verify the identity.

56. cos 3A = 4 $\cos^3$ A – 3 cos A 57. $\tan 3A = \frac{3\tan A-tan^3 A}{1-3tan^2 A}$

58. $\cot 3A = \frac{cot^3 A-3\cot A}{2cot^2 A-1}$ 59. $\frac{\sin 3A}{\sin A} - \frac{\cos 3A}{\cos A} = 2$

60. $\frac{3\cos A+\cos 3A}{3\sin A-\sin 3A} = cot^3 A$ 61. $\frac{\sin 3A+sin^2 A}{cos^3 A-\cos 3A} = \cot A$

19.5 Solving Trigonometric Equations

An equation which contains at least one trigonometric function is called a trigonometric equation. The method for solving some trigonometric equations is similar to the method for solving a linear equation or a quadratic equation as described earlier in this book.

Example 1

Solve the equation $1 - 2\cos x = 0$.

Solution

Begin by rewriting the equation so that $\cos x$ is isolated on one side of the equation.

$$1 - 2\cos x = 0$$

$$\cos x = \tfrac{1}{2}$$

First, find all solutions in the interval $[0, 2\pi]$. Cosine is positive in the first and the fourth quadrants, so $\cos x = \tfrac{1}{2}$ has solution $x = \pi/3$ and $x = 5\pi/3$. Because $\cos x$ has a period of 2π, add multiples of 2π to each of these solutions. So,

$$x = (\pi/3) + 2n\pi \qquad \text{and} \qquad x = (5\pi/3) + 2n\pi$$

where n is an integer .

These are the general solutions of the equation $1 - 2\cos x = 0$.

Example 2

Find all solutions of $\sin^2 x = \tfrac{3}{4}$ in the interval $[0, 2\pi)$.

Solution

Begin by taking the square root of each side.

$$\sin^2 x = \tfrac{3}{4} \qquad \text{Original equation}$$

$$\sin x = \pm\tfrac{\sqrt{3}}{2} \qquad \text{Take square root of each side}$$

First we consider the equation $\sin x = \sqrt{3}/2$

This equation has two solutions $x = \pi/3$ and $x = 2\pi/3$ in the interval $[0, 2\pi)$.

Next we consider the equation $\sin x = -\sqrt{3}/2$

This equation has two solutions $x = 4\pi/3$ and $x = 5\pi/3$ in the interval $[0, 2\pi)$.

So, in the interval $[0, 2\pi)$, the solutions are $x = \pi/3$, $x = 2\pi/3$, $x = 4\pi/3$ and $x = 5\pi/3$.

Example 3

Find all solutions of $\tan x + 2\sin x = 0$ in the interval $[0, 2\pi)$.

Solution

You can rewrite $\tan x$ as $\frac{\sin x}{\cos x}$

$$\frac{\sin x}{\cos x} + 2\sin x = 0$$

$\sin x + 2\sin x\cos x = 0$ Multiply each term by $\cos x$.

$\sin x\,(1 + 2\cos x) = 0$ Factor

By setting each of these factors equal to zero we get

$\sin x = 0$ and $1 + 2\cos x = 0$

$\cos x = -\frac{1}{2}$

The solutions of $\sin x = 0$ in the interval $[0, 2\pi)$ are $x = 0$ and π.

Now, the solutions of $\cos x = -\frac{1}{2}$ are $x = 2\pi/3, 4\pi/3$.

Therefore, the solutions of the equation in the interval $[0, 2\pi)$ are $x = 0, 2\pi/3, \pi$ and $4\pi/3$.

Example 4

Find all solutions of $2\cos^2 x + \cos x - 1 = 0$ in the interval $[-\pi, \pi]$.

Solution

This equation is a quadrant equation. Factor and solve as usual.

$$2\cos^2 x + \cos x - 1 = 0$$

$(2\cos x - 1)(\cos x + 1) = 0$ Factor

Setting each factor equal to zero gives

$2\cos x - 1 = 0$ and $\cos x + 1 = 0$

$\cos x = \frac{1}{2}$ $\cos x = -1$

$x = \pm(\pi/3)$ $x = \pm\pi$

The solutions of the equation in the interval $[-\pi, \pi]$ are $x = -\pi, -\pi/3, \pi/3, \pi$.

Example 5

Find all solutions of $2\cos^2 x + 3\sin x = 0$ in the interval $[0, 2\pi)$.

Solution

Begin by rewriting the equation in terms of sine only.

$2\cos^2 x + 3\sin x = 0$	Original equation
$2(1 - \sin^2 x) + 3\sin x = 0$	$\cos^2 x = 1 - \sin^2 x$
$2\sin^2 x - 3\sin x - 2 = 0$	Simplify
$(2\sin x + 1)(\sin x - 2) = 0$	Factor

Setting each factor equal to zero gives

$$2\sin x + 1 = 0 \quad \text{and} \quad \sin x - 2 = 0$$

$$\sin x = -\tfrac{1}{2} \qquad\qquad \sin x = 2$$

$$x = 7\pi/6,\ 11\pi/6$$

The equation $\sin x = 2$ has no solution because 2 is outside of the range of the sine function.

So, in the interval $[0, 2\pi)$, the only solutions are $x = 7\pi/6,\ 11\pi/6$.

Note: When an equation involves more than two functions, it will usually be best to rewrite the equation in terms of just one trigonometric function or in terms of sine and cosine only.

Exercise 19.5

Find all values of x from 0° to 360° inclusive that satisfy each of the following equations.

1. $2\cos x = 1$
2. $\tan x - 1 = 0$
3. $2\sin^2 x = 1$
4. $\tfrac{1}{3}\tan^2 x = 1$
5. $\csc^2 x = 4\cot^2 x$
6. $\sec^2 x + \tan^2 x = 7$
7. $\cos^2 x + \sin x + 1 = 0$
8. $2\sec^2 x + \tan^2 x - 3 = 0$
9. $2\sin^2 x = 2 + \cos x$

Find all values of x from 0 to 2π inclusive that satisfy each of the following equations.

10. $2\sin^2 x + \cos x = 1$
11. $2\cos^2 x = 5\cos x - 2$
12. $2\sin^2 x = 3\cos x$
13. $2\cos^2 x + 4\sin^2 x = 3$
14. $\sec^2 x + \tan x = 1$
15. $2\sin^2 x - 5\sin x + 2 = 0$

Find all values of x from $-\pi$ and $+\pi$ inclusive that satisfy each of the following equations.

16. $\cos x = -\frac{1}{2}$

17. $\tan x = 1$

18. $\tan x = -\sqrt{3}$

19. $\cos (x - 60)° = \frac{1}{2}$

20. $\sin(x + 30)° = \sqrt{3}/2$

21. $\tan x = 2 \sin x$

22. $\sec^2 x = 2 \tan^2 x$

23. $\sec x = 4 \cos x$

24. $4 \sin x = 3 \csc x$

25. $\sin^2 x + \sin x = 0$

26. $\tan^2 x = \tan x$

27. $2 \cos^2 x - \cos x - 1 = 0$

28. $2 \sin^2 x + 3 \sin x + 1 = 0$

29. $\sin x = 4 \csc x + 3$

30. $\tan x + \cot x + 2 = 0$

31. $\cos^2 x - \sin^2 x = \sin x$

Find all values of x from 0 to 2π inclusive that satisfy each of the following equations.

32. $\cos 2x + \sin x + 2 = 0$

33. $3 \sin x = \sin 2x$

33. $\cos 2x = \sin x$

35. $\cos 2x - 3 \cos x + 2 = 0$

36. $\tan 2x = \cot x$

37. $\tan x + 3 \cot 2x = 0$

38. $\cos 2x - \cos x = 0$

39. $\sin 2x - \sin x = 0$

19.6 Graphs of Trigonometric Functions

Graph of the sine function

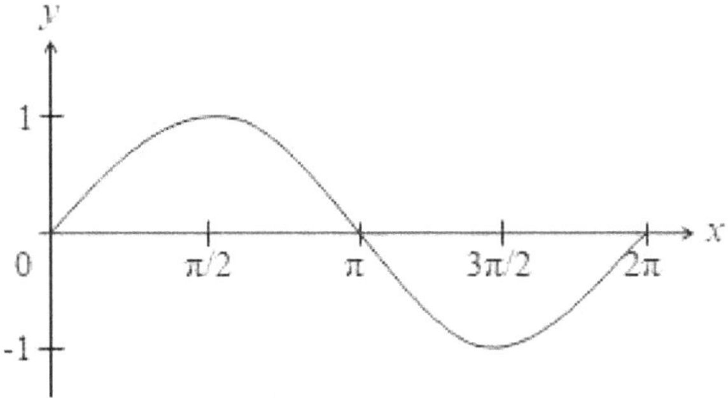

Figure 19.17

Figure 19.17 shows one cycle of the sine curve. The sine curve repeats indefinitely in the positive and negative directions, and is symmetric with respect to the origin. The values of the sine function for angles that differ by multiples of 2π radians (or 360°) are the same. The sine function is therefore periodic with period 2π (or 360°). The range of the sine function is the interval $[-1, 1]$, and the domain is $(-\infty, \infty)$.

Graph of the cosine function

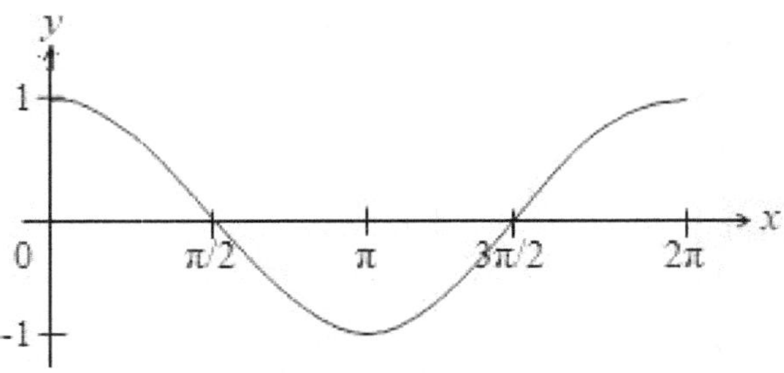

Figure 19.18

Figure 19.18 shows one cycle of the cosine curve. The cosine curve repeats indefinitely in the positive and negative directions, and is symmetric with respect to the *y*-axis. The values of the cosine function for angles that differ by multiples of 2π radians (or 360°) are the same. The cosine function is therefore periodic with period 2π (or 360°). The range of the cosine function is the interval [− 1, 1], and the domain is (− ∞, ∞).

Graph of the tangent function

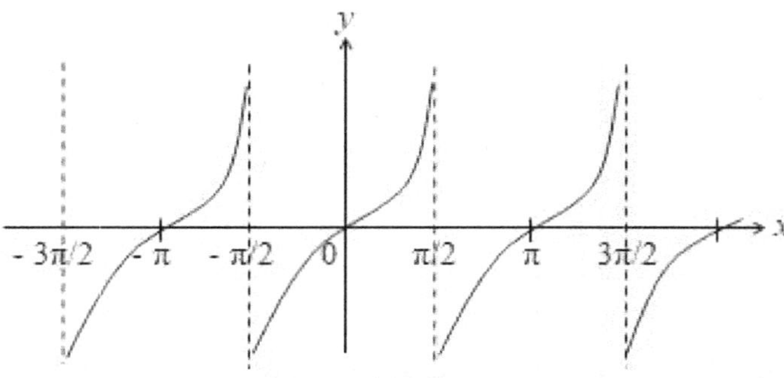

Figure 19.19

Figure 19.19 shows the graph of $y = \tan x$. Since $\tan x = \sin x/\cos x$, $\tan x$ is undefined when $\cos x = 0$. Because $\tan x$ is undefined for $x = -3\pi/2, -\pi/2, \pi/2$ and $3\pi/2$ and so on, the graph has vertical asymptotes at these values. You can see from the graph that the tangent is periodic, with a period of π (or 180°). The range of the tangent function is the set of all real numbers.

Amplitude and Period

Amplitude

The amplitude of a periodic function is the absolute value of half the difference between its maximum value and its minimum value. For instance, both the sine and cosine functions have a maximum value of 1 and a minimum value of − 1, so the amplitude of their graphs is $|1 + 1|/2$ or 1.The tangent function has no maximum or minimum, so it has no amplitude.

The amplitude represents half the distance between the maximum and minimum values of the function. The maximum and minimum values of $y = a \sin x$ (or $y = a \cos x$) could be found by multiplying the maximum and minimum values of $y = \sin x$ (or $y = \cos x$) by a. For instance, the maximum and minimum values of $y = -3 \cos x$ are 3 and -3 respectively, so the amplitude of $y = -3 \cos x$ is $|3 + 3|/2$ or 3. In general, the amplitude of $y = a \sin x$ (or $y = a \cos x$) is $|a|$.

The amplitude corresponds to the number of units that the graph of $y = \sin x$ or $y = \cos x$ is stretched vertically. For instance, the graph of $y = 2 \sin x$ is the graph of $y = \sin x$ stretched vertically by a factor of 2. The graphs of $y = \sin x$ and $y = 2 \sin x$ are shown in Figure 19.20.

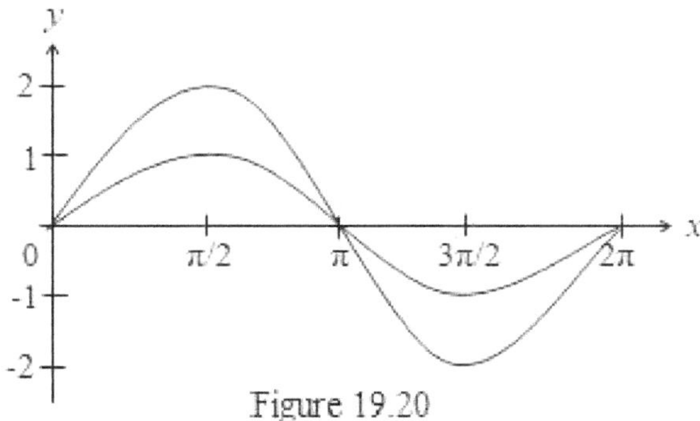

Figure 19.20

Observe that the graph of y = 2 sin x has the same shape as the graph of $y = \sin x$ except that each value of $y = 2 \sin x$ is twice the value of $y = \sin x$. Similarly, the graph of $y = 3 \cos x$ is the graph of $y = \cos x$ stretched vertically by a factor of 3. The examples show that you can sketch the graph of $y = a \sin x$ (or $y = a \cos x$) by stretching the graph of $y = \sin x$ (or $y = \cos x$) by a factor of a.

Period

Recall that the graph of $y = \sin x$ is a periodic function, and goes through one full cycle on the interval $[0, 2\pi]$, so its period is 2π. The graphs of $y = \sin x$ and $y = \sin 2x$ are shown in Figure 19.21.

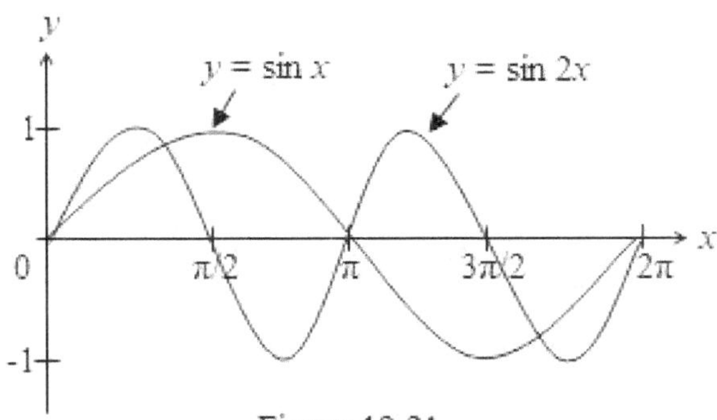

Figure 19.21

You can see from the graph in Figure 19.21 that the function $y = \sin 2x$ is a periodic function with period of π, and goes through one full cycle on the interval $[0, \pi]$ and two complete cycle in the interval $[0, 2\pi]$.

In general the period of $\sin ax$ (or cos ax) is given by $2\pi/|a|$, and the period of $y = \tan x$ is $\pi/|a|$. For instance, the period of $\cos 6\theta$ goes through one full cycle on the interval $[0, \pi/3]$ and has six complete cycles on the interval $[0, 2\pi]$.

The period of $\sin \frac{1}{2}\theta$ is 4π, so the graph of $y = \sin \frac{1}{2}\theta$ goes through one full cycle on the interval $[0, 4\pi]$. Notice that $y = \sin \frac{1}{2}\theta$ has half of a full cycle on the interval $[0, 2\pi]$.

Notice that, the value of a in $y = \sin ax$ or $y = \cos ax$ either stretches or constricts the graph but it does not change the maximum value or the minimum value.

Phase Shift

The graphs of $y = \sin x$ and $y = \sin (x - \pi/3)$ are shown in Figure 19.22.

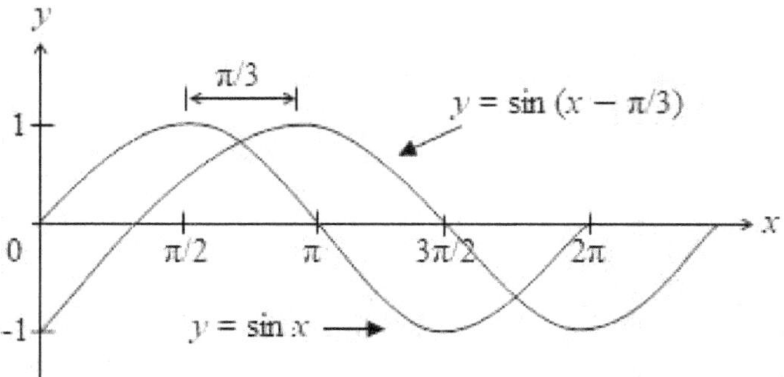

Figure 19.22

The graph of $y = \sin (x - \pi/3)$ has the same shape as $y = \sin x$ but is shifted to the right by $\pi/3$ radians. A horizontal translation of a trigonometric function is called a phase shift.

The phase shift of the functions $y = a \sin b(x - c)$ and $y = a \cos b(x - c)$ is c, where $b > 0$. If $c > 0$, the shift is to the right, and if $c < 0$, the shift is to the left.

The phase shift of the function $y = a \sin (bx - c)$ is not c. To find the phase shift first rewrite this function by factoring b to get $y = a \sin b\,(x - c/b)$. Now you can see that the phase shift will be c/b. Likewise, the phase shift of $y = a \cos (bx - c)$ is c/b.

It can be shown that the graph of $y = a \sin(bx - c)$ has one complete cycle in the interval $c/b \le x \le (c/b + 2\pi/b)$. It is important to note that the starting point of the graphs is determined by the phase shift.

Vertical Shifts

As you learned earlier, adding a constant to a function move the basic graph of the function up or down. Similarly, if a constant is added to a trigonometric function the graph is shifted upward or downward. For instance, the graph of $y = \sin x + 1$ has the same shape as the graph of $y = \sin x$ but shifted upward 1 unit.

A vertical shift occurs when the entire graph shifts up or down along the y-axis. A new horizontal axis called the midline becomes the reference line about which the graph oscillates. For the graph of $y = \sin x + c$, the midline is the graph of $y = c$.

Given the functions $y = a \sin (bx - c) + d$ and $y = a \cos (bx - c) + d$, where $b > 0$, you can see that the amplitude of these functions is $|a|$, the period is $2\pi/|b|$, the phase shift is c/b and the vertical shift is d.

Sketching Trigonometric Functions

The discussion up to now provides needed aid for sketching the graph of sine or cosine functions. Because sine and cosine functions are periodic, the graph is found by first finding the graph over one period and then repeating as many times as required.

To sketch the graph of a sine (or a cosine) function use the following steps.

1. Identify the amplitude
2. Identify the phase shift. The phase shift determines the starting point of the graph.
3. Find the period
4. Identify the vertical shift and the horizontal shift.
5. Determine five key points in one period of the graph: the intercepts, maximum points and minimum points.

Example 1

Sketch the graph of the function $y = -3 \cos 2x$ over one period.

Solution

Here, the amplitude is 3, period is $2\pi/2 = \pi$, phase shift is 0, and vertical shift is 0.

The graph would have one complete cycle on the interval $[0, \pi]$. The starting point is 0. Now divide π by 4 to get $\pi/4$. Add this to successive values beginning with 0, which gives the following values $\pi/4$, $\pi/2$, $3\pi/4$ and π. Evaluate the equation $y = -3 \cos 2x$ at $x = 0, \pi/4, \pi/2, 3\pi/4$ and π to obtain the following five key points in the interval $[0, \pi]$.:

$$(0, -3), (\pi/4, 0), (\pi/2, 3), (3\pi/4, 0), (\pi, -3).$$

Plotting these points and connecting them with a smooth curve gives the graph in Figure 19.23.

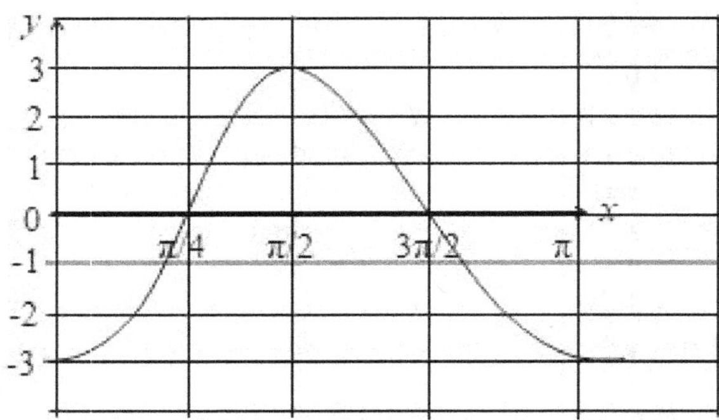

Figure 19.23

Noticed that the graph of $y = -3 \cos 2x$ is a reflection in the x-axis of the graph of $y = 3 \cos 2x$.

Example 2

Sketch the graph of the function $y = 2 \sin (3x - \pi)$ over one period.

Solution

Here, the amplitude is 2, period is $2\pi/3$, phase shift is $\pi/3$ and vertical shift is 0.

The graph would have one complete cycle on the interval $[\pi/3, \pi]$. Divide the period $2\pi/3$ by 4 to get $\pi/6$. Starting with the left endpoint of the interval add successively to get $\pi/2$, $2\pi/3$, $5\pi/6$ and π. Evaluate $y = 2 \cos (3x - \pi)$ at $\pi/3$, $\pi/2$, $2\pi/3$, $5\pi/6$ and π to obtain the following five key points in the interval $[\pi/3, \pi]$.

$$(\pi/3, 0), (\pi/2, 2), (2\pi/3, 0), (5\pi/6, -2), (\pi, 0)$$

Plotting these points and connecting them with a smooth curve gives the graph in Figure 19.24.

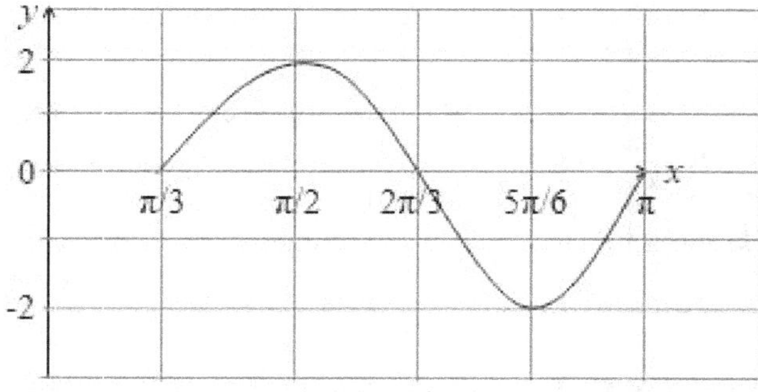

Figure 19.24

Example 3

Sketch the graph of the function $y = \sin 4x + 2$ over one period.

Solution

The amplitude is 1, phase shift is 0, period is $2\pi/4 = \pi/2$ and vertical shift is 2.

The graph would have one complete cycle on the interval $[0, \pi/2]$. Dividing this interval into four equal parts produces the following key points in the interval $[0, \pi/2]$.

$$(0, 2), (\pi/8, 3), (\pi/4, 2), (3\pi/8, 1)\ (\pi/2, 2)$$

The graph is shown in Figure 19.25.

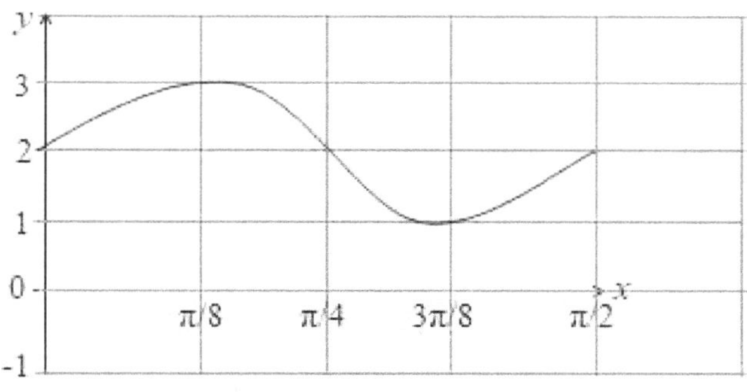

Figure 19.25

The graphs of the secant, cosecant, and cotangent are not covered in this textbook.

Exercise 19.6

Find the maximum and minimum values of the following expressions, giving the smallest positive or zero value of θ for which they occur.

1. $3 \sin \frac{1}{2} \theta$
2. $-2 \cos \theta$
3. $3 + 2 \cos \theta$
4. $1 - 2 \sin 3\theta$
5. $2 \csc \theta$
6. $\frac{1}{2} \sec \theta - 1$
7. $-3 \sec \theta + 1$
8. $2 - \frac{1}{3} \csc \theta$
9. $1/(3 - 2\sin \theta)$
10. $3/(2\cos \theta + 1)$
11. $2/(4 - 3\csc \theta)$
12. $4/(3 - 2\sec \theta)$

Find the amplitude and period of the following:

13. $y = 4 \sin 6x$
14. $y = -6 \cos 4x$
15. $y = 3 \sin 5x$
16. $y = -5 \sin \frac{1}{2} x$
17. $y = -\cos 3x$
18. $y = 2 \cos \frac{3}{4} x$

Graph each function over a two- period interval.

19. $y = 2 \cos x$
20. $y = \frac{1}{2} \sin x$
21. $y = 3 \cos 2x$
22. $y = 2 \sin \frac{1}{3} x$
23. $y = 3 \cos \frac{1}{2} x$
24. $y = -2 \sin \frac{1}{2} x$

Find the amplitude, period, phase shift and vertical shift and graph each function over a two- period interval.

25. $y = -2 \cos x$
26. $y = \sin (3/2)\, x$
27. $y = \frac{1}{2} \cos (x - \pi)$
28. $y = 4 \sin (\frac{1}{2} x + \pi) + 2$
29. $y = 4 \sin (x + \pi/4)$
30. $y = -2 + 3\cos (2x - \pi/3)$

19.7 Inverse Trigonometric Functions

For each of the six functions there is an inverse function that is defined in a restricted domain.

Recall from earlier section that the inverse of a function is the relation in which all values of x and y are interchanged. The graphs of $y = \sin x$ and its inverse $x = \sin y$ are shown in Figure 19.26(a) and Figure 19.26(b) respectively.

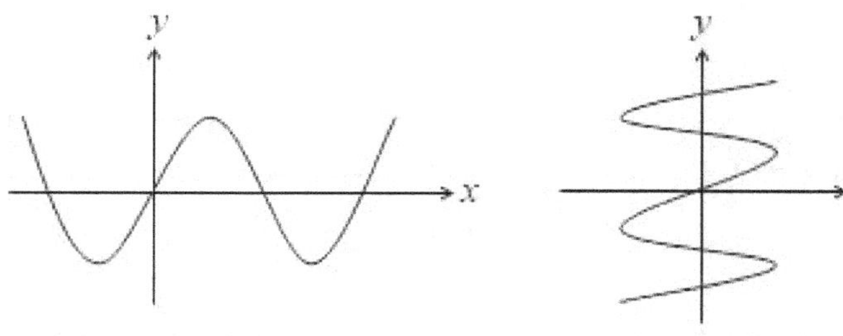

Figure 19.26(a) Figure 19.26(b)

You can see that the inverse is not a function, since it fails the vertical line test. However, if you restrict the domain to the interval $[-\pi/2, \pi/2]$ (see Figure 19.27 (a)) you obtain a unique inverse function of $y = \sin x$, called the inverse sine function. The graph of the inverse sine function, shown in Figure 19.27 (b) is obtained by reflecting the graph of $y = \sin x$ on the interval $[-\pi/2, \pi/2]$ in the line $y = x$.

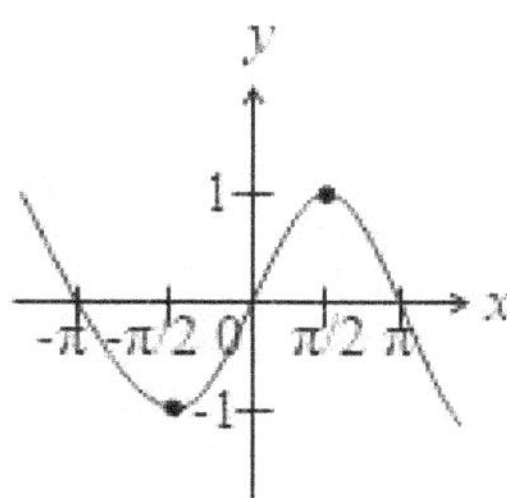

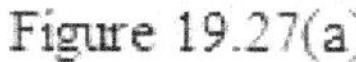
Figure 19.27(a)

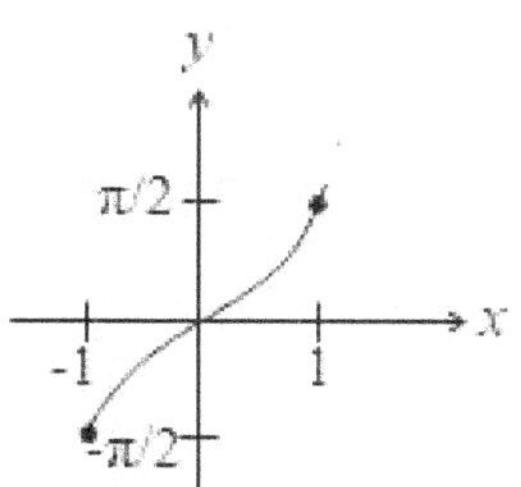

Figure 19.27(b)

As you can see from the graph in Figure 19.27(b), the domain of the inverse sine function is $[-1, 1]$ and the range is $[-\pi/2, \pi/2]$.

We denote the inverse function as $y = \sin^{-1}x$ or $y = \arcsin x$, read " y is the inverse of $\sin x$". The inverse sine function of x is the angle between $-\pi/2$ and $\pi/2$ whose sin is x. Hence, if $y = \sin^{-1}x$, then $\sin y = x$, and $-\pi/2 \leq y \leq \pi/2$. For example, $\sin(\pi/3) = \sqrt{3}/2$ so $arc\sin(\sqrt{3}/2) = \pi/3$.

Inverse Cosine Function

In the case of the inverse cosine function we restrict the domain of $y = \cos x$ to the interval $[0, \pi]$, (see Figure 19.28(a)). In a similar manner, the graph of the inverse cosine function, shown in Figure 19.28(b), is obtained by reflecting the graph of the cosine function in the restricted domain in the line $y = x$.

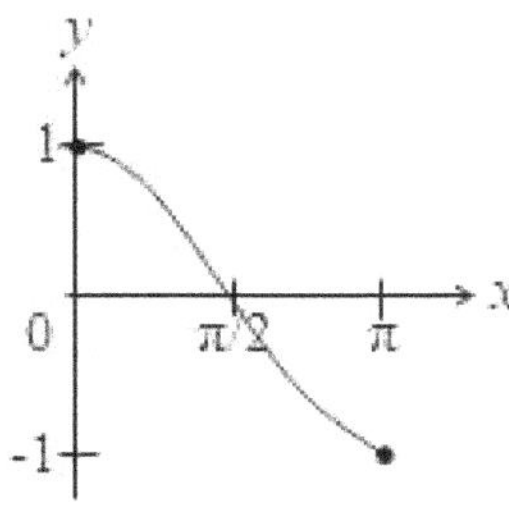

Figure 19.28(a)

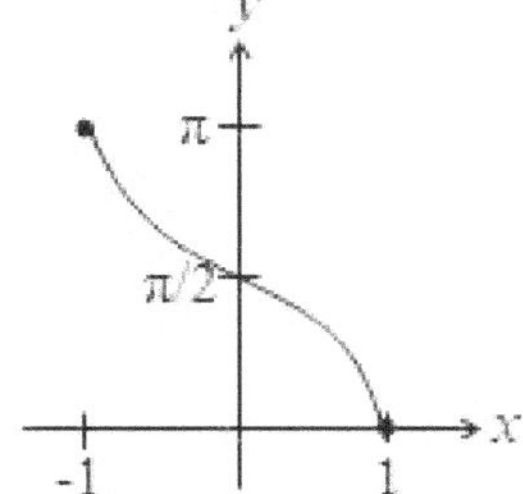

Figure 19.28(b)

You can see from Figure 19.28(b) that the domain of the inverse cosine function is $[-1, 1]$ and the range is $[0, \pi]$.

Similarly, we denote the inverse cosine function as $\cos^{-1}x$ or arcos x. By definition, inverse cosine function of x is the angle between 0 and π whose cosine is x. For example,

$$cos\left(\frac{3\pi}{4}\right) = -\frac{\sqrt{2}}{2}, \text{ so } arccos\left(-\frac{\sqrt{2}}{2}\right) = \frac{3\pi}{4}.$$

Inverse Tangent Function

We can define an inverse tangent function by restricting the domain of y = tan x to the interval [$-\pi/2$, $\pi/2$] (see Figure 19.29(a)). The graph of the inverse tangent function is shown in Figure 19.29(b).

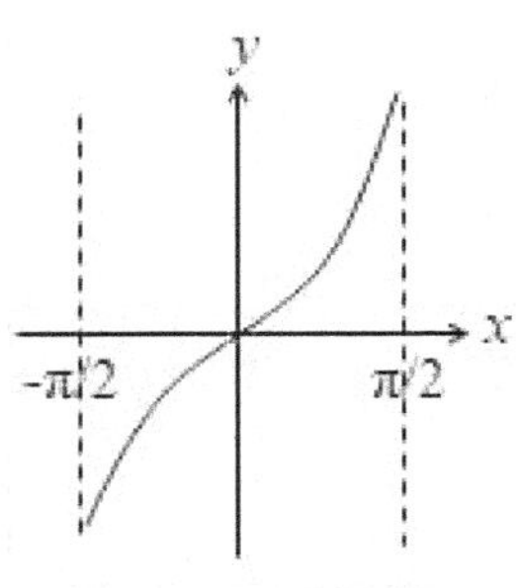

Figure 19.29(a)

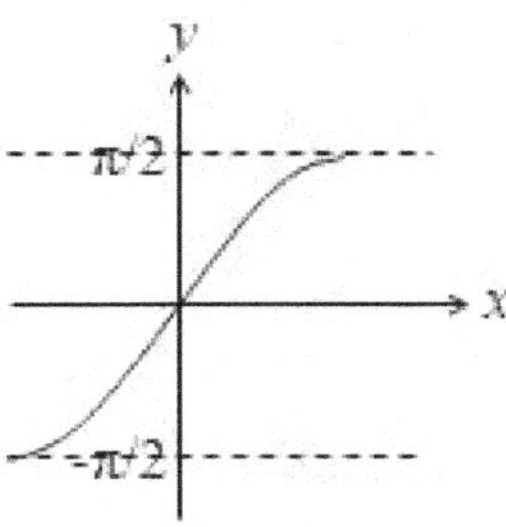

Figure 19.29(b)

You can see from Figure 19.29(b) that the domain of the inverse tangent function is $(-\infty, \infty)$ and the range is $(-\pi/2, \pi/2)$.

Evaluating Inverse Trigonometric Functions

Example 1

Evaluating the following without using a calculator:

(a) $\sin^{-1}(\sqrt{2}/2)$ (b) $\cos^{-1}(\sqrt{3}/3)$ (c) $\tan^{-1}(-1)$

Solution

(a) You must look for an angle x in the interval $-\pi/2 \le x \le \pi/2$ for which $\sin x = -\sqrt{2}/2$. Because $\sin(-\pi/4) = \sqrt{2}/2$ and $-\pi/4 \in \{-\pi/2 \le x \le \pi/2\}$, we have $\sin^{-1}(-\sqrt{2}/2) = -\pi/4$.

(b) In this case, you must look for an angle x in the interval $0 \le x \le \pi$ for which $\cos x = \sqrt{3}/3$. Because $\cos \pi/6 = \sqrt{3}/3$ and $\pi/6 \in \{0 \le x \le \pi\}$, $\cos^{-1}(\sqrt{3}/3) = \pi/6$.

(c) You must look for an angle x in the interval $-\pi/2 \le x \le \pi/2$ for which $\tan x = -1$. Because $\tan(-\pi/4) = -1$ and $-\pi/4 \in \{-\pi/2 \le x \le \pi/2\}$, $\tan^{-1}(-1) = -\pi/4$.

In the following examples you will need to recognise the relationship between the trigonometric function and its inverse. In such cases it will be helpful to draw a diagram.

Example 2

Simplify each of the following without using a calculator:

(a) sin [arccos (3/5)] (b) tan [arcsin (− 5/13)]

Solution

(a) Let θ be the angle for which $\cos\theta = 3/5$. Because $\cos\theta$ is positive, θ is in the first quadrant. Draw a right triangle with sides as shown in Figure 19.30.

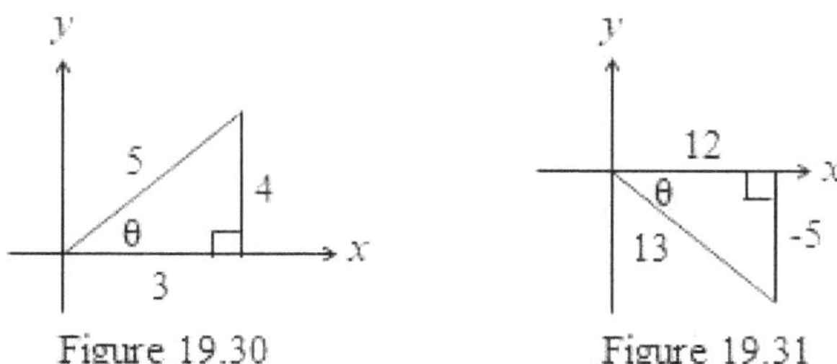

Figure 19.30 Figure 19.31

From Figure 19.30, we see that

$$\sin[\text{arc}\cos(3/5)] = \sin\theta = 4/5$$

(b) Let θ be the angle for which $\sin\theta = -5/13$. Because $\sin\theta$ is negative, θ is in the fourth quadrant. Draw a right triangle with sides as shown in Figure 19.31. From Figure 19.31, we see that

$$\tan[\text{arc}\sin(-5/13)] = \tan\theta = -5/12.$$

Example 3

Find $\sin\theta$ if $\sin^{-1}(5/13) - \cos^{-1}(3/5)$

Solution

Let $\alpha = \sin^{-1}(5/13)$ and $\beta = \cos^{-1}(3/5)$

So, $\sin\alpha = 5/13$ and $\cos\beta = 3/5$.

Draw right triangles with sides as shown in Figure 19.32

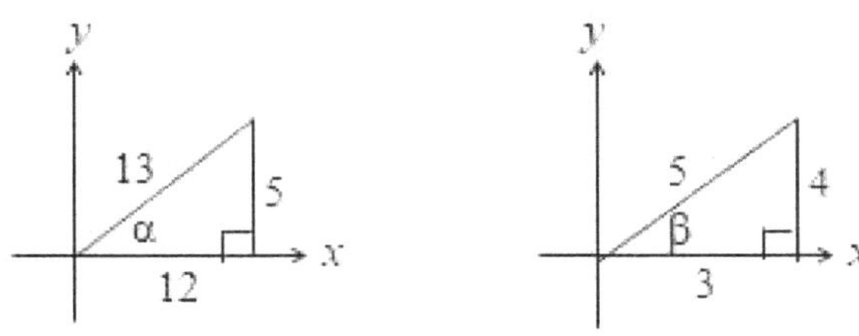

Figure 19.32

Now $\theta = \alpha - \beta$

Hence, $\sin\theta = \sin(\alpha - \beta)$

$$= \sin\alpha\cos\beta - \cos\alpha\sin\beta$$

$$= \left(\frac{5}{13}\right)\left(\frac{3}{5}\right) - \left(\frac{12}{13}\right)\left(\frac{4}{5}\right)$$

$$= \frac{15}{65} - \frac{48}{65}$$

$$= -\frac{33}{65}$$

Exercise 19.7

Evaluate each of the following without using a calculator:

1. arcsin ½ 2. arccos ½ 3. arctan 1

4. arcsin 0 5. arccos 0 6. arcsin (− 1)

7. arctan(−√3) 8. arcos(−√3/3) 9. arctan(−√3/3)

Simplify each of the following without using a calculator:

10. cos [arcsin (12/13)]

11. sin [arccos (8/17)]

12. sec [arctan (− ¾)]

13. Find sin θ if $\theta = \sin^{-1}(5/13) - \cos^{-1}(8/17)$

14. Find tan θ if $\theta = \tan^{-1}(3/4) + \tan^{-1}(4/5)$

Review Exercises

Convert the following degree measures to radians.

1. 75° 2. −45° 3. −105° 4. 330°

Convert the following radian measures to degrees.

5. π/3 6. 11π/4 7. −π/3 8. 5π

Find the reference angle for the given angles.

9. 165° 10. 310° 11. −218° 12. − 640°

State the quadrant in which the angle lies.

13. 183° 14. –208° 15. 697° 16. –305°

Evaluate the following trigonometric functions without using a calculator.

17. tan 135° 18. sin 240° 19. cos (– 150°) 20. sin (– 570°)

Simplify the following expressions.

21. $\cos\theta \tan\theta$

22. $(\csc\theta - 1)(\csc\theta + 1)$

23. $\frac{\sin^2\theta}{1+\cos\theta} + \frac{\sin^2\theta}{1-\cos\theta}$

24. $\frac{1}{\cos^2\theta} - \frac{1}{\cot^2\theta}$

Verify the identity.

25. $(1 - \sin^2\theta)\sec^2\theta = 1$

26. $(1 - \sin^2\theta)\csc^2\theta = \cot^2\theta$

27. $\frac{1}{1-\cos^2\theta} - \frac{1}{\sec^2\theta - 1} = 1$

28. $\frac{\tan\theta}{\sec\theta} - \frac{\tan\theta}{\sec\theta + 1} = 2\cot\theta$

Find all values of x from 0 to 2π inclusive that satisfy each of the following equations.

29. $\sec^2 x + \tan x = 1$

30. $3 \csc x = 4 \sin x$

31. $2\sin^2 x - 5\sin x - 3 = 0$

32. $2\cos^2 x = 3 - 5\cos x$

33. $\sec^2 x + \tan^2 x = 7$

34. $\cot x + \tan x = 2\sec x$

Find all values of x from $-\pi$ to π inclusive that satisfy the following equations.

35. $2\cos^2 x + 3\cos x + 1 = 0$

36. $\cos x = 4\sec x + 3$

37. $\sin^2 x - \cos^2 x = \cos x$

38. $2\sin^2 x = 3\cos x$

39. $2\cos^2 x + 4\sin^2 x = 3$

40. $\sec^2 x = 3\tan^2 x - 1$

Find all values of x from 0 to 2π inclusive that satisfy each of the following equations.

41. $\sin 2x + \cos x = 0$

42. $2\cos 2x + 8\sin^2 x - 5 = 0$

43. $\cos 2x - 7\cos x = 3$

44. $2\cos x - \sin 2x = 0$

Find the maximum and minimum values of the following expressions, giving the smallest positive or zero value of θ for which they occur:

45. $2\cos\frac{1}{2}\theta°$ 46. $-2\sin\theta°$

47. $3 - 2\sin\theta°$ 48. $1 - 2\cos 3\theta°$

Find the amplitude and period.

49. $y = 5 \sin 2x$
50. $y = -3 \cos 2x$
51. $y = 2 \sin 3x$
52. $y = -2 \cos 4x$

Find the amplitude, period, phase shift and vertical shift of each of the following and graph each function over a two-period interval.

53. $y = 5 \sin 2x$
54. $y = -3 \cos 4x$
55. $y = 2 \sin (3x - \pi)$
56. $y = -\cos (x + \pi/4)$
57. $y = 2 + 3 \sin 2x$
58. $y = 1 - 2 \cos (3x - 2\pi)$

Sketch for $0° \leq x \leq 360°$ the graph of each of the following functions:

59. $y = 2 + 3 \sin 2x$
60. $y = 1 - 2 \cos 3x$

Evaluate each of the following without using a calculator:

61. $\arcsin(-\frac{1}{2})$
62. $\arccos(-\frac{1}{2})$
63. $\arctan(\sqrt{3}/3)$
64. $\arcsin(-\sqrt{3}/2)$
65. $\arccos(\sqrt{2}/2)$
66. $\arctan(\sqrt{3})$

Simplify each of the following without using a calculator.

67. $\cos [\arcsin (5/13)]$
68. $\sin [\arctan (-5/12)]$
69. $\tan [\arccos (8/17)]$
70. $\sin[\arccos(\sqrt{3}/3)]$

20 Vectors

20.1 Vector Algebra

Quantities such as displacement, force and velocity can be described by stating both a magnitude (size or length) and a direction. Such quantities are called vectors. Other quantities, such as area, time and temperature can be represented by a single real number. Such quantities have only magnitude, and are called scalar quantities.

A vector can be represented graphically as a directed line segment, that is a line of a given length with a specific direction, as shown in Figure 20.1.

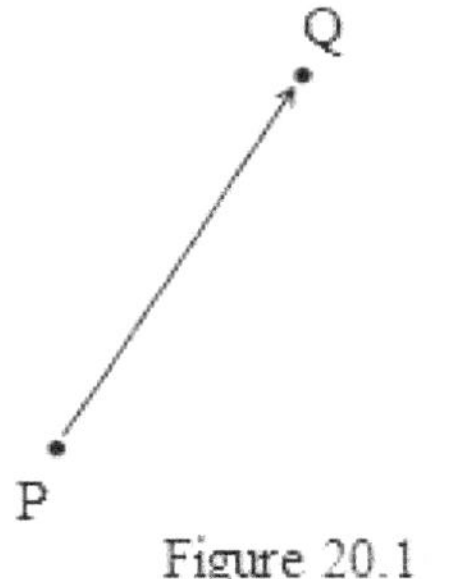

Figure 20.1

Figure 20.1 shows a vector which has its initial point at P and terminal point at Q. This vector is denoted as $\overrightarrow{PQ}$. The arrow indicates the direction of the vector. The length of the directed line segment $\overrightarrow{PQ}$ is called the magnitude of the vector, and is denoted by $|\overrightarrow{PQ}|$ or PQ. In print, we usually denote vectors by boldface characters such as **PQ** or ***a***. In handwriting you can indicate a vector as a character with an arrow on it or a character with a line under it, such as $\vec{A}$, $\vec{a}$ or $\underline{a}$.

Addition of Vectors

The addition of two vectors can be defined in two ways.

The Triangle Law

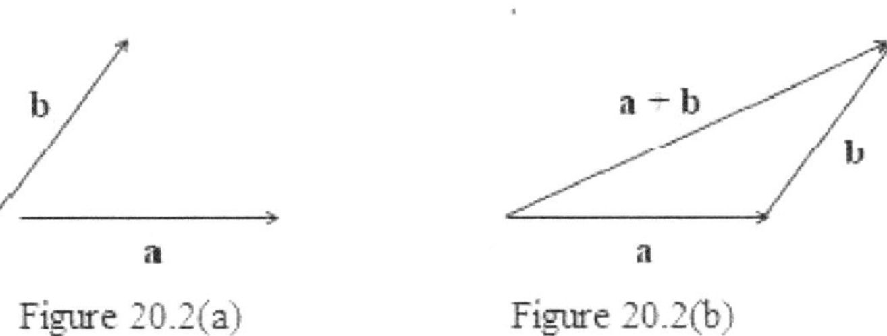

Figure 20.2(a) Figure 20.2(b)

To add the two vectors ***a*** and ***b*** (see Figure 20.2(a)), we place the initial point of ***b*** at the terminal point of ***a*** as shown in Figure 20.2(b). The sum ***a*** + ***b*** is formed by joining the initial point of $\boldsymbol{a}$ with the terminal point of $\boldsymbol{b}$. This method is called the triangle law for vector addition.

The Parallelogram Law

Given two vectors $\boldsymbol{a}$ and $\boldsymbol{b}$ with initial point at the same point (see Figure 20.3(a)) to add, we translate the initial point of vector $\boldsymbol{b}$ to the terminal point of vector $\boldsymbol{a}$ and complete the parallelogram. The vector lying along the diagonal, as shown in Figure 20.3(b) is the sum $\boldsymbol{a} + \boldsymbol{b}$ of the two vectors.

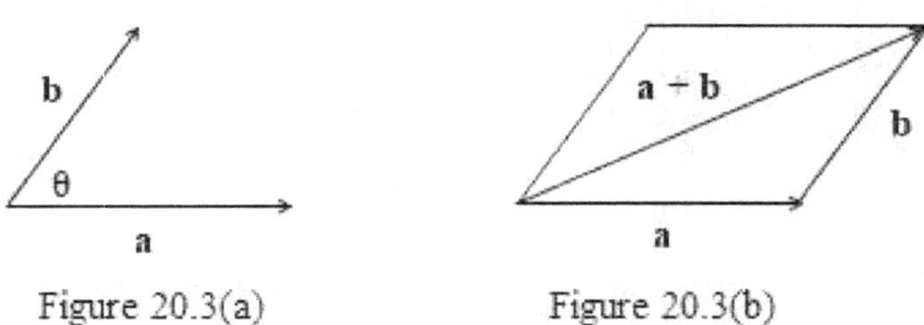

Figure 20.3(a) Figure 20.3(b)

This method is called the parallelogram law for vector addition.

Zero Vectors

A vector with a zero magnitude and has no specific direction is called a zero vector, denoted **O**.

Equal Vectors

Two nonzero vectors are equal if they have the same magnitude and the same direction regardless of the position of their initial points.

Component Form of a Vector

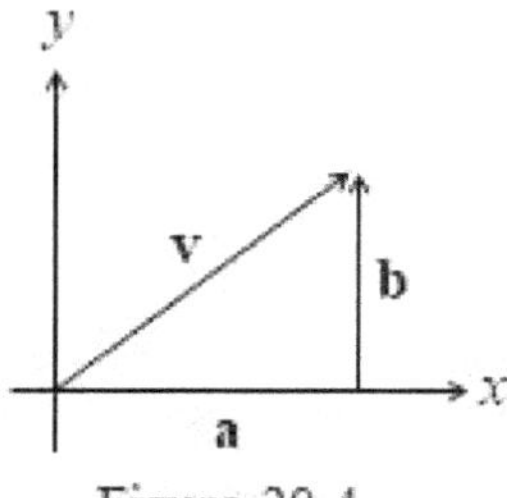

Figure 20.4

The vector $\boldsymbol{v}$, shown in Figure 20,4, can be written in terms of its movement in the x- and y-directions as $\boldsymbol{v} = \langle a, b \rangle$. In this form the vector is said to be written in component form. We call a the horizontal component of $\boldsymbol{v}$ and b the vertical component. In the x-direction, movement to the right is indicated as positive and movement to the left is indicated as negative. In the y-direction, movement up is positive and movement down is negative.

Adding Vectors in Component Form

The sum of two vectors in component form is obtained by adding their corresponding components. If $\boldsymbol{a} = < a_1, a_2 >$ and $\boldsymbol{b} = < b_1, b_2 >$, then $\boldsymbol{a} + \boldsymbol{b} = < a_1 + b_1, a_2 + b_2 >$.

Example 1

Given $\boldsymbol{a} = < -3, 2 >$ and $\boldsymbol{b} = < 5, -3 >$ find:

(a) $\boldsymbol{a} + \boldsymbol{b}$ (b) $\boldsymbol{b} + \boldsymbol{a}$

Solution

(a) $\boldsymbol{a} + \boldsymbol{b} = < -3, 2 > + < 5, -3 > = < 2, -1 >$

(b) $\boldsymbol{b} + \boldsymbol{a} = < 5, -3 > + < -3, 2 > = < 2, -1 >$

Negative Vectors

A vector having direction opposite to that of vector $\boldsymbol{v}$ but having the same magnitude is called a negative vector of $\boldsymbol{v}$, denoted by $-\boldsymbol{v}$. If $\boldsymbol{v} = < v_1, v_2 >$, then $-\boldsymbol{v} = < -v_1, -v_2 >$.

Example 2

Given $\boldsymbol{a} = < -3, 4 >$ and $\boldsymbol{b} = < -2, -1 >$ find:

(a) $-\boldsymbol{a}$ (b) $-\boldsymbol{b}$

Solution

(a) $-\boldsymbol{a} = -< -3, 4 > = < 3, -4 >$

(b) $-\boldsymbol{b} = -< -2, -1 > = < 2, 1 >$

Subtraction of Vectors

The difference of $\boldsymbol{a}$ and $\boldsymbol{b}$ is $\boldsymbol{a} - \boldsymbol{b}$, and is defined as $\boldsymbol{a} - \boldsymbol{b} = \boldsymbol{a} + (-\boldsymbol{b})$. If $\boldsymbol{a} = < a_1, a_2 >$ and $\boldsymbol{b} = < b_1, b_2 >$ then

$$\begin{aligned} \boldsymbol{a} - \boldsymbol{b} &= \boldsymbol{a} + (-\boldsymbol{b}) \\ &= < a_1, a_2 > + < -b_1, -b_2 > \\ &= < a_1 - b_1, a_2 - b_2 > \end{aligned}$$

Example 3

Given $\boldsymbol{a} = < -3, 2 >$ and $\boldsymbol{b} = < -5, 6 >$ find;

(a) $\boldsymbol{a} - \boldsymbol{b}$ (b) $\boldsymbol{b} - \boldsymbol{a}$

Solution

(a) $\boldsymbol{a} - \boldsymbol{b} = <-3, 2> - <-5, 6> = <2, -4>$

(b) $\boldsymbol{b} - \boldsymbol{a} = <-5, 6> - <-3, 2> = <-2, 4>$

Multiplying a Vector by a Scalar

If we multiply a vector $\boldsymbol{a}$ by a scalar (a real number) $k > 0$, the product $k\boldsymbol{a}$ has the same direction as $\boldsymbol{a}$, and is k times as long as $\boldsymbol{a}$, as shown in Figure 20.5.

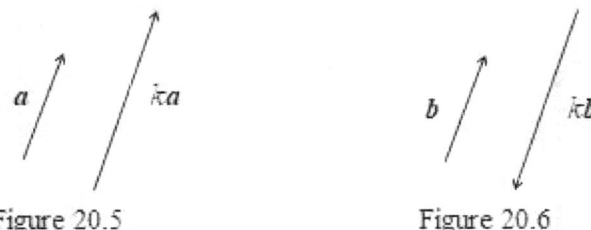

Figure 20.5 Figure 20.6

If we multiply a vector $\boldsymbol{b}$ by a scalar $k < 0$, the product $k\boldsymbol{b}$ has the direction opposite that of $\boldsymbol{b}$, and is $|k|$ times as long as $\boldsymbol{b}$, as shown in Figure 20.6.

To multiply the vector $\boldsymbol{a} = < a_1, a_2 >$ by a scalar k, multiply each component by k. That is, $k\boldsymbol{a} = < ka_1, ka_2 >$.

Example 4

Given $\boldsymbol{a} = < 1, -2 >$ and $\boldsymbol{b} = < 4, -6 >$ find:

(a) $-3\boldsymbol{a}$ (b) $\frac{1}{2}\boldsymbol{b}$

Solution

(a) $-3\boldsymbol{a} = -3 < 1, -2 > = < -3, 6 >$

(b) $\frac{1}{2}\boldsymbol{b} = \frac{1}{2} < 4, -6 > = < 2, -3 >$

Parallel Vectors

Two vectors $\boldsymbol{a}$ and $\boldsymbol{b}$ are parallel if $\boldsymbol{a} = k\boldsymbol{b}$, where k is any scalar. Parallel vectors may have the same direction or opposite direction.

Example 5

Determine whether or not $\boldsymbol{a} = < 3, 2 >$ and $\boldsymbol{b} = < -9, -6 >$ are parallel

Solution

If $\langle -9, -6 \rangle = k\langle 3, 2 \rangle$, then $-9 = 3k$, so $k = -3$ and $-6 = 2k$ giving $k = -3$.

Since $\langle -9, -6 \rangle = -3\langle 3, 2 \rangle$ vector ***a*** and ***b*** are parallel.

Magnitude of Vectors

The magnitude of a vector is represented geometrically by the length of a line segment.

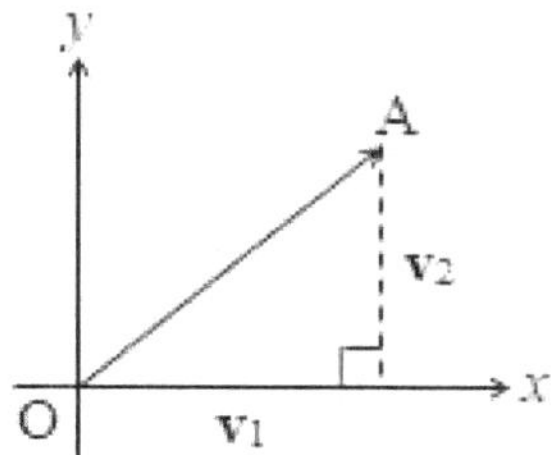

Figure 20.7

Using the Pythagorean Theorem, we have

$$|\boldsymbol{OA}| = \sqrt{v_1{}^2 + v_2{}^2}$$

Example 6

Find the magnitude of the vector

(a) $\boldsymbol{a} = < -8, 15 >$ (b) $\boldsymbol{b} = < 1, -7 >$

Solution

(a)

$$|\boldsymbol{a}| = \sqrt{(-8)^2 + 15^2}$$
$$= \sqrt{289}$$
$$= 17$$

(b)

$$|\boldsymbol{b}| = \sqrt{1^2 + (-7)^2}$$
$$= \sqrt{50}$$
$$= 5\sqrt{2}$$

Unit Vectors

A unit vector is a vector of unit length. If vector ***a*** has magnitude $|\boldsymbol{a}| > 0$. then the unit vector in the direction of ***a***, is obtained by dividing ***a*** by its magnitude.

$$\text{Unit vector} = \frac{1}{|\boldsymbol{a}|}\boldsymbol{a}$$

The unit vector in the direction of $\boldsymbol{a} = < a_1, a_2 >$, is

$$\text{unit vector} = \frac{1}{\sqrt{a_1 + a_2}} \langle a_1, a_2 \rangle.$$

Example 7

Find a unit vector in the same direction as $\boldsymbol{a} = < 3, -4 >$.

Solution

Begin by finding the magnitude of the vector.

$$|\boldsymbol{a}| = \sqrt{3^2 + (-4)^2}$$
$$= \sqrt{25}$$
$$= 5$$

So, the unit vector in direction of $\boldsymbol{a}$ is

$$\frac{1}{5} \langle 3, -4 \rangle = \langle 3/5, -4/5 \rangle$$

Unit Vectors along the Axes

Occasionally, we find it convenient to write vectors in terms of two standard unit vectors in the x- and y- directions, denoted by $\boldsymbol{i}$ and $\boldsymbol{j}$ respectively, as shown in Figure 20.8. We define the standard basis vector $\boldsymbol{i}$ and $\boldsymbol{j}$ by $\boldsymbol{i} = < 1, 0 >$ and $\boldsymbol{j} = < 0, 1 >$.

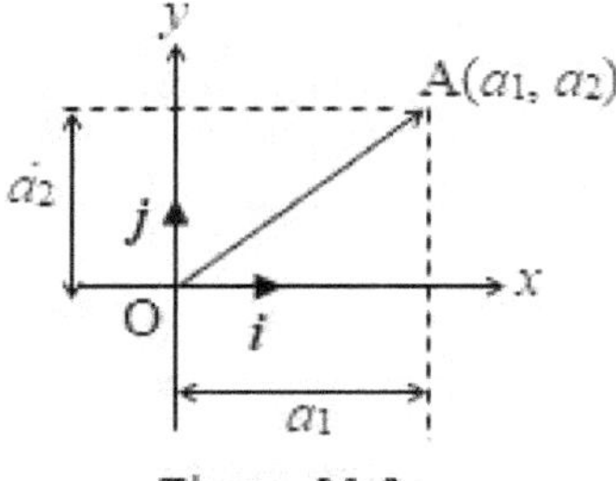

Figure 20.8

We can represent vector **OA** in terms of $\boldsymbol{i}$ and $\boldsymbol{j}$ as follows

$$\mathbf{OA} = a_1\boldsymbol{i} + a_2\boldsymbol{j}$$

This is called a linear combination of the vectors $\boldsymbol{i}$ and $\boldsymbol{j}$. The real number a_1 is called the horizontal component of vector **OA**, and a_2 is the vertical component.

Any vector in a plane can be written as a linear combination of the standard unit vectors $\boldsymbol{i}$ and $\boldsymbol{j}$. Note that, you can rewrite $a_1\boldsymbol{i} + a_2\boldsymbol{j}$ as $< a_1, a_2 >$.

Using Vectors in the form $a_1\boldsymbol{i} + a_2\boldsymbol{j}$

Example 8

Given that $\boldsymbol{a} = 3\boldsymbol{i} + 2\boldsymbol{j}$ and $\boldsymbol{b} = -5\boldsymbol{i} + \boldsymbol{j}$, find $\boldsymbol{a} + \boldsymbol{b}$.

Solution

$$\begin{aligned} \boldsymbol{a} + \boldsymbol{b} &= (3\boldsymbol{i} + 2\boldsymbol{j}) + (-5\boldsymbol{i} + \boldsymbol{j}) \\ &= (3 - 5)\boldsymbol{i} + (2 + 1)\boldsymbol{j} \\ &= -2\boldsymbol{i} + 3\boldsymbol{j} \end{aligned}$$

Example 9

Given that $\boldsymbol{a} = 3\boldsymbol{i} - 4\boldsymbol{j}$, find $|-3\boldsymbol{a}|$.

Solution

$$\begin{aligned} -3\boldsymbol{a} &= -3(3\boldsymbol{i} - 4\boldsymbol{j}) \\ &= -9\boldsymbol{i} + 12\boldsymbol{j} \\ |-3\boldsymbol{a}| &= \sqrt{(-9)^2 + 12^2} \\ &= \sqrt{225} \\ &= 15 \end{aligned}$$

Example 10

Find a unit vector in the same direction as $\boldsymbol{a} = -5\boldsymbol{i} + 12\boldsymbol{j}$.

Solution

Begin by finding the magnitude of the vector.

$$\begin{aligned} |\boldsymbol{a}| &= \sqrt{(-5)^2 + 12^2} \\ &= \sqrt{169} \\ &= 13 \end{aligned}$$

So, the unit vector is $-\frac{5}{13}\boldsymbol{i} + \frac{12}{13}\boldsymbol{j}$

Example 11

Given that $\boldsymbol{a} = -\boldsymbol{i} + 3\boldsymbol{j}$ and $\boldsymbol{b} = -2\boldsymbol{i} + \boldsymbol{j}$, find $|2\boldsymbol{a} - 3\boldsymbol{b}|$.

Solution

$$2\boldsymbol{a} - 3\boldsymbol{b} = 2(-\boldsymbol{i} + 3\boldsymbol{j}) - 3(-2\boldsymbol{i} + \boldsymbol{j})$$

$$= -2\boldsymbol{i} + 6\boldsymbol{j} + 6\boldsymbol{i} - 3\boldsymbol{j}$$

$$= 4\boldsymbol{i} + 3\boldsymbol{j}$$

$$|4\boldsymbol{i} + 3\boldsymbol{j}| = \sqrt{4^2 + 3^2}$$

$$= \sqrt{25}$$

$$= 5$$

Properties of Vectors

If $\boldsymbol{a}$, $\boldsymbol{b}$ and $\boldsymbol{c}$ are vectors, and k and m are scalars then the following holds:

1. $\boldsymbol{a} + \boldsymbol{b} = \boldsymbol{b} + \boldsymbol{a}$ — Commutative property
2. $(\boldsymbol{a} + \boldsymbol{b}) + \boldsymbol{c} = \boldsymbol{a} + (\boldsymbol{b} + \boldsymbol{c})$ — Associative property
3. $k(m\boldsymbol{a}) = (km)\boldsymbol{a}$ — Associative property
4. $(k + m)\boldsymbol{a} = k\boldsymbol{a} + m\boldsymbol{a}$ — Distributive property
5. $k(\boldsymbol{a} + \boldsymbol{b}) = k\boldsymbol{a} + m\boldsymbol{b}$ — Distributive property

Position Vectors

A vector $\boldsymbol{a}$ with its initial point located at the origin and its terminal point at (a_1, a_2) is called the position vector of the point (a_1, a_2), denoted by $< a_1, a_2 >$. Position vectors are uniquely represented by the coordinates of their terminal points.

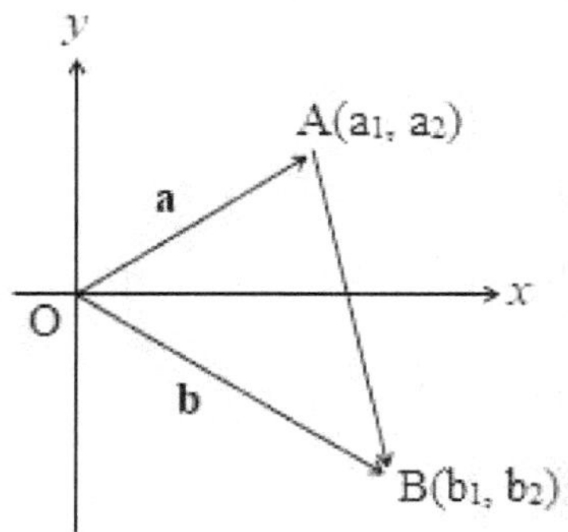

Figure 20.9

You can see from Figure 20.9 that the position vector of A is $\boldsymbol{a} = < a_1, a_2 >$, and the position vector of B is $\boldsymbol{b} = < b_1, b_2 >$. The vector with initial point $A(a_1, a_2)$ and terminal $B(b_1, b_2)$ can be obtained by using the position vectors of the points A and B. From Figure 20.9 you can see that

$$\mathbf{AB} = \mathbf{AO} + \mathbf{OB}$$

$$= \mathbf{OB} + (-\mathbf{OA})$$

$$= < b_1, b_2 > - < a_1, a_2 >$$

$$= < b_1 - a_1, b_2 - a_2 >$$

Example 12

Given that A (−7, 5) and B (−2, 3), find **AB**.

Solution

$$\mathbf{AB} = \mathbf{OB} - \mathbf{OA}$$

$$= < -2, 3 > - < -7, 5 >$$

$$= < 5, -2 >$$

Example 13

Given that A (−2, 3) and B (5, 2), find |**AB**|.

Solution

$$\mathbf{AB} = \mathbf{OB} - \mathbf{OA}$$

$$= < 5, 2 > - < -2, 3 >$$

$$= < 7, -1 >$$

$$|\mathbf{AB}| = \sqrt{7^2 + (-1)^2}$$

$$= \sqrt{50}$$

$$= 2\sqrt{5}$$

Example 14

Given that P (−3, 2) and **PQ** = < 5, 3 >, find the coordinates of Q.

Solution

$$\mathbf{PQ} = \mathbf{OQ} - \mathbf{OP}$$

$$\mathbf{OQ} = \mathbf{PQ} + \mathbf{OP}$$

$$= < 5, 3 > + < -3, 2 >$$

$$= < 2, 5 >$$

The coordinates of Q is (2, 5)

Applications of Vectors

Some proves in geometry can be simplified by using vector algebra.

Example 15

In the triangle ABC, M and N are the midpoint of the sides AB and AC respectively. Show that MN is parallel to BC and that MN = ½ BC.

Solution

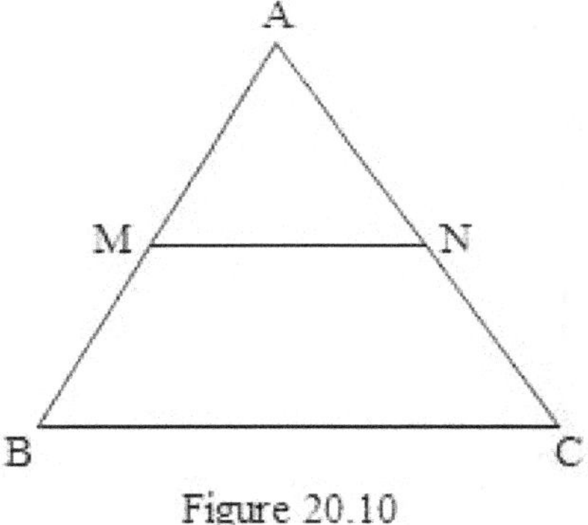

Figure 20.10

Figure 20.10 shows a triangle ABC.

Let $\mathbf{AB} = \boldsymbol{b}$ and $\mathbf{AC} = \boldsymbol{a}$

Then $\mathbf{AM} = \frac{1}{2}\boldsymbol{b}$ and $\mathbf{AN} = \frac{1}{2}\boldsymbol{a}$

$$\mathbf{MN} = \mathbf{MA} + \mathbf{AN}$$

$$= \mathbf{AN} - \mathbf{AM}$$

$$= \tfrac{1}{2}(\boldsymbol{a} - \boldsymbol{b})$$

$$= \tfrac{1}{2}\mathbf{BC}$$

∴ MN is parallel to BC and MN = ½ BC.

Example 16

Prove that the diagonals of a parallelogram bisect each other.

Solution

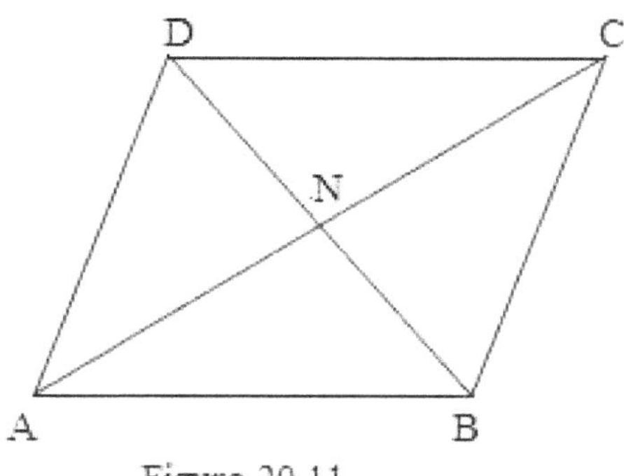

Figure 20.11

Figure 20.11 shows a parallelogram ABCD.

Let **AD** = $\boldsymbol{a}$ and **AB** = $\boldsymbol{b}$.

Now **BD** = **BA** + **AD** = $\boldsymbol{a} - \boldsymbol{b}$

and **AC** = **AB** + **BC** = $\boldsymbol{b} + \boldsymbol{a}$

If **BN** = k **BD and AN** = m**AC**, where k and m are scalars, then

$$\mathbf{AB} = \mathbf{AN} - \mathbf{BN}$$

$$= m\,\mathbf{AC} - k\,\mathbf{BD}$$

$$= m(\boldsymbol{b} + \boldsymbol{a}) - k(\boldsymbol{a} - \boldsymbol{b})$$

$$\boldsymbol{b} = (m - k)\boldsymbol{a} + (m + k)\boldsymbol{b}$$

From this equation, we see that

$$m - k = 0 \qquad (1)$$

$$m + k = 1 \qquad (2)$$

Adding Equation (1) and Equation (2) gives $m = ½$.

Also, substituting $m = ½$ into Equation (1) gives $k = ½$.

So, N is the midpoint of both diagonals, hence the diagonals bisect each other.

Composition and Decomposition of Vectors

Composition of Vectors

We can compose, that is add, two or more vectors to a single called the resultant vector. Consider two vectors $\boldsymbol{v}_1$ and $\boldsymbol{v}_2$ shown in Figure 20.12(a).

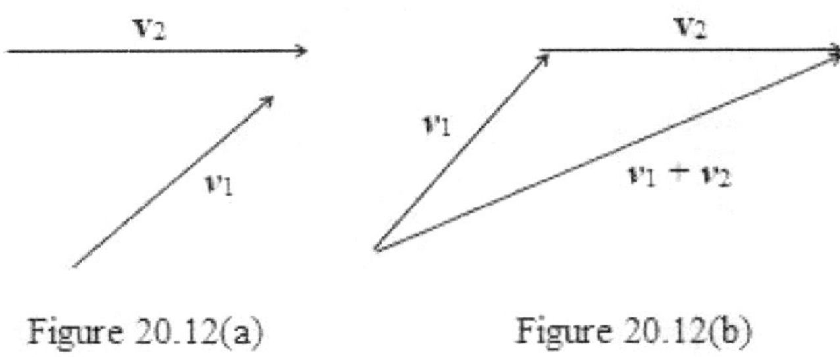

Figure 20.12(a) Figure 20.12(b)

You can see that the resultant vector is the vector joining the initial point of $\boldsymbol{v}_1$ to the terminal point of $\boldsymbol{v}_2$, as shown in Figure 20.12(b). In general, the resultant of two or more vectors can be found by drawing directed line segments representing the vectors in magnitude and direction to form a polygon. The line segment which completes the polygon represents the resultant vector. This result apply to vector such as velocity and others.

Decomposing a Vector into Components

Occasionally, it may be useful to decompose a vector into the sum of two perpendicular vector components as shown in Figure 20.13.

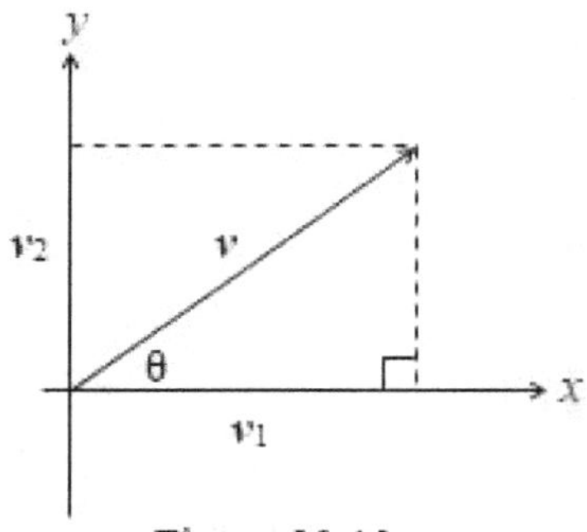

Figure 20.13

Figure 20.13 shows a vector $\boldsymbol{v}$ decomposed into orthogonal components $\boldsymbol{v}_1$ and $\boldsymbol{v}_2$ along the x- direction and the y-direction respectively. Vector $\boldsymbol{v}$ can now be written as $\boldsymbol{v} = \boldsymbol{v}_1 + \boldsymbol{v}_2$.

From elementary trigonometry, we have $\boldsymbol{v}_1 = |\boldsymbol{v}| \cos \theta$ and $\boldsymbol{v}_2 = |\boldsymbol{v}| \sin \theta$, so the component form of the vector $\boldsymbol{v}$ is written as $\boldsymbol{v} = < |\boldsymbol{v}| \cos \theta, |\boldsymbol{v}| \sin \theta >$

The magnitude of $\boldsymbol{v}$ is given by $|\boldsymbol{v}| = \sqrt{v_1{}^2 + v_2{}^2}$. From Figure 20.13, you can see that $\tan \theta = v_2/v_1$. Hence, the angle which $\boldsymbol{v}$ makes with the horizontal direction is $\tan^{-1}(v_2/v_1)$.

The resultant of two or more vectors can be obtained by decomposing each vector and then adding the corresponding components.

Example 17

Two particles travel with velocities $\boldsymbol{u}$ = (6 ms^{-1}, 060°) and $\boldsymbol{v}$ = (12 ms^{-1}, 120°), calculate the magnitude and direction of the resultant velocity.

Solution

Begin by finding the component form of the vectors $\boldsymbol{u}$ and $\boldsymbol{v}$. Then add the vectors.

$$\boldsymbol{u} = \,< 6\cos 30°, 6\sin 30° > \,=\, < 3\sqrt{3}, 3 >$$

$$\boldsymbol{v} = \,< 12\cos 30°, -12\sin 30° > \,=\, < 6\sqrt{36}, -6 >$$

$$\boldsymbol{u} + \boldsymbol{v} = \,< 3\sqrt{3}, 3 > + < 6\sqrt{3}, -6 >$$

$$= \,< 9\sqrt{3}, -3 >$$

The magnitude of the resultant velocity is

$$|\boldsymbol{v}| = \sqrt{(9\sqrt{3})^2 + (-3)^2}$$

$$= \sqrt{252}$$

$$= 15.9$$

You can use the diagram shown in Figure 20.14 to find direction of the resultant velocity.

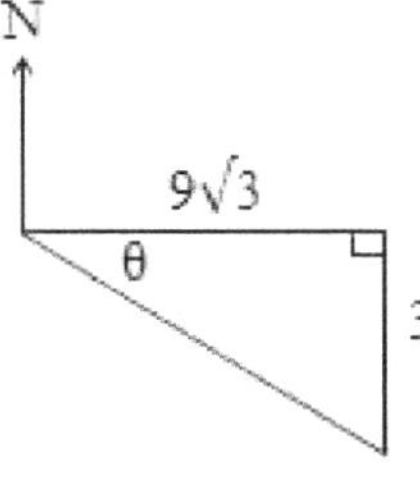

Figure 20.14

Let θ represents the directional angle of the resultant velocity. Using Figure 20.14, we have

$$\tan\theta = \frac{3}{9\sqrt{3}} = 0.1925$$

$$\theta = 10.9°$$

The direction of the resultant velocity is (90° + 10.9°) = 100.9°

Relative Velocity

There are many situations where a velocity is made up of two other velocities. Relative velocity is the velocity of one body as seen from another.

Suppose an airplane travelling with velocity ***a*** north encounter a wind blowing from the west with velocity ***b*** as illustrated in Figure 20.15(a).

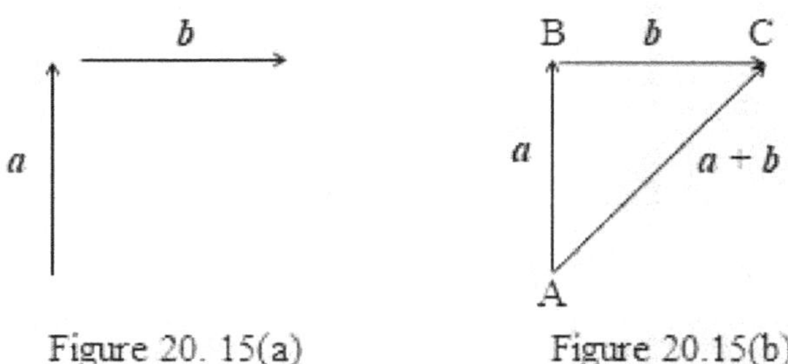

Figure 20. 15(a) Figure 20.15(b)

An observer on the ground would see the airplane as moving along **AC**, as shown in Figure 20.15 (b). The vector **AC** represents the velocity of the plane relative to the ground, called the ground speed, and the direction of the plane as seen from the ground, called the track. Figure 20.15(b) shows that the velocity of the airplane relative to the ground is the vector sum of the velocity of the airplane relative to the air, called the airspeed and the velocity of the wind relative to the ground, called the wind speed. The direction in which the pilot steers an airplane is called its course. The various velocities and directions are summarized in the table below.

Velocity	Air speed	Wind speed	Ground speed
Direction	Course	Wind direction	Track

The effect of the river current upon a boat is similar to the effect of the wind upon a airplane. The velocity of a boat as seen by an observer on the ground is the vector sum of the velocity of the boat in the water and the velocity of the current.

Note that, a current is described as moving towards a certain direction whereas a wind is described as coming from a certain direction.

Example 18

An airplane moves on a bearing of 315^0 at 120 km h^{-1} relative to the ground due to the fact that there is a wind blowing from the west of 50 km h^{-1} relative to the ground. How fast and in what direction will the airplane have travelled in still air?

Solution

Let $\boldsymbol{v}_1$ = the velocity of the airplane relative to the ground, $\boldsymbol{v}_2$ = the velocity of the wind and $\boldsymbol{v}$ = the velocity of the airplane in still air.

The diagram in Figure 20.16, illustrates the velocity vectors for the airplane and the wind.

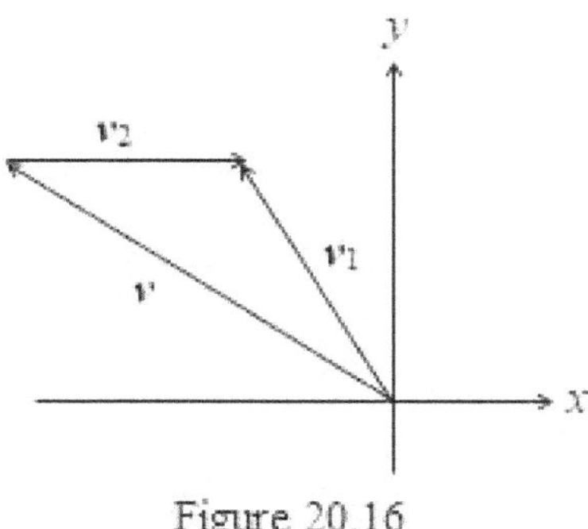

Figure 20.16

The velocity of the airplane relative to the ground is given by

$$\boldsymbol{v}_1 = <-\ 120 \cos 45°,\ 120 \sin 45° >$$

$$= < -60\sqrt{2},\ 60\sqrt{2} >$$

and the velocity of the wind is $\boldsymbol{v}_2 = < 50, 0 >$.

The velocity of the plane in still air is given by

$$\boldsymbol{v} = \boldsymbol{v}_1 + (-\ \boldsymbol{v}_2)$$

$$= < -60\sqrt{2},\ 60\sqrt{2} > + < -\ 50, 0 >$$

$$= < -60\sqrt{2} - 50, 60\sqrt{2} >$$

$$= < -\ 134.85,\ 84.85 >$$

So, the magnitude of $\boldsymbol{v}$ is

$$|\boldsymbol{v}| = \sqrt{(-134.85)^2 + 84.65^2}$$

$$= \sqrt{25384.8}$$

$$= 159.3$$

The speed of the airplane in still air is 159.3 kilometers.

You can use the diagram shown in Figure 20.17 to find the direction of the resultant velocity.

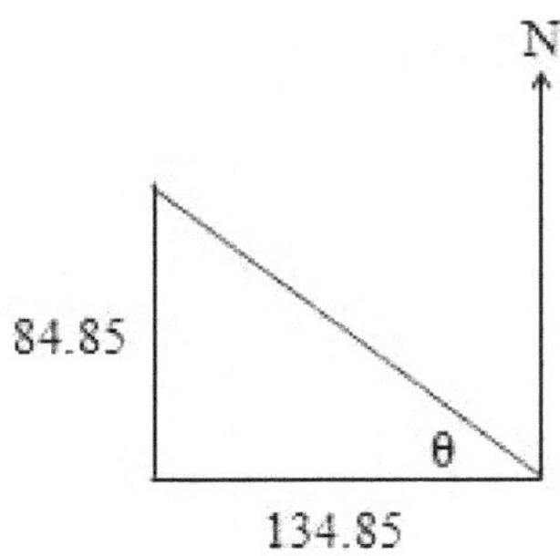

Figure 20.17

If θ is the direction angle of the flight path, we have

$$\tan\theta = \frac{84.85}{134.85}$$

$$= 0.6292$$

$$\theta = 32.2°$$

So, the direction of the plane in still air is (270° + 32.2°) = 302.2°

Exercise 20.1

Find:

1. $< 2, 3 > + < 1, -4 >$
2. $< 3, -1 > + < -4, 3 >$
3. $< 3, -4 > + < -3, 5 >$
4. $< 2, 4 > + < -3, -5 >$
5. $< -7, 2 > + < 1, 5 >$
6. $< -2, -3 > + < 5, 1 >$

Find the values of x and y from the following vector equations.

7. $< 3, x > + < y, 2 > = < 8, 7 >$
8. $< 1, 2 > + < y, -5 > = < 3, x >$
9. $< 2x + 3, x > + < -3, 3y - 2 > = < y, 5 >$
10. $< 3x, 2x - y > + < y, -7 > = < 2 - 3y, 2y >$
11. Given that $\boldsymbol{a} = < 2, -1 >$, $\boldsymbol{b} = < x, y >$, $\boldsymbol{c} = < -1, 3 >$ and $\boldsymbol{a} + \boldsymbol{b} = \boldsymbol{c}$, find the values of x and y.

Find:

12. $- < 1, 3 >$
13. $- < 3, -4 >$
14. $- < -2, 5 >$
15. $- < -2, 0 >$

Given that the vectors $\boldsymbol{a} = < 2, 3 >$, $\boldsymbol{b} = < -2, 4 >$ and $\boldsymbol{c} = < 4, 1 >$, find

16. $-\boldsymbol{a}$ 17. $-\boldsymbol{b}$ 18. $-\boldsymbol{c} + \boldsymbol{b}$ 19. $-\boldsymbol{b} + \boldsymbol{a}$ 20. $\boldsymbol{b} - \boldsymbol{a}$

Find:

21. $< 3, 5 > - < 1, 2 >$ 22. $< 5, 2 > - < -2, 3 >$

23. $< 1, 2 > - < 1, -2 >$ 24. $< -7, -6 > - < -5, 3 >$

25. $< -2, 1 > - < 1, 2 >$ 26. $< -2, 1 > - < -8, -4 >$

Given that $\boldsymbol{a} = < 2, -3 >$, $\boldsymbol{b} = < 3 - 4 >$ and $\boldsymbol{c} = < 4, 1 >$, find:

27. $\boldsymbol{b} - \boldsymbol{a}$ 28. $\boldsymbol{c} - \boldsymbol{a}$ 29. $\boldsymbol{b} - \boldsymbol{c}$ 30. $\boldsymbol{a} - \boldsymbol{b}$

Find:

31. $2< 3, -1 >$ 32. $-3< -1, 2 >$ 33. $\frac{1}{2} < -2, 6 >$

34. $-\frac{2}{3} < 3, -6 >$ 35. $-2< -3, 0 >$

Given that $\boldsymbol{a} = < 1, 2 >$, $\boldsymbol{b} = < 2, -3 >$ and $\boldsymbol{c} = < 3, 4 >$, find:

36. $-2\boldsymbol{a}$ 37. $3\boldsymbol{b}$ 38. $3\boldsymbol{a} + 2\boldsymbol{b}$

39. $2\boldsymbol{b} - \boldsymbol{c}$ 40. $3\boldsymbol{c} - 2\boldsymbol{a}$

41. Given that $\boldsymbol{a} = < 2, 3 >$ and $\boldsymbol{b} = < 3, -1 >$, solve the equation $x\boldsymbol{a} + y\boldsymbol{b} = < 12, 7 >$.

42. Given that $\boldsymbol{a} = < 3, 4 >$, $\boldsymbol{b} = < 2, -3 >$ and $\boldsymbol{c} = < 5, 1 >$, solve the equation $x\boldsymbol{a} + y\boldsymbol{b} = \boldsymbol{c}$.

43. Given that $\boldsymbol{a} = < 3, 2 >$, $\boldsymbol{b} = < 2, -3 >$ and $\boldsymbol{c} = < 12, -5 >$, find k and m such that $k\boldsymbol{a} + m\boldsymbol{b} = \boldsymbol{c}$.

Determine whether the vectors $\boldsymbol{a}$ and $\boldsymbol{b}$ are parallel

44. $\boldsymbol{a} = < -2, -4 >$, $\boldsymbol{b} = < 1, 2 >$ 45. $\boldsymbol{a} = < 1, 2>$, $\boldsymbol{b} = < -2, 1 >$

46. $\boldsymbol{a} = < 6, 4 >$, $\boldsymbol{b} = < 3, -2 >$ 47. $\boldsymbol{a} = < -2, 1 >$, $\boldsymbol{b} = < 8, -4 >$

Find the magnitude of each of the following vectors:

48. $< 4, 3 >$ 49. $< -2, 4 >$

50. $< -3, 5 >$ 51. $< -5, -12 >$

Given that $\boldsymbol{a} = < -7, 2 >$ and $\boldsymbol{b} = < 1, 5 >$, evaluate:

52. $|\boldsymbol{a} + \boldsymbol{b}|$ 53. $|\boldsymbol{b} - \boldsymbol{a}|$

54. $|2\boldsymbol{a} + \boldsymbol{b}|$ 55. $|\boldsymbol{a} - 2\boldsymbol{b}|$

Find the unit vector in the direction of each of the following vectors

56. $< -8, 6 >$ 57. $< 15, 8 >$

58. $< -12, -5 >$ 59. $< 7, -24 >$

Given that $\boldsymbol{a} = < 4, -3 >$ and $\boldsymbol{b} = < -1, 2 >$, find the unit vector in the direction of $\boldsymbol{c}$ in each case:

60. $\boldsymbol{c} = \boldsymbol{a} + \boldsymbol{b}$ 61. $\boldsymbol{c} = \boldsymbol{b} - \boldsymbol{a}$

62. $\boldsymbol{c} = \boldsymbol{a} + 2\boldsymbol{b}$ 63. $\boldsymbol{c} = 2\boldsymbol{a} + 3\boldsymbol{b}$

Write each of the following vectors as a linear combination of the standard unit vectors of $\boldsymbol{i}$ and $\boldsymbol{j}$

64. $< 4, 3 >$ 65. $< 3, -2 >$ 66. $< -5, -2 >$

67. $< 0, 5 >$ 68. $< -7, 0 >$

Given that $\boldsymbol{a} = -3\boldsymbol{i} + 2\boldsymbol{j}$, $\boldsymbol{b} = 2\boldsymbol{i} - 3\boldsymbol{j}$ and $\boldsymbol{c} = 5\boldsymbol{i} + 4\boldsymbol{j}$, find:

69. $\boldsymbol{a} + \boldsymbol{b}$ 70. $\boldsymbol{b} - \boldsymbol{c}$ 71. $2\boldsymbol{c} + \boldsymbol{b}$ 72. $3\boldsymbol{a} - 2\boldsymbol{b}$

Find the magnitude of each of the following vectors

73. $-3\boldsymbol{i} + \boldsymbol{j}$ 74. $4\boldsymbol{i} - 3\boldsymbol{j}$ 75. $-12\boldsymbol{j}$ 76. $-15\boldsymbol{i} - 8\boldsymbol{j}$

Find the unit vector in the direction of the following vectors

77. $\boldsymbol{i} + \boldsymbol{j}$ 78. $-3\boldsymbol{i} + 2\boldsymbol{j}$ 79. $2\boldsymbol{i} - \boldsymbol{j}$ 80. $-\boldsymbol{i} - 7\boldsymbol{j}$

Find the component form of the vector with the given initial point and terminal point.

81. Initial point: (4, 2); Terminal point: (−5, 3)

82. Initial point: (0, 4); Terminal point: (−3, −2)

83. Initial point: (−5, 0); Terminal point: (3, 2)

84. Initial point: (1, 3); Terminal point: (− 2, 0)

Find the vector with the given initial point and terminal point. Write the vector as a linear combination of the standard unit vectors $\boldsymbol{i}$ and $\boldsymbol{j}$.

85. Initial point: (7, 3); Terminal point: (9, 4)

86. Initial point: (1, −2); Terminal point: (−3, −5)

87. Initial point: (3, 4); Terminal point: (2, −1)

88. Initial point: (2, 3); Terminal point: (−2, 4)

89. Given A (x, y), B (3, 4) and **AB** = < − 2, 1 >, find the values of x and y.

90. Given A (4, −5) and **AB** = < 2, 3 >, find the coordinates of B.

91. Given Q (− 2, − 3) and **PQ** = < − 3, 4 >, find the coordinates of P.

92. The position vectors relative to the origin O, of three points A, B and C are $2\boldsymbol{i} + 3\boldsymbol{j}$, $4\boldsymbol{i} + 7\boldsymbol{j}$ and $\boldsymbol{i} + 2\boldsymbol{j}$ respectively. Given that **OB** = m **OA** + n **OC**, where m and n are scalar constants, find the values of m and n.

93. The position vectors of A and B are $-3\boldsymbol{i} + 2\boldsymbol{j}$ and $-\boldsymbol{i} - \boldsymbol{j}$. Given that **AC** = $5\boldsymbol{i} + \boldsymbol{j}$, find **BC**.

94. A and B have position vectors $5\boldsymbol{i} + 2\boldsymbol{j}$ and $-\boldsymbol{i} + 4\boldsymbol{j}$ respectively. Find the position vector of **C** if 3 **OA** = 2 **OB** + **OC**.

95 .The coordinates of A and B are (2, 3) and (−2, 5) respectively. Find the position vector of **C** if 2 **OA** = 2 **OB** + **BC**.

96. PQRS is a parallelogram in which PR is perpendicular to QS. Prove that PQRS is a rhombus.

97. ABC is a triangle and **AB** = $\boldsymbol{b}$, **AC** = $\boldsymbol{c}$, X and Y are the midpoint of AB and AC respectively. Express in terms of $\boldsymbol{b}$ and $\boldsymbol{c}$:

a. **BC** b. **AX** c. **AY** d. **XY**

What can you conclude about BC and XY?

If Δ ABC has area 6, what is the area of Δ AXY?

98. M and N divides the sides PQ and PR respectively of triangle PQR in the ratio 1: 2. Show that **MN** is parallel to **QR**.

99. ABCD is any quadrilateral where E, F, G, H are the midpoints of the sides

a. Express **EF** in terms of **AC**.

b. Express **HG** in terms of **AC**.

c. Prove that EFGH is a parallelogram

100. In a given quadrilateral ABCD, k is the midpoint of AC and also the midpoint of BD. Prove that ABCD is a parallelogram.

101. Using vector method, prove that a quadrilateral with two sides equal and parallel is a parallelogram.

102. OAB is a triangle with **OA** = $\boldsymbol{a}$ and **OB** = $\boldsymbol{b}$. X and Y are such that **OX** = $k\boldsymbol{a}$ and **OY** = $m\boldsymbol{b}$. If XY is parallel to AB, show that $k = m$.

103. Using vector method, prove that if ABCD is a parallelogram and X and Y are the midpoint of AB and CD respectively, then CXAY is a parallelogram.

104. Given a trapezium ABCD with AD//BC, if X and Y are the midpoint of AB and DC respectively prove that **XY** = (**AD** + **BC**). [Hint: Let **XA** = $\boldsymbol{a}$, **DY** = $\boldsymbol{b}$]

105. OABC is a parallelogram with the position vectors of A and C relative to O being a and c. X is the midpoint of OA, and OB and XC meet at Y. Find the position vector of Y.

106. A pilot is steering an aircraft due east and its airspeed is 600 km h $^{-1}$. There is a wind blowing from the south at a speed of 50 km h $^{-1}$. Find the direction in which the aircraft travels and its speed over the ground.

107. A pilot was steering his aircraft on a bearing of 136^0 and his airspeed indicator showed 200 km h $^{-1}$. However there was a wind blowing at 50 km h $^{-1}$ from the west. What was its speed over the ground and its direction?

108. A helicopter leaves an airfield A to fly to B 500 km away on a bearing of 140°. There is a steady wind of 30 km h $^{-1}$ from NE. The helicopter has airspeed of 150 km h $^{-1}$. Find the course the pilot must take and the time taken for him to reach B.

109. A current flows at 5 km h $^{-1}$ on a bearing of 150^0. A boat which can travel at 12 km h $^{-1}$ is to sail 60 km due east. In what direction should it be steered, and how long will the journey take?

110. A pilot steered an aircraft on a bearing of 060°. There was a 45 km h $^{-1}$ wind blowing from the south-east. If the airspeed of the aircraft was 235 km h $^{-1}$, find the aircraft's track and its ground speed

111. A ship sets off on a bearing of 147° at a speed of 30 km h $^{-1}$ through the water. The current flows at 5 km h $^{-1}$ in the direction 083°. Find the direction the ship travels and the distance of the ship after 3 hours?

112. An airplane whose speed in still air is 450 km h $^{-1}$ travels directly from A to B, a distance of 1200 km. The bearing of B from A is 215^0 and there is a wind of 60 km h $^{-1}$ from the east.

a. Find the bearing on which the plane was steered.

b. Find the time taken for the journey.

20.2 Dot Product

The dot product $\boldsymbol{a} \cdot \boldsymbol{b}$ of two vectors $\boldsymbol{a}$ and $\boldsymbol{b}$ is a real number, defined by

$$\boldsymbol{a} \cdot \boldsymbol{b} = |\boldsymbol{a}||\boldsymbol{b}| \cos \theta$$

where θ is the angle between $\boldsymbol{a}$ and $\boldsymbol{b}$.

Since $\cos(\pi/2) = 0$, $\boldsymbol{a} \cdot \boldsymbol{b} = 0$ when $\theta = \pi/2$. Hence, two vectors $\boldsymbol{a}$ and $\boldsymbol{b}$ are orthogonal or perpendicular if and only if $\boldsymbol{a} \cdot \boldsymbol{b} = 0$.

If $\boldsymbol{a}$ and $\boldsymbol{b}$ are parallel but in the same direction, then $\theta = 0°$. Since $\cos 0° = 1$.

$$\boldsymbol{a} \cdot \boldsymbol{b} = |\boldsymbol{a}||\boldsymbol{b}|$$

Note that if $\boldsymbol{a}$ and $\boldsymbol{b}$ are parallel but have opposite direction then $\theta = \pi$, and

$$\boldsymbol{a} \cdot \boldsymbol{b} = -|\boldsymbol{a}||\boldsymbol{b}|$$

The dot product is defined for vectors in component form as follows:

If $\boldsymbol{a} = < a_1, a_2 >$ and $\boldsymbol{b} = < b_1, b_2 >$ are two vectors then the dot product is defined by

$$\boldsymbol{a} \cdot \boldsymbol{b} = a_1b_1 + a_2b_2$$

The dot product satisfies the following properties.

Properties of Scalar Product

For vectors $\boldsymbol{a}$, $\boldsymbol{b}$, $\boldsymbol{c}$ and any scalar k, the following hold:

1. $\boldsymbol{a} \cdot \boldsymbol{b} = \boldsymbol{b} \cdot \boldsymbol{a}$ Commutative property
2. $\boldsymbol{a} \cdot (\boldsymbol{b} + \boldsymbol{c}) = \boldsymbol{a} \cdot \boldsymbol{b} + \boldsymbol{a} \cdot \boldsymbol{c}$ Distributive property
3. $(k\boldsymbol{a}) \cdot \boldsymbol{b} = k(\boldsymbol{a} \cdot \boldsymbol{b}) = \boldsymbol{a} \cdot (k\boldsymbol{b})$
4. $\boldsymbol{a} \cdot \boldsymbol{a} = |\boldsymbol{a}|^2$

Example 1

Find dot product of:

(a) $\boldsymbol{a} = < 3, 2 >$ and $\boldsymbol{b} = < -2, 4 >$ (b) $\boldsymbol{a} = 3\boldsymbol{i} - 6\boldsymbol{j}$ and $\boldsymbol{b} = 2\boldsymbol{i} + 5\boldsymbol{j}$

Solution

(a) $\boldsymbol{a} \cdot \boldsymbol{b} = < 3, 2 > \cdot < -2, 4 > = 3(-2) + 2(4) = 2$

(b) $\boldsymbol{a} \cdot \boldsymbol{b} = (3\boldsymbol{i} - 6\boldsymbol{j}) \cdot (2\boldsymbol{i} + 5\boldsymbol{j})$

$= (3\boldsymbol{i} - 6\boldsymbol{j}) \cdot 2\boldsymbol{i} + (3\boldsymbol{i} - 6\boldsymbol{j}) \cdot 5\boldsymbol{j}$

$= 6\boldsymbol{i} \cdot \boldsymbol{i} - 12\boldsymbol{j} \cdot \boldsymbol{i} + 15\boldsymbol{i} \cdot \boldsymbol{j} - 30\boldsymbol{j} \cdot \boldsymbol{j}$

$= 6|\boldsymbol{i}|^2 - 30|\boldsymbol{j}|^2$

$= 6 - 30$

$= -24$

Because the standard unit vectors $\boldsymbol{i}$ and $\boldsymbol{j}$ are orthogonal $\boldsymbol{i} \cdot \boldsymbol{j} = \boldsymbol{j} \cdot \boldsymbol{i} = 0$. Note also that, $\boldsymbol{i} \cdot \boldsymbol{i} = \boldsymbol{j} \cdot \boldsymbol{j} = 1$.

Example 2

Verify that $\boldsymbol{a} = 3\boldsymbol{i} - 4\boldsymbol{j}$ and $\boldsymbol{b} = 4\boldsymbol{i} + 3\boldsymbol{j}$ are orthogonal.

Solution

Begin by finding the dot product of the two vectors.

$$\boldsymbol{a} \cdot \boldsymbol{b} = 3(4) + (-4)(3) = 0$$

Because the dot product is 0, the two vectors are orthogonal.

Example 3

Let $\boldsymbol{a} = <3, 2>$, $\boldsymbol{b} = <2, -1>$ and $\boldsymbol{c} = <4, 3>$. Find each dot product

(a) $\boldsymbol{a} \cdot (\boldsymbol{b} + \boldsymbol{c})$ (b) $\boldsymbol{a} \cdot \boldsymbol{b} + \boldsymbol{a} \cdot \boldsymbol{c}$

Solution

(a) Begin by adding the two vectors

$$\begin{aligned} \boldsymbol{b} + \boldsymbol{c} &= <2, -1> + <4, 3> \\ &= <6, 2> \end{aligned}$$

So,

$$\begin{aligned} \boldsymbol{a} \cdot (\boldsymbol{b} + \boldsymbol{c}) &= <3, 2> \cdot <6, 2> \\ &= 3(6) + 2(2) \\ &= 18 + 4 \\ &= 22 \end{aligned}$$

(b)

$$\begin{aligned} \boldsymbol{a} \cdot \boldsymbol{b} + \boldsymbol{a} \cdot \boldsymbol{c} &= <3, 2> \cdot <2, -1> + <3, 2> \cdot <4, 3> \\ &= 3(2) + 2(-1) + 3(4) + 2(3) \\ &= 6 - 2 + 12 + 6 \\ &= 22 \end{aligned}$$

Notice that, the distributive property is also valid for the dot product.

The Angle between Two Vectors

Recall, that the dot product of two vectors $\boldsymbol{a}$ and $\boldsymbol{b}$ is defined by $\boldsymbol{a} \cdot \boldsymbol{b} = |\boldsymbol{a}||\boldsymbol{b}| \cos \theta$.

Thus, if θ is the angle between two nonzero vectors $\boldsymbol{a}$ and $\boldsymbol{b}$, where $0 \leq \theta \leq \pi$, then

$$\cos\theta = \frac{\boldsymbol{a}\cdot\boldsymbol{b}}{|a||b|}$$

Given $\boldsymbol{a} = < a_1, a_2 >$ and $\boldsymbol{b} = < b_1, b_2 >$, we have

$$\cos\theta = \frac{a_1 b_1 + a_2 b_2}{\left(\sqrt{a_1{}^2 + a_2{}^2}\right)\left(\sqrt{b_1{}^2 + b_2{}^2}\right)}$$

Example 4

Find the acute angle between $\boldsymbol{a} = < 2, 3 >$ and $\boldsymbol{b} = < -1, 5 >$.

Solution

We have $\cos\theta = \frac{2(-1)+3(5)}{(\sqrt{2^2+3^2})(\sqrt{(-1)^2+5^2})}$

$$= \frac{13}{(\sqrt{13})(\sqrt{26})}$$

$$= \frac{1}{\sqrt{2}}$$

$$\theta = 45°$$

Components and Projections

In a previous discussion we learned to decompose a vector into the sum of two orthogonal vector components.

Consider a particle at rest on a plane inclined at an angle θ to the horizontal, as shown in Figure 20.18.

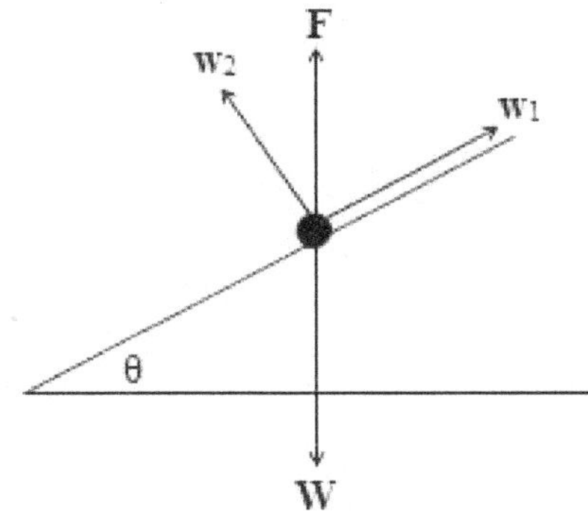

Figure 20.18

The particle has a weight **W** vertically downwards so the plane must be supporting it with a force **F** acting vertically upwards. The force **F** can be decomposed into two orthogonal components $\mathbf{w}_1$ and $\mathbf{w}_2$, so

$$\mathbf{F} = \mathbf{w}_1 + \mathbf{w}_2$$

The component $\boldsymbol{w}_1$ which is acting up the plane keeps the particle from rolling down the plane.

Now, we consider a procedure for finding $\boldsymbol{w}_1$ and $\boldsymbol{w}_2$.

Figure 20.19 shows two nonzero vectors $\boldsymbol{u}$ and $\boldsymbol{v}$.

Let the angle θ be the angle between the two vectors.

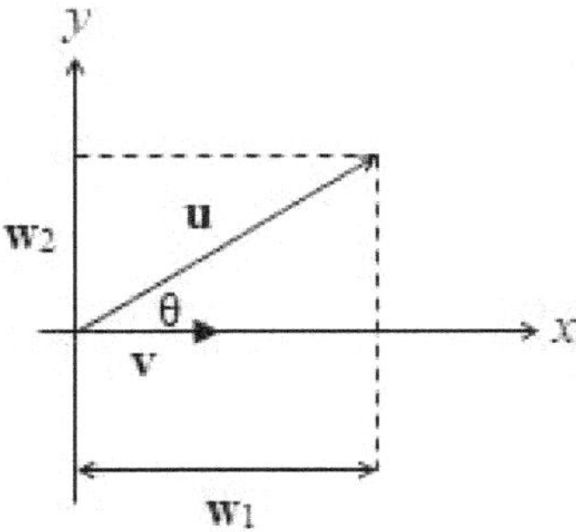

Figure 20.19

The vector $\boldsymbol{u}$ is decomposed into orthogonal components $\mathbf{w}_1$ and $\mathbf{w}_2$. The component $\mathbf{w}_1$ is parallel to $\boldsymbol{v}$, and is a scalar multiple of $\boldsymbol{v}$.

The vector component $\mathbf{w}_1$ is called the projection of $\boldsymbol{u}$ onto $\boldsymbol{v}$, denoted by $\mathbf{w}_1 = \text{proj}_v\boldsymbol{u}$. The vector $\boldsymbol{u}$ is given by

$$\boldsymbol{u} = \mathbf{w}_1 + \mathbf{w}_2$$

Because $\mathbf{w}_1$ is a scalar product $\boldsymbol{v}$ we have $\mathbf{w}_1 = k\boldsymbol{v}$, where k is a scalar.

So,

$$\begin{aligned}\boldsymbol{u} &= \mathbf{w}_1 + \mathbf{w}_2 \\ &= k\boldsymbol{v} + \mathbf{w}_2\end{aligned}$$

Taking dot product of each side with $\boldsymbol{v}$ gives

$$\begin{aligned}\boldsymbol{u} \cdot \boldsymbol{v} &= (k\boldsymbol{v} + \mathbf{w}_2) \cdot \boldsymbol{v} \\ &= k\boldsymbol{v} \cdot \boldsymbol{v} + \mathbf{w}_2 \cdot \boldsymbol{v}\end{aligned}$$

Because $\mathbf{w}_2$ and $\boldsymbol{v}$ are orthogonal $\mathbf{w}_2 \cdot \boldsymbol{v} = 0$, so

$$\boldsymbol{u} \cdot \boldsymbol{v} = k\,|\boldsymbol{v}|^2$$

Hence $$k = \frac{u \cdot v}{|v|^2}$$

Because $\mathbf{w}_1 = k\boldsymbol{v}$, we have $\boldsymbol{w}_1 = proj_{\boldsymbol{v}}\boldsymbol{u} = \left(\frac{u \cdot v}{|v|^2}\right)\boldsymbol{v}$.

Example 5

Let $\boldsymbol{u} = < 3, 2 >$ and $\boldsymbol{v} = < 4, -3 >$. Write vector $\boldsymbol{u}$ as the sum of two orthogonal vectors one of which is projection of $\boldsymbol{u}$ onto $\boldsymbol{v}$.

Solution

Begin by finding the projection of $\boldsymbol{u}$ onto $\boldsymbol{v}$.

$$\begin{aligned}
\boldsymbol{w}_1 &= proj_{\boldsymbol{v}}\boldsymbol{u} = \left(\frac{u \cdot v}{|v|^2}\right) v \\
&= \left(\frac{3\cdot 4 + 2\cdot(-3)}{\sqrt{4^2 + (-3)^2}}\right)\langle 4, -3\rangle \\
&= \left(\frac{6}{5}\right)\langle 4, -3\rangle \\
&= \langle \frac{24}{5}, \frac{-18}{5}\rangle
\end{aligned}$$

Because $\boldsymbol{u} = \mathbf{w}_1 + \mathbf{w}_2$, we have

$$\begin{aligned}
\mathbf{w}_2 &= \boldsymbol{u} - \mathbf{w}_1 \\
&= \langle 3, 2 \rangle - \langle \frac{24}{5}, \frac{-18}{5} \rangle \\
&= \langle \frac{-9}{5}, \frac{28}{5} \rangle
\end{aligned}$$

Now

$$\begin{aligned}
\boldsymbol{u} &= \mathbf{w}_1 + \mathbf{w}_2 \\
&= \langle \frac{24}{5}, \frac{-18}{5} \rangle + \langle \frac{-9}{5}, \frac{28}{5} \rangle \\
&= \langle 3, 2 \rangle
\end{aligned}$$

Example 6

A body of mass 5 kilograms rests on a smooth plane inclined at 30^0 to the horizontal. Find the least value of the force required to keep the body from rolling down the plane. [Take g = 10 ms^{-2}]

Solution

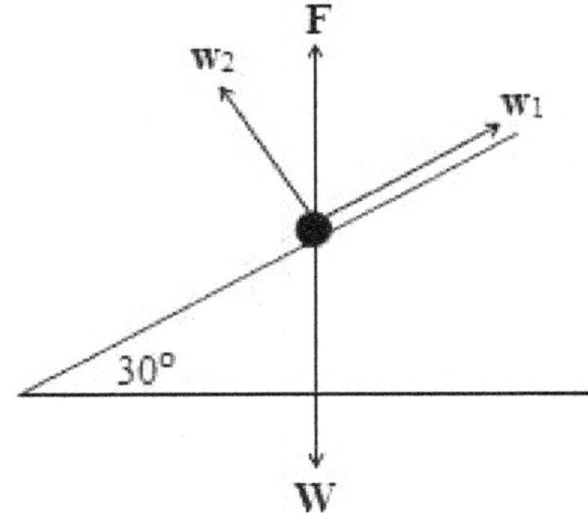

Because the weight W is 50 N we represent **F** by the vector

$$\mathbf{F} = -50\boldsymbol{j}$$

To find the force required to keep the body from rolling down, we project **F** onto a vector $\boldsymbol{v}$ in the direction of the plane, as follows.

$$\boldsymbol{v} = (\cos 30°)\boldsymbol{i} + (\sin 30°)\boldsymbol{j}$$

$$= \frac{\sqrt{3}}{2}\boldsymbol{i} + \frac{1}{2}\boldsymbol{j}$$

So, the projection of **F** onto $\boldsymbol{v}$ is

$$\mathbf{w}_1 = \text{proj}_v\mathbf{F}$$

$$= \left(\frac{\boldsymbol{F}\cdot\boldsymbol{v}}{|\boldsymbol{v}|^2}\right)\boldsymbol{v} = (-50)\left(\frac{1}{2}\right)\boldsymbol{v} = (-25)\left(\frac{\sqrt{3}}{2}\boldsymbol{i} + \frac{1}{2}\boldsymbol{j}\right)$$

$$= (-50)\left(\frac{1}{2}\right)\boldsymbol{v}$$

$$= (-25)\left(\frac{\sqrt{3}}{2}\boldsymbol{i} + \frac{1}{2}\boldsymbol{j}\right)$$

The magnitude of this force is 25 N, and so a force of 25 N is required to keep the body from rolling down the plane.

Work

Work is done upon an object when a force acting upon it cause a displacement of the object. The work W done by a constant force F acting in the direction of motion is defined as the product of the magnitude of the force and the distance d. That is, the work done is given by

$$W = F \cdot d.$$

When a force is exerted on an object at an angle to the horizontal, as shown in Figure 20.20, it is only the horizontal component of the force that directly causes the displacement of the object.

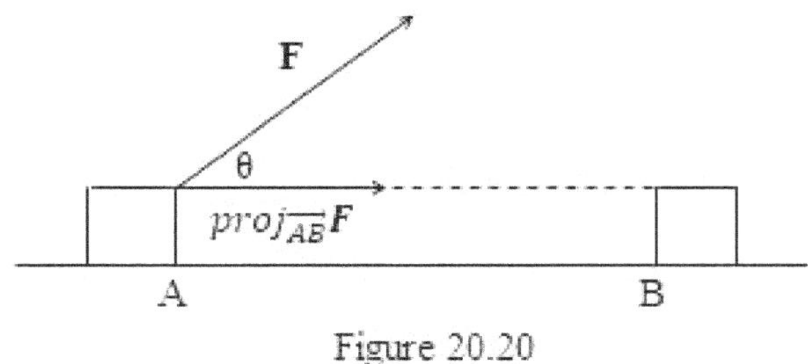

Figure 20.20

The force exerted in the direction of motion is the projection of the force, **F**, $proj_{\overrightarrow{AB}}\boldsymbol{F}$. The work W done by the force is given by

$$W = |proj_{\overrightarrow{AB}}\boldsymbol{F}||\overrightarrow{AB}|$$

$$= |\boldsymbol{F}cos\theta||\overrightarrow{AB}|$$

$$= \boldsymbol{F} \cdot \overrightarrow{AB}$$

Example 7

A toy wagon is pulled by exerting a force of 5 Newtons on a handle that makes an angle of 30° with the horizontal. Find the work done in pulling the wagon 40 meters.

Solution

The force exerted in the direction of motion is the horizontal component. The work W done is

$$\mathrm{W} = \mathbf{F} \cdot d$$

$$= |\mathbf{F}||d| \cos \theta$$

$$= 5 \cdot 40 \cdot \cos 30°$$

$$= 200 \left(\tfrac{\sqrt{3}}{2}\right)$$

$$= 173.2 \text{ Newton meter}$$

Exercise 20.2

Find the dot product of $\boldsymbol{a}$ and $\boldsymbol{b}$.

1. $\boldsymbol{a} = < 3, 1 >, \boldsymbol{b} = < 2, 5 >$

2. $\boldsymbol{a} = < 2, 5 >, \boldsymbol{b} = < 2, 1 >$

3. $\boldsymbol{a} = < 0, -2 >, \boldsymbol{b} = < -2, 4 >$

4. $\boldsymbol{a} = 3\boldsymbol{i} + \boldsymbol{j}, \boldsymbol{b} = -2\boldsymbol{i} + 3\boldsymbol{j}$

5. $\boldsymbol{a} = -2\boldsymbol{i} - 3\boldsymbol{j},\ \boldsymbol{b} = 2\boldsymbol{i} - 3\boldsymbol{j}$

6. $\boldsymbol{a} = 4\boldsymbol{i} + 3\boldsymbol{j},\ \boldsymbol{b} = -3\boldsymbol{i} + 4\boldsymbol{j}$

Use the vectors $\boldsymbol{a} = < 2, 3 >$, $\boldsymbol{b} = < -2, 1 >$ and $\boldsymbol{c} = < 4, -3 >$ to find each of the following:

7. $\boldsymbol{a} \cdot \boldsymbol{c}$

8. $2\boldsymbol{a} \cdot \boldsymbol{b}$

9. $(\boldsymbol{a} \cdot \boldsymbol{b})\boldsymbol{c}$

10. $(\boldsymbol{a} \cdot 3\boldsymbol{c})\boldsymbol{b}$

11. $(\boldsymbol{a} \cdot \boldsymbol{c}) - (\boldsymbol{a} \cdot \boldsymbol{b})$

12. $\boldsymbol{a} \cdot (\boldsymbol{b} + \boldsymbol{c})$

Verify that the vectors $\boldsymbol{a}$ and $\boldsymbol{b}$ are orthogonal.

13. $\boldsymbol{a} = < 1, 4 >,\ \boldsymbol{b} = < 8, -2 >$

14. $\boldsymbol{a} = <2, 3 >,\ \boldsymbol{b} = < -6, 4 >$

15. $\boldsymbol{a} = < 2, -1 >,\ \boldsymbol{b} = < 2, 4 >$

16. $\boldsymbol{a} = 6\boldsymbol{i} + 2\boldsymbol{j},\ \boldsymbol{b} = -\boldsymbol{i} + 3\boldsymbol{j}$

17. $\boldsymbol{a} = 2\boldsymbol{i} - 3\boldsymbol{j},\ \boldsymbol{b} = 3\boldsymbol{i} + 2\boldsymbol{j}$

18. $\boldsymbol{a} = -2\boldsymbol{i} + 7\boldsymbol{j},\ \boldsymbol{b} = 7\boldsymbol{i} + 2\boldsymbol{j}$

Find the angle between the vectors

19. $\boldsymbol{a} = < 1, 4 >,\ \boldsymbol{b} = < -2, 3 >$

20. $\boldsymbol{a} = < 3, -4 >,\ \boldsymbol{b} = < 12, 5 >$

21. $\boldsymbol{a} = < -3, 2 >,\ \boldsymbol{b} = < 2, 4 >$

22. $\boldsymbol{a} = 3\boldsymbol{i} - 2\boldsymbol{j},\ \boldsymbol{b} = \boldsymbol{i} + \boldsymbol{j}$

23. $\boldsymbol{a} = -5\boldsymbol{i} + 3\boldsymbol{j},\ \boldsymbol{b} = 2\boldsymbol{i} + 7\boldsymbol{j}$

24. $\boldsymbol{a} = 4\boldsymbol{i} - \boldsymbol{j},\ \boldsymbol{b} = 2\boldsymbol{i} + 3\boldsymbol{j}$

Find the projection of $\boldsymbol{u}$ onto $\boldsymbol{v}$. Then write $\boldsymbol{u}$ as the sum of two orthogonal vectors, one of which is $\text{proj}_{\boldsymbol{v}}\boldsymbol{u}$.

25. $\boldsymbol{u} = < 3, 2 >,\ \boldsymbol{v} = < 4, 1 >$

26. $\boldsymbol{u} = < 1, 3 >,\ \boldsymbol{v} = < -2, 4 >$

27. $\boldsymbol{u} = < -2, -3 >,\ \boldsymbol{v} = < -1, -4 >$

28. $\boldsymbol{u} = < -2, 3 >,\ \boldsymbol{v} = < -1, 2 >$

29. A body of mass 12 kilograms rests on a smooth plane inclined at 30° to the horizontal. Find the least value of the force required to keep the body from rolling down the plane. [Take $g = 10\ \text{ms}^{-2}$]

30. A toy wagon is pulled by exerting a force of 15 Newtons on a handle that makes an angle of 60° with the horizontal. Find the work done in pulling the wagon 50 meters.

Review Exercises

Given that $\boldsymbol{a} = 3\boldsymbol{i} + 2\boldsymbol{j}$, $\boldsymbol{b} = -\boldsymbol{i} + 3\boldsymbol{j}$ and $\boldsymbol{c} = 2\boldsymbol{i} - 3\boldsymbol{j}$, express in terms of $\boldsymbol{i}$ and $\boldsymbol{j}$.

1. $\boldsymbol{a} + 2\boldsymbol{b}$

2. $2\boldsymbol{c} - 3\boldsymbol{b}$

3. $3\boldsymbol{a} + 2\boldsymbol{c}$

4. Given $\boldsymbol{p} = 3\boldsymbol{i} + 2\boldsymbol{j}$ and $\boldsymbol{q} = 2\boldsymbol{i} - 3\boldsymbol{j}$, find numbers x and y such that $x\boldsymbol{p} + y\boldsymbol{q} = 3\boldsymbol{i} - 11\boldsymbol{j}$.

5. If $\boldsymbol{a} = \boldsymbol{i} - 3\boldsymbol{j}$, $\boldsymbol{b} = 2\boldsymbol{i} + 3\boldsymbol{j}$ and $\boldsymbol{c} = 2\boldsymbol{i} + 4\boldsymbol{j}$, find the values of the constants m and n such that $2\boldsymbol{a} = 5m\boldsymbol{b} - 3n\boldsymbol{c}$.

6. Given that $\boldsymbol{a} = -12\boldsymbol{i} + 4\boldsymbol{j}$ and that $\boldsymbol{b} = \boldsymbol{i} + p\boldsymbol{j}$, find the values of the constants p and q such that $\boldsymbol{a} + q\boldsymbol{b} = -27\boldsymbol{i} + 19\boldsymbol{j}$.

7. Given that $\boldsymbol{a} = \langle 3, 1 \rangle$, $\boldsymbol{b} = \langle -4, 3 \rangle$ and $\boldsymbol{c} = \langle 17, -3 \rangle$, find m and n such that $m\boldsymbol{a} + n\boldsymbol{b} = \boldsymbol{c}$.

Given $\boldsymbol{a} = -2\boldsymbol{i} + \boldsymbol{j}$, $\boldsymbol{b} = 3\boldsymbol{i} + 2\boldsymbol{j}$ and $\boldsymbol{c} = -\boldsymbol{i} - 2\boldsymbol{j}$, find:

8. $|\boldsymbol{a} + 2\boldsymbol{b}|$ 9. $|2\boldsymbol{c} - 3\boldsymbol{b}|$ 10. $|3\boldsymbol{a} - 2\boldsymbol{c}|$

Find the magnitude of each vector and the acute angle it makes with the x-axis.

11. $\langle -3, 4 \rangle$ 12. $\langle 5, -12 \rangle$ 13. $-2\boldsymbol{i} + 3\boldsymbol{j}$ 14. $4\boldsymbol{i} - 5\boldsymbol{j}$

Find the unit vectors in the direction of the following vectors:

15. $\boldsymbol{a} = -3\boldsymbol{i} + 4\boldsymbol{j}$ 16. $\boldsymbol{b} = 2\boldsymbol{i} - \boldsymbol{j}$ 17. $\boldsymbol{a} = 2\boldsymbol{i} + \boldsymbol{j}$ 18. $\boldsymbol{b} = -5\boldsymbol{i} + 12\boldsymbol{j}$

19. A $(-3, 2)$, B $(1, -2)$ and C $(2, 3)$ are points in the $x-y$ plane. Find **BA** and **BC** in the form $\langle x, y \rangle$.

20. A, B and C are points with position vectors $3\boldsymbol{i} - 2\boldsymbol{j}$, $2\boldsymbol{i} + \boldsymbol{j}$ and $-\boldsymbol{i} + 4\boldsymbol{j}$ respectively. Find in terms of $\boldsymbol{i}$ and $\boldsymbol{j}$, the vectors **AB**, **BC** and **CA**.

21. The coordinates of A are $(2, -3)$ and the position vector of B is $4\boldsymbol{i} + 2\boldsymbol{j}$. Find the vector **BA**.

22. If the coordinates of A are $(4, 2)$ and **AB** $= 2\boldsymbol{i} + \boldsymbol{j}$, find the position vector of B.

23. A and B have position vectors $2\boldsymbol{i} + 5\boldsymbol{j}$ and $4\boldsymbol{i} - \boldsymbol{j}$ respectively. Find the position vector of **C** if 3 **OA** = 2**OB** + **OC**.

24. The coordinates of A and B are $(3, 2)$ and $(5, -2)$ respectively. Find the position vector of C if 2**OA** = 2 **OB** + **BC.**

25. The position vectors relative to an origin O, of three points A, B and C are $3\boldsymbol{i} + 2\boldsymbol{j}$, $7\boldsymbol{i} + 4\boldsymbol{j}$ and $2\boldsymbol{i} + \boldsymbol{j}$ respectively. Given that **OB** = m **OA** + n **OC** where m and n are scalar constants, find the value of m and n.

26. The position vector of A and B are $2\boldsymbol{i} - 3\boldsymbol{j}$ and $-\boldsymbol{i} - \boldsymbol{j}$. Given that **AC** $= \boldsymbol{i} + 5\boldsymbol{j}$, find the position vector of C. Find |**BC**| and the angle **BC** makes with the x-axis.

27. A quadrilateral has vertices A $(4, 0)$, B $(7, -3)$, C $(-2, -2)$ and D $(-5, 1)$. Show that ABCD is a parallelogram.

28. Show that the triangle whose vertices have position vectors $4\boldsymbol{i} + 2\boldsymbol{j}$, $2\boldsymbol{i} + 5\boldsymbol{j}$ and $5\boldsymbol{i} + 3\boldsymbol{j}$ is isosceles.

29. A $(1, 2)$, B $(3, 5)$, C $(3, -6)$ and D(x, y) are the vertices of the parallelogram ABCD. Find the values of x and y.

30. P and Q divides the sides BC and AC respectively of triangle ABC in the ratio 2 : 1. If **AB** = $\boldsymbol{a}$ and **AC** = $\boldsymbol{b}$, find (a) **OP** and (b) show that **QP** is parallel to **AB** and one-third its length.

31. OABC is a parallelogram in which the position vectors of A, B and C relative to O are $\boldsymbol{a}$, $\boldsymbol{b}$ and $\boldsymbol{c}$ respectively. Let X and Y be the midpoints of the diagonals OB and AC respectively.

a. Express $\boldsymbol{b}$ in terms of $\boldsymbol{a}$ and $\boldsymbol{c}$

b. Find expressions for **AC, AY, OX, OY**.

c. What can you say about the points X and Y?

Given that $\boldsymbol{a} = 3\boldsymbol{i} + 2\boldsymbol{j}$, $\boldsymbol{b} = 2\boldsymbol{i} - \boldsymbol{j}$ and $\boldsymbol{c} = -2\boldsymbol{i} + 3\boldsymbol{j}$ find the scalar product.

32. $\boldsymbol{a} \cdot \boldsymbol{c}$ 33. $\boldsymbol{a} \cdot \boldsymbol{b}$ 34. $\boldsymbol{b} \cdot \boldsymbol{c}$

35. Given that $\boldsymbol{a} = 3\boldsymbol{i} + 2\boldsymbol{j}$ and $\boldsymbol{b} = 2\boldsymbol{i} - 3\boldsymbol{j}$ show that $\boldsymbol{a}$ is perpendicular to $\boldsymbol{b}$.

36. Given that $\boldsymbol{a} = \boldsymbol{i} + 3\boldsymbol{j}$ and $\boldsymbol{b} = 6\boldsymbol{i} + 2\boldsymbol{j}$ find the angle between $\boldsymbol{a}$ and $\boldsymbol{b}$.

37. A (2, 3), B (−1, 4) and C (5, −2) are three points. Evaluate **BA** · **BC** and hence find ∠ ABC.

38. Given $\boldsymbol{a} = -2\boldsymbol{i} + 3\boldsymbol{j}$, $\boldsymbol{b} = \boldsymbol{i} + 4\boldsymbol{j}$ and $\boldsymbol{c} = 7\boldsymbol{i} + 2\boldsymbol{j}$ verify that $\boldsymbol{a} \cdot (\boldsymbol{b} + \boldsymbol{c}) = \boldsymbol{a} \cdot \boldsymbol{b} + \boldsymbol{a} \cdot \boldsymbol{c}$.

39. If $\boldsymbol{p} = 4\boldsymbol{i} + 3\boldsymbol{j}$, $\boldsymbol{q} = -3\boldsymbol{i} + 2\boldsymbol{j}$ and $\boldsymbol{r} = \boldsymbol{i} + 5\boldsymbol{j}$ show that $\boldsymbol{p} \cdot (\boldsymbol{q} - \boldsymbol{r}) = \boldsymbol{p} \cdot \boldsymbol{q} - \boldsymbol{p} \cdot \boldsymbol{r}$ and find the angle between $\boldsymbol{p}$ and $\boldsymbol{q}$.

40. A light aircraft flies on a course of S 40° E, and its airspeed is 120 km h^{-1}. The speed of a wind blowing from north-west is 35 km h^{-1}. Find the ground speed and the direction of the aircraft.

41. An airplane moves on a bearing of 315° at 130 km h^{-1} relative to the ground, due to the fact that there is a wind blowing from the west at 45 km h^{-1}. Find the speed and the direction the plane will have traveled if there was no wind.

42. A river is flowing at 2 ms^{-1}. A man can row at 2.5 ms^{-1}. In which direction should he row if he is to cross the river directly? If the river is 60 meters wide how long will it take him?

43. A body of mass 8 kilograms rests on a smooth plane inclined at 30° to the horizontal. Find the least value of the force required to keep the body from rolling down the plane. [Take g = 10 ms^{-2}]

44. A toy wagon is pulled by exerting a force of 6 Newtons on a handle that makes an angle of 30° with the horizontal. Find the work done in pulling the toy wagon 20 meters.

21 Mathematical Induction

Mathematical induction is used to establish that a mathematical statement involving integers such as

$$1 + 2 + 3 + \cdots + n = \frac{n(n+1)}{2} \qquad (1)$$

is true for any positive integer.

To prove that Equation (1) is true for all positive integers we may substitute possible values of n. For example, we can substitute the first three positive integers.

For $n = 1$, we have

$$1 = \frac{1(1+1)}{2} = \frac{1(2)}{2} = 1$$

For $n = 2$, we have

$$1 + 2 = \frac{2(1+2)}{2} = \frac{2(3)}{2} = 3$$

For $n = 3$, we have

$$1 + 2 + 3 = \frac{3(3+1)}{2} = \frac{3(4)}{2} = 6$$

Equation (1) is true for n = 1, 2 and 3. It would not be possible to substitute all possible values of n because there is an infinite number of them. Although Equation (1) is true for some values of n, we could never be sure that it would be true for all values of n. A general proof of the statement that covers all values of n is to use mathematical induction.

Proof by Induction

To prove that a mathematical statement involving positive integers is true for all positive integers, $n \geq 1$, we use the following 3 steps.

First, we show that the statement is true for a specific value of n, usually $n = 1$. We then assume that the statement is true for a particular value of n, say $n = k$ (This is called the induction assumption). If we can show that the statement is also true for $n = k + 1$ then the statement must be true for all positive integers.

Example

Show that each of the following statements is true for all positive integers n.

(a) $1 + 2 + 3 + \cdots + n = \frac{n(n+1)}{2}$

(b) $2 + 5 + 8 + \cdots + (3n - 1) = \frac{n(3n+1)}{2}$

Solution

(a) $$1 + 2 + 3 + \cdots + n = \frac{n(n+1)}{2}$$

For $n = 1$, we have

$$1 = \frac{1(1+1)}{2} = \frac{1(2)}{2} = 1$$

So, the statement is true for $n = 1$.

Next, we assume that the statement is true for some positive integer, $n = k$. So,

$$1 + 2 + 3 + \cdots + k = \frac{k(k+1)}{2}$$

which is the Induction assumption.

Finally, we show that the statement is true for $n = k + 1$.

Add $(k + 1)$ to the left hand side.

$$1 + 2 + 3 + \cdots + k + (k + 1)$$

Using the induction assumption, we have

$$1 + 2 + 3 + \cdots + k + (k + 1) = \frac{k(k+1)}{2} + (k + 1)$$

$$= \frac{k(k+1)+2(k+1)}{2}$$

$$= \frac{(k+1)(k+2)}{2}$$

$$= \frac{(k+1)[(k+1)+1]}{2}$$

Thus, the statement is true for $n = k + 1$, and so it is true for all positive integers.

(b) $$2 + 5 + 8 + \cdots + (3n - 1) = \frac{n(3n+1)}{2}$$

For $n = 1$, we have

$$2 = \frac{1(3+1)}{2} = \frac{1(4)}{2} = 2$$

So, the statement is true for $n = 1$.

We assume that the statement is true for $n = k$. So,

$$2 + 5 + 8 + \cdots + (3k - 1) = \frac{k(3k+1)}{2}$$

Add [3(k + 1) – 1] = 3*k* +2 to the left hand side.

$$2+5+8+\cdots+(3k-1)+(3k+2)$$

Using the induction assumption, we have

$$2+5+8+\cdots+(3k-1)+(3k+2)=\frac{k(3k+1)}{2}+(3k+2)$$

$$=\frac{k(3k+1)+2(3k+2)}{2}$$

$$=\frac{3k^2+k+6k+4}{2}$$

$$=\frac{3k^2+6k+3+k+1}{2}$$

$$=\frac{3(k^2+2k+1)+k+1}{2}$$

$$=\frac{3(k+1)^2+k+1}{2}$$

$$=\frac{(k+1)[3(k+1)+1]}{2}$$

Thus, the statement is true for $n = k + 1$, and so it is true for all positive integers.

Exercise 21

Prove the following results by induction.

1. $1+3+5+\cdots+(2n-1)=n^2$

2. $1^3+2^3+3^3+\cdots+n^3=\frac{1}{4}n^2(n+1)^2$

3. $1^2+3^2+5^2+\cdots+(2n-1)^2=\frac{1}{3}n(4n^2-1)$

4. $1^2+2^2+3^2+\cdots+n^2=\frac{1}{6}n(n+1)(2n+1)$

5. $1+3+3^2+3^3+\cdots+3^{n-1}=\frac{1}{2}(3^n-1)$

Answers to Exercises

Chapter 1

Exercise 1.1

1. {3, 4, 5, 6, 7, 8} 2. {11, 13, 17, 19} 3. {31, 33, 35, 37, 39} 4. {15, 18, 21, 24}

5. {-2, -1, 0, 1, 2} 6. {8, 9, 10, 11, 12, 13, 14} 7. $\notin$ 8. $\in$ 9. $\notin$ 10. $\in$

11. {2, 4, 6, 8, 10} 12. {1, 3, 5, 7, 9} 13. {2, 3, 5, 7} 14. {5, 10] 15. True

16. False 17. True 18. False 19. False 20. True 21. Equal 22. Equivalent

23. Equivalent 24. Equal 25. Equal 26. Empty 27. Not empty 28. Empty

29. Not empty 30. Not empty 31. Infinite 32. Finite 33. Finite 34. Infinite

35. $\subset$ 36. $\supset$ 37. $\subset$ 38. $\supset$ 39. $\subset$ 40. $\emptyset$, {3}, {4},

{5}, {3, 4}, {3, 5}, {4, 5}, {3, 4, 5} 41. 128 42. 9

Exercise 1.2

1. {1, 2, 3, 4, 5, 6} 2. {1, 2, 3, 4, 5, 6, 8} 3. {1,2, 4, 5, 6, 8, 10, 20}

4. {1, 2, 3, 4, 5, 6, 7, 8} 5. {2, 3, 5, 7, 9} 6. {2, 3, 5} 7. {3, 6}

8. {3, 6} 9. $\emptyset$ 10. {3, 6} 11. {1, 2, 3, 4, 5, 6, 7} 12. {1, 2, 3, 4, 5, 6, 7}

13. {1, 2, 3, 4, 6, 8} 14. {1} 15. {2} 16. {3} 17. {1, 2, 3, 4, 5, 6, 7, 8}

18. {1, 2, 3, 4, 5, 6, 7, 8} 19. $\emptyset$ 20. $\emptyset$ 21. {1, 2, 3, 4, 5, 6, 8} 22. {1, 2, 3, 4, 5, 6}

23. {3} 24. {1, 6} 25. {1. 2. 3, 4, 5, 6} 26. {1, 3, 6} 27. {1, 2, 3, 4, 5, 6}

28. {1, 3, 6} 29. {2, 4, 6, 8} 30. {1, 3, 6, 8} 31. {1, 2, 4, 6, 7} 32. {1, 2, 4, 5, 6}

33. $\emptyset$ 34. {1, 2, 4, 6} 35. {1, 2, 4, 5, 6, 7} 36. {7}

Review Exercises

1. {9, 12, 15, 18} 2. {2, 3, 4, 5, 6, 7, 8} 3. {-1, 0, 1, 2, 3, 4, 5, 6} 4. {2, 4, 6, 8, 10}

5. Equal 6. Equivalent 7. Equivalent 8. Equal 9. Finite 10. Infinite

11. Infinite 12. Finite 13. False 14. True 15. False 16. True 17. False

18. False 19. True 20. False 21. $\in$ 22. $\notin$ 23. $\subset$ 24. $\not\subset$

25. {1, 2, 4, 5, 7, 8, 9, 10} 26. {1, 2, 4, 5, 7, 8, 9, 10} 27. {1, 2, 4, 5, 7, 8, 9}

28. {1, 4, 7, 8, 9, 10} 29. {8, 9} 30. {2, 5, 8, 9} 31. {1, 2, 4, 5, 7, 8, 9, 10}

32. {1, 2, 4, 5, 7, 8, 9, 10} 33. {8, 9} 34. {8, 9} 35. {1, 2, 3, 4, 5, 6, 8}

36. {1, 2, 3, 5, 6, 8} 37. {3, 6} 38. {2, 5} 39. {1, 2, 3, 5, 6, 8}

40. {2, 3, 5, 6} 41. {1, 2, 3, 5, 6, 8} 42. {2, 3, 5, 6} 43. {1, 2, 5, 7, 9}

44. {1, 2, 4, 6, 9} 45. {1, 2, 4, 5, 6, 7, 9} 46. {1, 2, 9} 47. {1, 2, 9}

48. {1, 2, 4, 5, 6, 7, 9} 49. {1, 2, 3, 4, 6, 8, 9} 50. {4, 6} 51. {3, 8}

52. $\emptyset$ 53. $\cup$ 54. {2, 5, 8} 55. {1, 3, 4, 6, 7} 56. {6} 57. {1, 2, 3, 5, 6, 7, 8}

58. {1, 3, 5} 59. $\emptyset$ 60. {4, 7} 61. {1, 2, 3, 5}

Chapter 2

Exercise 2.1

1. (a) { 1. 8} (b) { - 12, 0, 1, 8} (c) $\{-12, -\frac{3}{4}, -\frac{1}{4}, 0, 3/8, 1, 8\}$ (d) $\{-\sqrt{7}, 4\pi\}$

2. (a) {8, 12} (b) {0, 8, 12} (c) $\{-9/2, -3/8, 0, 10/3, 8, 12\}$ (d) $\{-\sqrt{6}, -\pi/2, \sqrt{13}\}$

3. (a) {10} (b) {−1, 10} (c) $\{-3.6, -\sqrt{4}, -1, -\frac{1}{2}, -0.\dot{3}, -0.3, 10, 26.4\}$ (d) $\{\sqrt{5}, 3\pi\}$

4. (a) {3, 20} (b) {-6, 0, 3, 20} (c) $\{-6, -\sqrt{25}, -5/2, -0.11, 0, 0.75, 3, 20\}$ (d) $\{-\sqrt{6}\}$

13. < 14. > 15. < 16. > 17. < 19. > 20. < 21. > 22. > 23. <

24. > 25. 6 26. 55 27. 21 28. 66 29. 50 30. 20 31. 9

32. 3 33. 12 34. 85 35. 12 36. 15 37. 17 38. 18 39. – 16 40. – 25

41. -20 42. -31 43. > 44. < 45. = 46. = 47. < 48. >

Exercise 2.2

1. 2 2. 3 3. 12 4. 7 5. 13 6. – 5 7. – 8 8. – 19 9. – 16

10. – 28 11. – 14 12. – 9 13. 15 14. 25 15. – 10 16. – 25 17. 3

18. – 2 19. 5 20. – 7 21. – 4 22. – 7 23. 4 24. – 5 25. – 8

26. – 30 27. – 7 28. – 3 29. 10 30. 90 31. – 30 32. – 44

33. 42 34. 72 35. – 36 36. – 70 37. – 56 38. – 70 39. 60 40. – 45

41. 56 42. – 165 43. – 5 44. 5 45. 2 46. – 6 47. 4 48. – 3 49. 3

50. – 5 51. – 4 52. – 8 53. – 7 54. 4 55. 100 56. 100 57. 1200

58. 160 59. 2500 60. 1080 61. 693 62. 336 63. 1182 64. 16983

65. 741 66. 1192

Exercise 2.3

1. 2^7 2. 3^5 3. 5^8 4. 6^{10} 5. 2^{-5} 6. 7^3 7. 8^5 8. 2^{-2} 9. 3^5

10. 5^2 11. 6^{-8} 12. 4^6 13. 2^{12} 14. 3^{10} 15. 7^{20} 16. 2^{-6} 17. 3^{-8}

18. 4^6 19. 5^{-3} 20. 3^{-6} 21. 2 22. 5 23. 4 24. 2 25. 27

26. 4 27. 125 28. 32 29. $\frac{1}{2}$ 30. $\frac{1}{3}$ 31. $\frac{1}{25}$ 32. $\frac{1}{8}$ 33. 1 34. 16

35. 2 36. 3^4 37. 2^3 38. 5^3 39. 4^2 40. 5^2 41. 3^2 42. 4.7×10^6

43. 8.6×10^4 44. 2.3×10^2 45. 7.6×10^1 46. 3.64×10^9 47. 1.24×10^5

48. 2.75×10^1 49. 5.72×10^2 50. 6.27×10^2 51. 9.5 52. 1.05×10^4

53. 4.156×10^3 54. 5.6×10^{-3} 55. 2.4×10^{-1} 56. 4.31×10^{-5}

57. 7.03×10^{-4} 58. 6.43×10^{-2} 59. 8.5×10^{-6} 60. 9×10^{-3} 61. 1.26×10^{-1}

62. 3.164×10^{-2} 63. 2×10^{-5} 64. 6.05×10^{-2} 65. 5.7×10^{-7} 66. 4×10^9

67. 4.05×10^{-6} 68. 2.25×10^2 69. 5.445×10^1 70. 2.448×10^3 71. 5.82×10^1

72. 6×10^3 73. 2.4×10^7 74. 1×10^{-1} 75. 2.8×10^2 76. 6.5×10^{-2}

77. 6×10^1 78. – 57 79. 20 80. – 48 81. 21 82. – 2 83. 3 84. – 8

85. 22 86. 2 87. 18 88. – 1 89. 22 90. – 5 91. – 3 92. – 56

93. 1 94. – 31

Review Exercises

1. (a) {12} (b) { - 1, 0, 12} (c) {– 3.5, – √9, – 2.3, – 1, – 0.75, 0, 19/2, 12} (d) [√3, 2π}

2. (a) {9, 23} (b) {- 3, 0, 9, 23} (c) {– 9/2, – 3, – ¾ , 0, √36, 15/2, 9, 23} (d) {– √5}

7. $<$ 8. $>$ 9. $<$ 10. $>$ 11. 8 12. 7.3 13. – 32 14. – 15 15. $<$ 16. $>$

17. = 18. 11 19. 2 20. 0 21. – 4 22. 24 23. – 8 24. – 10 25. – 24

26. – 36 27. 78 28. – 120 29. – 8 30. 8 31. – 3 32. 4 33. 300

34. 4150 35. 1379 36. 1488 37. 1 ¼ 38. 3/7 39. 14/15 40. 13/36

41. 1 ¼ 42. 7(11/12) 43. 2(11/12) 44. 6(8/15) 45. 4/9 46. 25 ½ 48. 9

49. 5/6 50. 1 ½ 51. 2(1/14) 52. 3^2 53. 2^{-1} 54. 5^{-1} 55. 5^{-4} 56. 3^3

57. 3^{-2} 58. 5^2 60. 3^6 61. 7^{-2} 62. 1 63. 2^2 64. 2 65. 6 66. 4

67. 9 68. 1/27 69. 1/16 70. 5.97×10^8 71. 6.4×10^2 72. 6.75×10^1

73. 7.582×10^{3} 74. 6.5×10^{-1} 75. 8.04×10^{-4} 76. 2.178×10^{-2} 77. 9×10^{-8}

78. 5.395×10^{8} 79. 4.836×10^{-3} 80. 2×10^{6} 81. 9×10^{2} 82. 2×10^{1}

83. 2.8×10^{-2} 84. 7.8×10^{3} 85. 6×10^{-1} 86. 26 87. 7 88. 8 89. 21

90. 22 91. – 14 92. 7 93. 11

Chapter 3

Exercise 3.1

1. $6x$ 2. $8y$ 3. $12x^2$ 4. $5x$ 5. $4y$ 6. $-a$ 7. $5x$ 8. $3y$ 9. $2x$ 10. $-9x$

11. $-4y$ 12. $-5ab$ 13. $8x + 7y$ 14. $4a + 5b$ 15. $-6a$ 16. $3y^2 - 16y$

17. $9xy + 7yz$ 18. $2x^3 + 3x^2$ 19. $2x^2 + 3x$ 20. $3ab^2 + 2a^2b$ 21. $12ab$ 22. $10xy$

23. $5st$ 24. $-6rs$ 25. $-8pq$ 26. $-24xy$ 27. $4ab$ 28. $42pq$ 29. $24rs$

30. $2y^2$ 31. $9x^4y^2$ 32. $-8x^3y^6$ 33. $3a^5b^4$ 34. $-8a^2b^2$ 35. $6a^3b^5$ 36. $30a^4b^2$

37. $8x^5y^3$ 38. $2a^3b^2c$ 39. $-15x^3y^3$ 40. $6x^3y^5$ 41. $-6a^3b^4c^2$ 42. $20a^5b^7$

43. $15x^3y^3$ 44. $42a^3b^4$ 45. $-6x^5y^6$ 46. $10a^7b^3$ 47. $-7b^5c^7$ 48. $4b$

49. $3b$ 50. $3y$ 51. $2y$ 52. $4xy$ 53. $\frac{5xz}{6y}$ 54. $-\frac{3}{2y}$ 55. $-\frac{5y}{z}$ 56. $\frac{4a^2z}{b}$

57. $\frac{2ab}{3}$ 58. $\frac{6}{b}$ 59. $\frac{4y}{x}$ 60. – 42 61. 120 62. 162 63. 100 64. 14 65. 9

66. 13 67. 11 68. 3 69. 42 70. 24 71. – 4 72. 20 73. – 10 74. 4

Exercise 3.2

1. $x^2 + 9x + 20$ 2. $x^2 + 13x + 42$ 3. $x^2 + 10x + 9$ 4. $x^2 + 17x + 60$

5. $x^2 + 12x + 20$ 6. $x^2 + 18x + 45$ 7. $2x^2 + 11x + 12$ 8. $6x^2 + 7x + 2$

9. $10x^2 + 43x + 28$ 10. $18x^2 + 29x + 20$ 11. $x^2 + 2x - 15$ 12. $x^2 + 3x - 70$

13. $x^2 + 4x - 32$ 14. $2x^2 + x - 15$ 15. $x^2 + x - 72$ 16. $3x^2 - x - 30$

17. $7x^2 - x - 30$ 18. $x^2 - 4x - 45$ 19. $x^2 - 3x - 70$ 20. $x^2 - 6x - 72$

21. $x^2 + 7x - 18$ 22. $x^2 + 3x - 40$ 23. $x^2 + 5x - 14$ 24. $x^2 + 8x - 65$

25. $x^2 - 3x - 40$ 26. $x^2 - 7x - 44$ 27. $x^2 - 3x - 54$ 28. $6x^2 + 5x - 6$

29. $8x^2 - 22x - 21$ 30. $5x^2 + 38x - 16$ 31. $x^2 - 7x + 10$ 32. $x^2 - 12x + 32$

33. $x^2 - 10x + 21$ 34. $x^2 - 13x + 30$ 35. $x^2 - 9x + 18$ 36. $x^2 - 22x + 120$

37. $15x^2 - 29x + 12$ 38. $5x^2 - 31x + 6$ 39. $6x^2 - 31x + 35$ 40. $12x^2 - 26x + 15$

41. $x^2 + 6x + 9$ 42. $x^2 + 14x + 49$ 43. $x^2 + 12x + 36$ 44. $x^2 + 24x + 144$

45. $4x^2 + 12x + 9$ 46. $25x^2 + 40x + 16$ 47. $x^2 - 10x + 25$ 48. $x^2 - 4x + 4$

49. $x^2 - 20x + 100$ 50. $x^2 - 6x + 9$ 51. $36x^2 - 12x + 1$ 52. $49x^2 - 28x + 4$

53. $x^2 - 16$ 54. $x^2 - 25$ 55. $x^2 - 9$ 56. $x^2 - 64$ 57. $x^2 - 49$ 58. $4x^2 - 9$

59. $16x^2 - 25$ 60. $9x^2 - 81$ 61. $49x^2 - 16$ 62. $36x^2 - 169$

Exercise 3.3

1. $3(a + 4)$ 2. $4(y - 2)$ 3. $5(p + 2q)$ 4. $7(x - 3y)$ 5. $4y(2x - 3z)$ 6. $3p(2p - 5r)$

7. $6pq(p - 3q)$ 8. $xy(7y + 4x)$ 9. $-2xy(2x + 3y)$ 10. $-4ab(3a - 2b)$

11. $-5xy(y + 2x)$ 12. $-7a^2b^2(a + 3b)$ 13. $(a + d)(b + c)$ 14. $(p - q)(r + s)$

15. $(x + y)(5a + 2b)$ 16. $(a + 2b)(3c + 4d)$ 17. $(y + 2z)(2x - 3y)$

18. $(2y - 3z)(3x - 4w)$ 19. $(a + b)(3x - y)$ 20. $2(5a - b)(y + 3)$

21. $(r - 2s)(4p - 5q)$ 22. $2(a + b)(2u + 3v)$ 23. $(x - 2y)(7x - 3z)$

24. $(5a + 2b)(2x + 3y)$ 25. $(x + 4)(x + 5)$ 26. $(x + 3)(x + 7)$ 27. $(x + 6)(x + 8)$

28. $(x + 2)(x + 10)$ 29. $(x + 9)(x + 5)$ 30. $(x + 3)(x + 12)$ 31. $(x + 4)(x + 8)$

32. $(x - 2)(x - 7)$ 33. $(x - 3)(x - 5)$ 34. $(x - 4)(x - 9)$ 35. $(x - 5)(x - 6)$

36. $(x - 7)(x - 10)$ 37. $(x - 5)(x - 8)$ 38. $(x - 3)(x - 9)$ 39. $(x - 6)(x - 9)$

40. $(x - 5)(x - 10)$ 41. $(x - 4)(x + 5)$ 42. $(x - 3)(x + 7)$ 43. $(x - 6)(x + 8)$

44. $(x - 5)(x + 10)$ 45. $(x - 7)(x + 10)$ 46. $(x - 5)(x + 8)$ 47. $(x - 3)(x + 9)$

48. $(x - 3y)(x + 5y)$ 49. $(x - 2y)(x + 3y)$ 50. $(x - 4y)(x + 8y)$ 51. $(x - 3y)(x + 10y)$

52. $(x - 4y)(x + 9y)$ 53. $(x - 7)(x + 2)$ 54. $(x - 10)(x + 3)$ 55. $(x - 12)(x + 8)$

56. $(x - 8)(x + 7)$ 57. $(x - 6)(x + 4)$ 58. $(x - 9)(x + 6)$ 59. $(x - 9)(x + 3)$

60. $(x - 8y)(x + 3y)$ 61. $(x - 7y)(x + 5y)$ 62. $(x - 5y)(x + 4y)$

63. $(x - 8y)(x + 2y)$ 64. $(x - 7y)(x + 6y)$ 65. $(2x + 1)(x + 4)$

66. $(3x + 2)(x + 3)$ 67. $(5x - 4)(x + 6)$ 68. $(2x - 3)(3x + 2)$ 69. $(2x - 3)(5x + 3)$

70. $(3x - 2)(7x + 4)$ 71. $(4x + 3)(6x - 7)$ 72. $(2x + 9)(3x - 5)$ 73. $(2x + 5)(x + 3)$

74. $(3x - 5)(5x + 2)$ 75. $(3x - 5)(4x - 3)$ 76. $(2x + 7)(4x - 5)$ 77. $(2x + 3)(3x - 5)$

78. $(5x - 3)(x + 1)$ 79. $(3x + 4)(x + 2)$ 80. $(x + 3)^2$ 81. $(x + 5)^2$ 82. $(x + 7)^2$

83. $(x+6)^2$ 84. $(x+9)^2$ 85. $(x+12)^2$ 86. $(x+15)^2$ 87. $(2x+4y)^2$

88. $(3x+5y)^2$ 89. $(4x+3y)^2$ 90. $(5x+4y)^2$ 91. $(2x+5y)^2$ 92. $(x-2)^2$

93. $(x-6)^2$ 94. $(x-8)^2$ 95. $(x-10)^2$ 96. $(x-11)^2$ 97. $(x-13)^2$ 98. $(x-6y)^2$

99. $(x-5y)^2$ 100. $(3x-2y)^2$ 101. $(2x-5y)^2$ 102. $(x-4)(x+4)$

103. $(x-8)(x+8)$ 104. $(x-12)(x+12)$ 105. $(3y-8)(3y+8)$

106. $(2x-5)(2x+5)$ 107. $(13-x)(13+x)$ 108. $(5x-4y)(5x+4y)$

109. $3(3x-2y)(3x+2y)$ 110. $8(x-3y)(x+3y)$ 111. $5x(x-3)(x+3)$

112. $3(x^3-4)(x^3+4)$ 113. $(1-7x)(1+7x)$ 114. $5x(2x-3y)(2x+3y)$

115. $2y(4x-3y)(4x+3y)$ 116. $2x(5x-2y)(5x+2y)$

Exercise 3.4

1. $\frac{2a}{3b}$ 2. $\frac{5y}{6x}$ 3. $\frac{4}{3x}$ 4. $\frac{3}{a}$ 5. $\frac{2x-1}{x}$ 6. $\frac{1}{y-4}$ 7. $\frac{x(x+4)}{x+2}$ 8. $\frac{x+5}{3x}$ 9. $\frac{x+2}{x+3}$

10. $\frac{x+4}{x+6}$ 11. $-\frac{2x-1}{x+3}$ 12. $\frac{x-3}{x+8}$ 13. $\frac{x-2y}{x+2y}$ 14. $\frac{2-x}{2+x}$ 15. $\frac{1}{x-3y}$ 16. $\frac{5y}{6x}$

17. $\frac{3x}{4}$ 18. $\frac{3y^2}{2x^3}$ 19. $\frac{y}{2x}$ 20. $\frac{5z}{3xy}$ 21. $\frac{x-y}{x}$ 22. $\frac{2x}{x+2y}$ 23. $\frac{3}{2}$ 24. $\frac{x-2y}{a}$

25. $\frac{a+b}{a(a-b)}$ 26. 1 27. $\frac{x+3}{x+1}$ 28. $\frac{xz}{y}$ 29. $3xy$ 30. $\frac{5}{6xy}$ 31. $\frac{x}{4a}$ 32. $\frac{x}{y}$

33. $\frac{4x}{x+y}$ 34. $\frac{x+2}{x}$ 35. $\frac{1}{x-4}$ 36. $\frac{x-3}{10x}$ 37. $\frac{b}{a}$ 38. $\frac{-2}{3y}$ 39. $\frac{1}{a+b}$ 40. $12x^2$

41. $6x^2y^2$ 42. $(a-b)(a+b)$ 43. $3(x-1)(x+1)$ 44. $(2x-y)(2x+y)$

45. $(x-2)(x+3)$ 46. $x(x+2)(x+3)$ 47. $(x-4)(x-3)(x-2)$

48. $(x+1)(x+2)$ 49. $x(x-y)(x+y)$ 50. x 51. $\frac{y}{4}$ 52. $\frac{13}{2x}$ 53. $\frac{1}{6y}$

54. $\frac{7x-y}{12}$ 55. $\frac{-5x}{12}$ 56. $\frac{11y}{12}$ 57. $\frac{x+5}{2}$ 58. $\frac{4(5x-3)}{(2x-3)(2x+3)}$ 59. $-\frac{2}{x+3}$ 60. $\frac{3}{x-y}$

61. $\frac{x+5y}{(x-y)(x+y)}$ 62. $\frac{6}{(x-6)(x+5)}$ 63. $\frac{2(x-5)}{(x-3)(x+1)(x+3)}$ 64. $\frac{3y+5}{(y-5)9y+3)}$ 65. $\frac{x+27}{(x-5)(x+2)(x+7)}$

66. $\frac{x+6}{(x-4)(x+3)^2}$ 67. $\frac{5(1-x)}{(x-3)(x+2)^2}$ 68. $\frac{19-7x}{9x-2)(x+2)}$ 69. $-\frac{5}{x+y}$ 70. $\frac{2(y+1)}{y-1}$ 71. $\frac{x(y+3)}{y(x-3)}$

72. $\frac{x}{x-1}$ 73. $\frac{x-y}{y}$ 74. $\frac{2y-x}{xy}$ 75. $\frac{y}{x}$ 76. $\frac{3x-1}{3-2x}$ 77. $\frac{x+3}{x-3}$ 78. $\frac{x-3y}{x=2y}$ 79. $\frac{1}{2y-x}$

80. $\frac{x-3}{x-2}$ 81. $\frac{x-3}{x-5}$

Review Exercises

1. $15x^4y^2$ 2. $-14x^5$ 3. $-12x^6$ 4. $21x^6y^5$ 5. $-6x^9y^3$ 6. $8x^5y^5$ 7. $4x^4y^6$

8. $-125a^3b^6$ 9. $81x^8y^{12}$ 10. $3x^4$ 11. $-4x^2$ 12. $3x^2$ 13. $-5mn$ 14. $-8xy$

15. $5x^3y$ 16. 32 17. – 4 18. – 26 19. – 40 20. 44 21. 14 22. $7x$ 23. $-2a$

24. $8y$ 25. $14y$ 26. $8x + y$ 27. $-9a + 3b$ 28. $5x^2 + 2x$ 29. $15x - 6x^2$

30. $-18xy + 12x$ 31. $6xy + 2x^2$ 32. $-2p^2 + 6pq - 8p$ 33. $3x^2 - 6xy + 3x$

34. $-3x + 3$ 35. $7y + 7$ 36. $y^3 + 6$ 37. $x^3 - 12$ 38. $a + 3$ 39. $5y - 5$

40. $2y^2 - y$ 41. $-2x - 6y$ 42. $x^2 + 9x + 18$ 43. $x^2 + 4x - 12$ 44. $a^2 - 13a + 42$

45. $6x^2 - 5x - 6$ 46. $y^2 - 12y + 32$ 47. $3x^2 - 7x - 6$ 48. $x^2 - 9$ 49. $x^2 - 36$

50. $4x^2 - 9$ 51. $x^2 + 6x + 9$ 52. $4x^2 - 12x + 9$ 53. $9x^2 + 12x + 4$ 54. $2x(x + 3)$

55. $5t(2t - 1)$ 56. $9pq(3q + 2p)$ 57. $4x(3x + 2)$ 58. $8xy(x - 3y)$ 59. $3xy(x + 2y)$

60. $(x + 2)(x + y)$ 61. $(x - 2)(2x + z)$ 62. $(x + y)(x + 1)$ 63. $(a - 3)(a + y)$

64. $(x - 2)(x + 2)(x + 3)$ 65. $(x + 6)(2x^2 - 5)$ 66. $(x + 1)(x + 5)$ 67. $(y + 4)(y + 7)$

68. $(y - 5)(y + 9)$ 69. $(x - 5)(x + 12)$ 70. $(x - 5)(x + 3)$ 71. $(x - 6)(x + 7)$

72. $(2x - 1)(x + 4)$ 73. $(2x - 7)(3x - 1)$ 74. $(7x + 1)(x + 2)$ 75. $(3a - 4)(3a + 2)$

76. $(2a + 3)(2a - 5)$ 77. $(3x + 1)(x - 2)$ 78. $(x - 9)^2$ 79. $(x + 7)^2$ 80. $(4x - 3)^2$

81. $(2x + 3)^2$ 82. $(8x + 1)^2$ 83. $(3x - 5)^2$ 84. $(x - 6)(x + 6)$ 85. $(x - 3)(x + 3)$

86. $(3y - 2)(3y + 2)$ 87. $(4a - 3)(4a + 3)$ 88. $(4a - 3b)(4a + 3b)$

89. $3x(5y^3 - x^2)(5y^3 + x^2)$ 90. $\frac{2x^3}{3}$ 91. $\frac{2y}{3x}$ 92. $\frac{3ab^3}{4}$ 93. $\frac{3x^2}{4y^2}$ 94. $\frac{3x}{4y}$ 95. $-\frac{2x^2}{3y}$

96. $-\frac{2x}{3y^3}$ 97. $\frac{b^2}{2a^2}$ 98. $\frac{x+1}{3}$ 99. $\frac{5y}{y-5}$ 100. $\frac{x-4}{x-5}$ 101. $x - 2$ 102. $\frac{x^2y^2}{6}$ 103. $\frac{3y^2}{2x^4}$

104. $\frac{5b}{12}$ 105. $\frac{2}{3y}$ 106. $\frac{2}{3}$ 107. $x(x - 2)$ 108. $\frac{3x}{x-8}$ 109. $\frac{5x}{x+2}$ 110. $-\frac{2(x+2)}{x}$ 111. $\frac{3+y^2}{2}$

112. $\frac{5}{6xy^2}$ 113. $\frac{9}{20}$ 114. $\frac{1}{3}$ 115. $\frac{a+3}{10a}$ 116. $\frac{2(p-2)}{9}$ 117. $\frac{4x}{2x+3}$ 118. $\frac{2x+1}{2x}$ 119. $\frac{x+2}{3x^2}$

120. $\frac{1}{xy}$ 121. $\frac{2}{ab}$ 122. 7 123. $\frac{x-y}{y}$ 124. $\frac{3-x}{(x+1)(x+3)}$ 125. $\frac{3(x-1)}{2(x-3)(x+3)}$

126. $\frac{3y^2+8y+10}{(y-3)(y+2)(y+5)}$ 127. $\frac{2a^2-7a+9}{(a-3)(a-2)(a+4)}$ 128. $\frac{x}{(x+2)(x+3)}$ 129. $\frac{4y^2+7y-15}{(y-5)(y-1)(y+1)}$

130. $-\frac{x(y+x)}{y}$ 131. 1 132. $\frac{x-y}{x+y}$ 133. $\frac{x-3}{x-1}$ 134. $\frac{a-2}{a}$ 135. x

Chapter 4

Exercise 4.1

1. 3 2. 2 3. 3 4. 5 5. 2 6. 2 7. – 6 8. 3 9. – 2 10. 2 11. $\frac{1}{4}$

12. $1\frac{1}{2}$ 13. 4 14. 1 15. 1 16. – 1 17. – 2 18. 2 19. 2 20. – 2

21. – 3 22. 3 23. 1 24. – 3 25. 3 26. 4 27. – 3 28. – 3 29. 3

30. 4 31. 2 32. 2 33. $1\frac{1}{3}$ 34. 2 35. – 5 36. 4 37. 7 38. 1

39. 7 40. $\frac{3}{4}$ 41. 6 42. 2 43. – 6 44. $-1\frac{1}{2}$ 45. 3 46. 18 47. 18

48. 12 49. 9 50. – 10 51. $7\frac{1}{3}$ 52. $-\frac{4}{5}$ 53. – 14 54. – 3 55. – 11

56. $\frac{7}{12}$ 57. 2 58. – 1

Exercise 4.2

1. $x = \frac{b+d}{c}$ 2. $x = \frac{c-a}{b}$ 3. $x = a(c-b)$ 4. $x = \frac{bc}{c-b}$ 5. $x = \frac{d}{f-e}$ 6. $x = \frac{2b-3a}{a+b}$

7. $x = \frac{bc}{a(c-b)}$ 8. $x = ac + b$ 9. $x = \frac{5a}{a+b}$ 10. $x = \frac{a-c}{b-c}$ 11. $x = \sqrt{\frac{b}{1-a}}$

12. $x = \sqrt[3]{\frac{b-1}{a}}$ 13. $x = \sqrt{\frac{3f-2g}{g+f}}$ 14. $x = \frac{b^3c}{a}$ 15. $x = \sqrt{\frac{c}{b-a}}$ 16. $t = \frac{v-u}{a}$

17. $n = \frac{2S}{a+l}$ 18. $R = \frac{E-rI}{I}$ 19. $r = \frac{s-a}{s}$ 20. $T = \frac{100I}{PR}$ 21. $v = \sqrt{\frac{2gk}{m}}$

22. $R = \sqrt{\frac{A+hr^2}{h}}$ 23. $r = \sqrt[3]{\frac{3V}{4\pi}}$ 24. $n = \frac{IR}{E-Ir}$ 25. $t = \sqrt{\frac{2s}{g}}$ 26. $g = \frac{4\pi^2 l}{T^2}$

27. $C_1 = \frac{CC_2}{C_2-C}$ 28. $v = \sqrt{\frac{2(E-V)}{m}}$ 29. $v = \frac{mg\cos\propto}{k^3}$ 30. $P = \frac{m+R^2Q}{R^2}$ 31. 10.5

32. 4 33. 200 34. 4 35(a) $l = g\frac{T^2}{4\pi^2}$ (b) 1.12 36. 60 ohms 37. 4.47 m s^{-1}

38. 45^0 39(a) $P = \frac{rk}{Q} - ms$ (b) 27 40(a) $v = \frac{fu}{u-f}$ (b) 4

Exercise 4.3

1. $x + y$ 2. $x + 3$ 3. $a + 2$ 4. $p + 5$ 5. $a - c$ 6. $s - 7$ 7. $7 - z$ 8. pq

9. $6b$ 10. $2x - 8$ 11. $3m + 1$ 12. $2(a + b)$ 13. $2(x + y)$ 14. $3(x - y)$

15. $3x + 2$ 16. $\frac{a-b}{3}$ 17. $\frac{2x-5}{3}$ 18. $\frac{2x+y}{5}$ 19. $2(x - 3)$ 20. $3(x - 4)$ 21. $2x + 3$

22. $3y-6$ 23. $2xy$ 24. $\frac{a+b}{5}$ 25. $x+5$ 26. $12-x$ 27. $x-3$ 28. $32-x$

29. $\frac{x}{4}$ 30. $x+5$ 31. $x+2$ 32. $3x+5$ 33. $15-9x$ 34. $2x+50$

35. $x+3x=12$ 36. $2x-x=5$ 37. $3x+10=19$ 38. $5x+3=2x+9$

39. $3x+6=30$ 40. $2(x+5)+3x=85$ 41. $6x=72$ 42. $x+4x=25$

43. $2x+3=5x-9$ 44. $3(x+1)=x$ 45. 3, 9 46. 3 47. 13, 18 48. 2

49. 9, 18 50. 26 51. 8, 10, 12 52. 32, 33, 34 53. 27, 29, 31 54. 6 cm

55. width 8 cm, length 22 cm 56. width 17 cm, length 25 cm

57. 7 of 35 Gp stamps 8 of15 Gp stamps 58. 35 59. 80 60. 4 at $1.50, 6 at $ 2

61. 40 years 62. 16 years, 48 years 63. 10 years 64. 9 years 65. 40 km h $^{-1}$, 30 km h $^{-1}$

66. 20 km h $^{-1}$, 25 km h $^{-1}$ 67. 7 p.m. 68. 5 pm 69. 4 pm 70. 4 pm 71. 3 hours

72. $2\frac{1}{2}$ hours 73. 25 km h $^{-1}$, 50 km h $^{-1}$ 74. 15 km h $^{-1}$, 20 km h $^{-1}$ 75. 160 km, 110 km

76. 12 noon

Exercise 4.4

1. 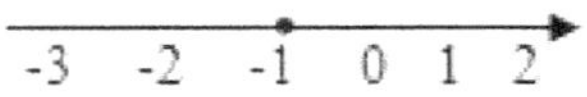 -3 -2 -1 0 1 2
2. -1 0 1 2 3 4 5
3. -5 -4 -3 -2 -1 0 1
4. -1 0 1 2 3 4
5. -2 -1 0 1 2 3 4
6. −2 ½ 0
7. 0 2 ¾
8.  0 ¾
9. 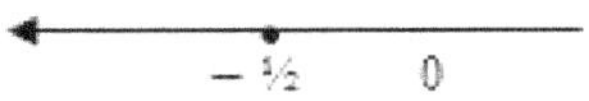 − ½ 0
10. 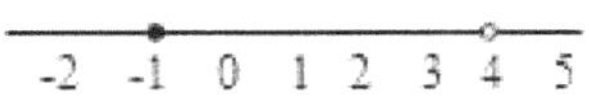 -2 -1 0 1 2 3 4 5
11. -2 -1 0 1 2 3
12. -3 -2 -1 0 1 2 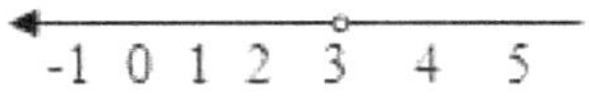 3 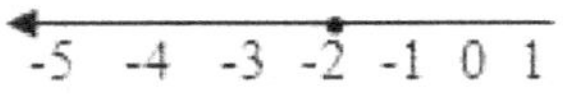 4 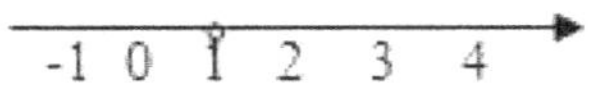 5 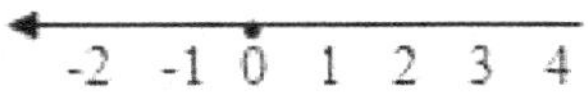 6

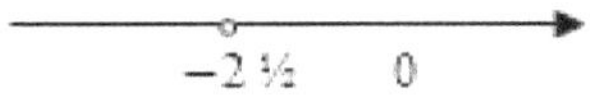

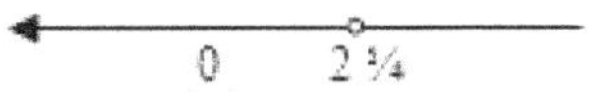

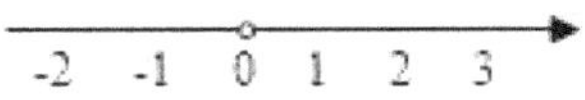

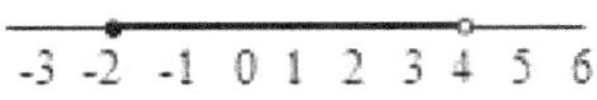

13. $x>3$ 14. $x<2$ 15. $x\le 2$ 16. $x\le 3$ 17. $x\ge -4$ 18. $x>5$

19. $x\ge 1\frac{1}{2}$ 20. $x\ge 1$ 21. $x>-4$ 22. $x\ge 2$ 23. $x>1\frac{1}{2}$ 24. $x<-20$

25. $x > -1$ 26. $x < 3$ 27. $x \geq 1\frac{1}{3}$ 28. $x < -2$ 29. $x \geq -9$ 30. $x > 1\frac{1}{2}$

31. $x \leq -5$ 32. $x < -3$ 33. $x > 3$ 34. $x \leq 2\frac{1}{2}$ 35. $x > -5$ 36. $x \geq -2$

37. $8 < x < 10$ 38. $-2 \leq x < -1$ 39. $-3 \leq x < 2$ 40. $-5 < x < 1$

41. $-5 < x < 11$ 42. $-4 < x < 1$ 43. $1 < x < 3$ 44. $1 \leq x \leq 4$

45. $-23 < x < -11$ 46. $2 \leq x < 5$ 47. $x < 2$ 48. $x \leq 7$ 49. $x \geq -4$

50. 5 51. At least 92 metres 52. at most 12 cm 53. $10 < x < 17$ 54. 4

55. 12 years 56. At least 97 57. At least 82 58. 20

Review Exercises

1. 3 2. 5 3. 2 4. 3 5. – 5 6. – 4 7. 5 8. $2\frac{1}{2}$ 9. 4 10. – 4

11. 10 12. 10 13. 2 14. 2 15. 2 16. 8 17. 17 18. 8 19. $a = \frac{2A-bh}{h}$

20. $P = \frac{A}{I+rt}$ 21. $c = \sqrt{\frac{E}{m}}$ 22. $h = \frac{A-2\pi r^2}{2\pi r}$ 23. $r = \sqrt[3]{\frac{3V}{4\pi}}$ 24. $d = \frac{ac}{2a+c}$

25. 3.75 26. 5 cm 27. 68^0 F 28. $x + 5 = 9$ 29. $x - 7 = 15$ 30. $3x - 4 = 2x$

31. $2(x + 3) = x + 12$ 32. 13 33. 16 34. 20 21 22 35. 33 35 37

36. 40 42 44 37. 1550, 1710 38. 9 years old 39. \$820

40. -5 -4 -3 -2 -1 0 1

41. -1 0 1 2 3 4 5 6

42. -5 -4 -3 -2 -1 0 1

43. -5 -4 -3 -2 -1 0 1

44. -5 -4 -3 -2 -1 0 1

45. – ¾ 0

46. $x \geq -4$ 47. $x < 7$ 48. $x < 8$ 49. $x > 8$ 50. $x > 9$ 51. $x \geq 8$ 52. $x < 1\frac{1}{2}$

53. $x > -\frac{2}{3}$ 54. $3x + 5 < 7$ 55. $x - 3 > 5$ 56. $2x + 7 \geq 12$ 57. $60 < x < 80$

58. $x \geq 45$ 59. $x \leq 1200$ 60. $x \leq 100$ 61. $x \geq 2.50$ 62. $2\frac{1}{2}$ hours 63. 6

64. At least 84 65. At most 12 m 66. at least 6 hours 67. at least 5 minutes

Chapter 5

Exercise 5.1

1. – 6, 2 2. 3, 5 3. – 9, 2 4. – 5, 6 5. – 4, 5 6. – 8, - 4 7. – 8, - 7

8. 5, 9 9. – 7, 4 10. – 6, 7 11. – 5, 3 12. – 7, 0 13. ± 4/3 14. ± 5/2

15. ± 4/3 16. – 2, 9 17. 4, 5 18. 3, 4 19. – 7, 2 20. – 3, 10 21. – 8, 2

22. – 1/3, 7 23. – 3/2, – ¼ 24. – ½ , 5/3 25. 4/3, 3 26. 2/7, 5 27. – 8, 3/2

28. – 5/3, 3 29. 7/2, 5 30. – 1, 5/3 31. – 1, 6/5 32. – 5/2, 4/3 33. – 5/2, 3/2

Exercise 5.2

1. – 3, 1 2. – 3, 4 3. – 1.16, 5.16 4. – 3.27, - 1.63 5. – 2.14, 5.14

6. – 2.37, 4.37 7. ½ , 1 8. 0.24, 2.76 9. – 4.44, - 0.56 10. – 3/5, 2

11. – 3/2, 5 12. – 0.72, 1.72

Exercise 5.3

1. 1, 3 2. – 2, 5 3. – 4.41, -1.59 4. 0.44, 4.56 5. – 10, 3 6. – 1/3, 2

7. – 2/7, 1 8. – 3.58, - 0.42 9. 0.44, 1.36 10. – 2, ¾ 11. – 1.09, 0.34

12. – 0.15, 2.15 13. – 1, 7/3 14. – 1.26, 2.42 15. – ½ , 3/2 16. – 2.12, 0.12

Exercise 5.4

1. 15, 16 2. 13, 15 3. 12, 14 4. 4, 8 5. 13 cm by 26 cm 6. 5, 8

7. 6 cm by 8 cm 8. 7 cm, 10 cm 9. 9 cm, 18 cm 10. 12 m 11. 6 m 12. 2 hours

Review Exercises

1. 1, 6 2. – 5, - 1 3. – 7, 3 4. – 9, 2 5. – 7, - 2 6. – 5, - 3 7. 0, 4

8. 1 twice 9. 4 twice 10. 3 twice 11. 1, 4 12. – 2, 3 ½ 13. – 5/3, 4

14. – 1, 5/3 15. – ¼ , 2/3 16. – 1, 2/3 17. – 1, 6/5 18. – 4, 5/2 19. – 7, 5/2

20. 1/5, 5 21. – 4, 3/2 22. – 4, 4/3 23. 14, 16 24. 5, 8 25. 12 cm by 16 cm

26. 9 cm, 12 cm 27. 5 cm, 12 cm 28. 8 cm 29. 5 hours 30. 378 km

Chapter 6

Exercise 6.1

1. (2, 3) 2. (1, -1) 3. (3, 2) 4. (-2, 3) 5. (1, - 3) 6. (1, -2) 7. (2, – ½)

8. (1, 1) 9. (2, -1) 10. (2, -2) 11. (0, -6) 12. (-12, -3) 13. (3, 2)

14. (3, 4) 15. (-3, -2)

Exercise 6.2

1. (-1, 2) 2. (3, 4) 3. (1, -2) 4. (-2, -3) 5. (1, 1) 6. (3, 2) 7. (4, -3)

8. (3, -1) 9. (2, -3) 10. (2, 3) 11. (1, 2) 12. (1, 4)

Exercise 6.3

1. 10, 15 2. 8, 9 3. 3, 4 4. 3, 7 5. $1, $0.50 6. 25¢, 40¢ 7. 18 m, 12m

8. 4 m, 9 m 9. 5 cm by 12cm 10. 12 kg, 18 kg 11. 12 years, 36 years

12. 11 years, 35 years 13. 13 of the 5¢ coin, 9 of the 20¢ coin

14. 250 student tickets, 150 adult tickets 15. 300 adult tickets, 400 children tickets

16. $100, $20 17. 120 ml, 80 ml 18. 20 ml, 10 ml 19. $7000, $5000

20. $5000, $3000

Exercise 6.4

1. (2, 6); (4, 3) 2. (− 3, − 10/3), (5, 2) 3. (− 6/5, 10/3), (− 2, 2) 4. (− 5/4, − 12), (3, 5)

5. (5, 6); (2, 15) 6. (4, -3); (-2, 6) 7. (2, 1); (8, -11) 8. (3, 1); (27/11, 29/11)

9. (-1, -4); (5, 8) 10. (4, 1); (-8, -3) 11. (2, 3); (− 7/5, − 19/5) 12. (2, -1); (6, 5)

13. (3, -3); (3/4, 3/2) 14. (3, 4) 15. (4, -1) 16. (-3, -1); (-1, 3)

17. (3, -5); (− 18/5, 19/5) 18. (2, 1); (-2, 3)

Exercise 6.5

1. (1, 2, 3) 2. (1, 2, -3) 3. (2, 1, -1) 4. (1, 2, 3)

5. (0, 5, 3) 6. (6, 4, -5)

Review Exercises

1. (1, 1) 2. (-2, 2) 3. (2, 1) 4. (-3, 3) 5. (7, -2) 6. (2, -5) 7. (3, 2)

8. (6, 3) 9. (-2, 3) 10. (2, -3) 11. (-3, -5) 12. (3, -1) 13. 13, 37

14. 35°, 145° 15. 10 of 5 ¢ coin, 12 of 10 ¢ 16. 300, 200 17. 5 cm by 13 cm

18. 5 cm, 7 cm, 7cm 19. $500, $300 20. 120 km h^{-1}, 60 km h^{-1} 21. (-1, 3); (-3, 1)

22. (0, 2); (2, 0) 23. (-1, -4); (3, 0) 24. (1, 0); (− ½ , 3/2) 25. (-2, 4); (4, -2)

26. (1, 2); (-3, 0) 27. (2, 3, 4) 28. (2, -1, 1) 29. (1, 3, -2) 30. (− 3, ½ , − 1)

Chapter 7

Exercise 7.1

1. $2\sqrt{5}$ 2. $2\sqrt{7}$ 3. $3\sqrt{3}$ 4. $2\sqrt{10}$ 5. $3\sqrt{5}$ 6. $4\sqrt{3}$ 7. $5\sqrt{2}$

8. $3\sqrt{6}$ 9. $2\sqrt{15}$ 10. $6\sqrt{2}$ 11. $5\sqrt{3}$ 12. $4\sqrt{5}$ 13. $4\sqrt{6}$ 14. $7\sqrt{2}$

15. $7\sqrt{3}$ 16. $5\sqrt{6}$ 17. $9\sqrt{2}$ 18. $6\sqrt{5}$ 19. $8\sqrt{3}$ 20. $7\sqrt{5}$ 21. $12\sqrt{2}$

22. $8\sqrt{5}$ 23. $9\sqrt{5}$ 24. $14\sqrt{3}$ 25. $11\sqrt{5}$ 26. $15\sqrt{3}$ 27. $11\sqrt{6}$ 28. $6\sqrt{15}$

29. $11\sqrt{7}$ 30. $12\sqrt{5}$ 31. $13\sqrt{2}$ 32. $9\sqrt{7}$

Exercise 7.2

1. $7\sqrt{3}$ 2. $8\sqrt{2}$ 3. $7\sqrt{5}$ 4. $2\sqrt{7}$ 5. $-3\sqrt{3}$ 6. $-2\sqrt{5}$ 7. $6\sqrt{2}$

8. $6\sqrt{3}$ 9. $4\sqrt{5}$ 10. $-5\sqrt{7}$ 11. $4\sqrt{3}$ 12. $8\sqrt{2}$ 13. $2\sqrt{3}$ 14. $2\sqrt{5}$

15. $16\sqrt{2}$ 16. $7\sqrt{2}$ 17. $\sqrt{5}$ 18. $3\sqrt{2}$ 19. 0 20. $-2\sqrt{3}$ 21. $2\sqrt{7}$

22. $15\sqrt{2}$ 23. $\sqrt{6}$ 24. $6\sqrt{2}$ 25. $\sqrt{5}$ 26. $7\sqrt{3}$ 27. $-\sqrt{2}$ 28. $4\sqrt{2}$

29. $-4\sqrt{5}$ 30. $5\sqrt{6}$ 31. $-9\sqrt{3}$ 32. $-6\sqrt{2}$ 33. $8\sqrt{2}$ 34. $3\sqrt{3}$

35. $3\sqrt{5}$ 36. $6\sqrt{2}$

Exercise 7.3

1. $2\sqrt{3}$ 2. $5\sqrt{3}$ 3. $3\sqrt{6}$ 4. $10\sqrt{15}$ 5. $6\sqrt{10}$ 6. $15\sqrt{6}$ 7. $6\sqrt{15}$

8. 6 9. $4\sqrt{5}$ 10. $4\sqrt{10}$ 11. $9\sqrt{5}$ 12. $6\sqrt{6}$ 13. $10\sqrt{15}$ 14. $35\sqrt{6}$

15. $4\sqrt{30}$ 16. $2\sqrt{15}$ 17. $12\sqrt{5}$ 18. $30\sqrt{2}$ 19. $18\sqrt{5}$ 20. $180\sqrt{3}$

21. $\sqrt{15}-3$ 22. $2\sqrt{15}-9$ 23. $3+\sqrt{15}$ 24. $2\sqrt{21}+21$ 25. $6+\sqrt{30}$

26. $2\sqrt{3}-2$ 27. $30-10\sqrt{6}$ 28. $12+4\sqrt{6}$ 29. $13+7\sqrt{3}$ 30. $7-3\sqrt{5}$

31. $6\sqrt{6}+6-6\sqrt{2}-2\sqrt{3}$ 32. $2\sqrt{6}-2$ 33. $2+3\sqrt{2}-2\sqrt{3}-3\sqrt{6}$ 34. $\sqrt{10}-\sqrt{3}$

35. $\sqrt{15}+5\sqrt{5}$ 36. $4\sqrt{5}+8+5\sqrt{2}-2\sqrt{10}$ 37. 1 38. – 18 39. – 15 40. 4

41. 3 42. – 20 43. $7+4\sqrt{3}$ 44. $14-6\sqrt{5}$ 45. $5+2\sqrt{6}$ 46. $11-4\sqrt{6}$

47. $22+12\sqrt{2}$ 48. $30-12\sqrt{6}$ 49. $18-12\sqrt{2}$ 50. $11+4\sqrt{6}$

Exercise 7.4

1. $\frac{2\sqrt{3}}{3}$ 2. $2\sqrt{5}$ 3. $2\sqrt{2}$ 4. $3\sqrt{3}$ 5. $\frac{\sqrt{30}}{2}$ 6. $4\sqrt{3}$ 7. $\frac{3\sqrt{10}}{2}$ 8. $3\sqrt{3}$

9. $2\sqrt{7}$ 10. $\frac{3\sqrt{2}}{8}$ 11. $\sqrt{2}$ 12. $\sqrt{10}$ 13. $\frac{3\sqrt{2}}{2}$ 14. $\frac{\sqrt{15}}{3}$ 15. $\frac{\sqrt{6}}{2}$ 16. $\frac{\sqrt{15}}{3}$

17. $\frac{\sqrt{6}}{2}$ 18. $\frac{2\sqrt{15}}{5}$ 19. $\frac{\sqrt{5}}{2}$ 20. $\frac{2\sqrt{3}}{3}$ 21. $\sqrt{2}-1$ 22. $\sqrt{3}+1$ 23. $2+\sqrt{3}$

24. $6-2\sqrt{7}$ 25. $\sqrt{3}+\sqrt{2}$ 26. $\frac{4-\sqrt{6}}{2}$ 27. $\frac{5\sqrt{7}+7}{6}$ 28. $\frac{12\sqrt{3}+2\sqrt{6}}{11}$ 29. $7-4\sqrt{3}$

30. $\frac{11+6\sqrt{2}}{7}$ 31. $2+\sqrt{3}$ 32. $\frac{17-4\sqrt{15}}{7}$ 33. $17-12\sqrt{2}$ 34. $-4-\sqrt{15}$

35. $\frac{1-2\sqrt{14}}{5}$ 36. $-7-4\sqrt{3}$ 37. $\frac{5\sqrt{5}+11}{-2}$ 38. $\frac{2\sqrt{6}+3\sqrt{3}}{3}$ 39. $\frac{13\sqrt{15}+45}{30}$ 40. $\frac{1+\sqrt{21}}{5}$

Exercise 7.5

1. 3 2. 3 3. – 5 4. 3 5. $\pm\frac{1}{2}$ 6. 7 7. 3 8. 3 9. – 4, - 2 10. 2, 5

11. 10 12. 4 13. 9 14. $\frac{9}{16}$ 15. 7 16. 7 17. 2 18. 3 19. 7 20. $\frac{3}{2}$, $\frac{7}{4}$

Review Exercises

1. $6\sqrt{3}$ 2. $3\sqrt{10}$ 3. $3\sqrt{2}$ 4. $3\sqrt{15}$ 5. $13\sqrt{3}$ 6. $17\sqrt{2}$ 7. $6\sqrt{6}$

8. $9\sqrt{3}$ 9. $8\sqrt{6}$ 10. $13\sqrt{5}$ 11. $-5\sqrt{6}$ 12. $4\sqrt{2}$ 13. $-6\sqrt{5}$ 14. $-2\sqrt{2}$

15. $\sqrt{2}$ 16. $4\sqrt{2}$ 17. $6\sqrt{3}$ 18. $4\sqrt{5}$ 19. $4\sqrt{7}$ 20. $6\sqrt{15}$ 21. $25\sqrt{6}$

22. $6\sqrt{15}$ 23. $18\sqrt{6}$ 24. $6\sqrt{6}$ 25. $60\sqrt{5}$ 26. $\frac{\sqrt{2}}{10}$ 27. $\frac{4\sqrt{2}}{3}$ 28. $\frac{4\sqrt{2}}{9}$

29. $\frac{3\sqrt{6}}{8}$ 30. $\frac{\sqrt{3}}{3}$ 31. $\frac{2\sqrt{3}+3}{3}$ 32. $23+7\sqrt{6}$ 33. $4+3\sqrt{14}$ 34. $5+2\sqrt{6}$

35. $79-20\sqrt{6}$ 36. $2-3\sqrt{6}$ 37. 2 38. $\frac{3\sqrt{2}+\sqrt{5}}{13}$ 39. $-12-9\sqrt{2}$ 40. $\frac{8+5\sqrt{3}}{11}$

41. $5+2\sqrt{6}$ 42. $\frac{3\sqrt{6}-5\sqrt{2}}{4}$ 43. $\frac{5\sqrt{6}-6}{19}$ 44. 3 45. 1, 3 46. 26 47. 15 48. – 3, 1

Chapter 8

Exercise 8.1

1. Not 2. Function 3. Not 4. Function 5. Function 6. Not 7. Function

8. Not 9. Domain = {- 1, 0, 1}; Range = {2, 3} 10. Domain = {- 3, -2}; Range = {0, 2}

11. Domain = {-1, 0, 1}; Range = {0} 12. Domain = {-2, -1, 0}; Range = {3, 4}

13. Domain = $\{x| -2 < x < 2\}$; Range = $\{y|y \geq -2\}$

14. Domain = $\{x| -3 < x < 2\}$; Range = $\{y|y \geq -3\}$

15. Domain = $\{x|0 < x \leq 2\}$; Range = $\{y| -1 < y \leq 2\}$

16. Domain = $\{x| -3 < x < 2\}$; Range = $\{y|y \leq 3\}$

17. Domain = $\{x| -2 \leq x \leq 3\}$; Range = $\{y| -2 \leq y \leq 2\}$

18. Domain = $\{x| -2 \leq x \leq 4\}$; Range = $\{y| -3 \leq y \leq 3\}$

Exercise 8.2

1. one – to –one 2. not 3. one –to – one 4. not 9. one –to –one

10. not 11. one – to – one 12. no 13. inverse 14. inverse

15. No 16. Inverse 17. inverse 18. $f^{-1}(x) = \frac{x+8}{3}$ 19. $g^{-1}(x) = \sqrt[3]{4-x}$

20. $h^{-1}(x) = \frac{5-x}{2x}, x \neq 0$ 21. $f^{-1}(x) = \frac{4x+3}{x-2}, x \neq 2$ 22. 5 23. $\frac{2}{5}$ 24. 1 25. – 2

Exercise 8.3

1. (a) $5x - 1$ (b) $x - 7$ (c) $6x^2 + x - 12$ (d) $\frac{3x-4}{2x+3}$

2. (a) $2x^2 + 5x + 3$ (b) $2x^2 + x - 3$ (c) $4x^3 + 12x^2 + 9x$ (d) x

3. (a) $x^2 + x - 6$ (b) $12 + x - x^2$ (c) $x^3 + 3x^2 - 9x - 2$ (d) $\frac{1}{x-3}$

4. (a) $\frac{2x+1}{3x+2}$ (b) $\frac{2x-7}{3x+2}$ (c) $\frac{8x-12}{(3x+2)^2}$ (d) $\frac{2x-3}{4}$

5. $3x^2 + 4x + 9$ 6. $1 + 9x - 2x^2$ 7. $x^5 - 8x^4 + 5x^3 + 50x^2$ 8. $\frac{x+2}{x^2}$

9. $25x + 18$ 10. $9x - 8$ 11. $15x - 7$ 12. $15x + 7$ 13. $2x^2 + 1$

14. $4x^2 + 12x + 8$ 15. $4x + 9$ 16. $x^4 - 2x^2$ 17. $\frac{3x-12}{x+4}, x \neq -4$

18. $\frac{6x-9}{2x+1}, x \neq -\frac{1}{2}$ 19. $2x - 6$ 23. $\frac{2x-4}{3}$. 24. $\frac{2x-14}{3}$ 29. $12x^2 + 36x + 25$

30. 1 31. $6x^2 - 1$ 32. 23 33. 26 34. $9x^2 - 6x + 1$ 35. 196 36. 11

37. $-\frac{19}{2}$ 38. 3 39. $-\frac{3}{22}$ 40. $\frac{23}{6}$ 41. – 15 42. 342 43. 5

44. – 9 45. $-\frac{1}{2}$

Exercise 8.4

5. – 8, 4 6. – 12, 2 7. 6, 9 8. $-\frac{3}{7}, 8$ 9. ∅ 10. – 2, 8 11. ∅

12. $\frac{1}{3}, 1$ 13. $\frac{2}{3}, 2$ 14. – 3, 7 15. 1, $\frac{5}{3}$ 16. – 3, 1 17. $\frac{1}{2}$ 18. 1

Review Exercises

1. function 2. Not 3. Function 4. Function 5. Not 6. Not 7. Function

8. Not 9. function 10. Not 11. Domain $= \{x|x \in R\}$; Range $= \{y|y \in R\}$

12. Domain $= \{x|x \in R\}$; Range $= \{y|y > 0\}$

13. Domain $= \{x|x \in R\}$; Range $= \{y|y \leq 8\}$

14. Domain $= \{x| -3 \leq x \leq 3\}$; Range $= \{y| -2 \leq y \leq 3\}$

15. Domain $= \{x| -4 \leq x \leq 3\}$; Range $= \{y| -4 \leq y \leq 3\}$

16. Domain $= \{x| -3 \leq x \leq 3\}$; Range $= \{y| -4 \leq y \leq 3\}$

17. $\{x|x \in R\}$ 18. $\{x|x \in R\}$ 19. $\{x|x \geq 3\}$ 20. $\{x| -6 \leq x \leq 6\}$

21. $\{x|x \neq -5\}$ 22. $\{x|x \neq 2\}$ 23. One – to – one 24. Not 25. One – to – one

26. One – to – one 27. not 28. One – to – one 32. No 33. Has inverse

34. No 35. Has inverse 36. no 37. Has inverse 38. 4 39. 5 40. 4

41. $5 - 4a + 2a^2$ 42. $1 - 4a + 4a^2$ 43. $3a^2 - 2$ 44. – 11 45. 9

46. $f^{-1}(x) = \frac{x+3}{2}$ 47. $f^{-1}(x) = \sqrt[3]{9-x}$ 48. $f^{-1}(x) = \frac{2x-8}{3}$ 49. $f^{-1}(x) = \frac{x-2}{3}$

50. $f^{-1}(x) = \frac{3x}{x-2}, x \neq 2$ 51. $f^{-1}(x) = \frac{5x+2}{2-x}, x \neq 2$

58. (a) $5x + 1$ (b) $x + 7$ (c) $6x^2 - x - 12$ (d) $\frac{3x+4}{2x-3}$

59.(a) $2x^2 - x - 3$ (b) $2x^2 - 5x + 3$ (c) $4x^3 - 12x^2 + 9x$ (d) x

60.(a) $x^2 + x + 6$ (b) $-3 + x - x^2$ (c) $x^3 - 3x^2 + 9x - 27$ (d) $\frac{x^2+9}{x-3}$

61.(a) $\frac{5x^2-8x}{(x-3)(x-2)}$ (b) $\frac{-12+10x-x^2}{(x-3)(x-2)}$ (c) $\frac{6x^2+5x-6}{x^2-5x+6}$ (d) $\frac{2x^2-x-6}{3x^2-11x+6}$

62. $x - 12$ 63. $11x + 1$ 64. $4x + 9$ 65. $-7x + 8$ 66. $2x^2 - 1$

67. $4x^2 - 12x + 10$ 68. $\frac{1}{16}x^2 - \frac{1}{2}x + 2$ 69. $\frac{1}{4}x^2 - \frac{3}{4}$ 70. $\frac{1}{2}x - 5$

71. $\frac{1}{2}x - \frac{7}{4}$ 72. $3x - 8$ 73. $\frac{x+8}{3}$ 74. $\frac{x+4}{3}$ 75. $3x - 4$ 76. $\frac{x+8}{9}$

77. $x - 4$ 82. $\frac{5}{2}$ 83. $-\frac{19}{8}$ 84. $\frac{13}{4}$ 85. $-\frac{5}{2}$ 86. $\frac{2}{3}$ 87. $\frac{3}{2}$ 88. $x + 3$

89. 3 90. – 1, 6 91. – 4, 6 92. – 12, 2 93. – 6, 3 94. $-\frac{19}{2}, \frac{5}{2}$

95. $-4, 9$ 96. $\frac{3}{2}, \frac{21}{2}$ 97. $-\frac{1}{4}, \frac{15}{4}$ 98. $-\frac{4}{5}, \frac{8}{5}$ 99. $\frac{2}{3}, 2$ 100. $-\frac{1}{4}, \frac{3}{2}$

101. $-3, 7$ 102. $-5, -\frac{1}{3}$

Chapter 9

Exercise 9.1

1. 13 **2**. 5 **3**. 10 **4**. 7.07 **5**. 3.61 **6**. 3.61 **7**. 2.83 **8**. 17

9. $3\sqrt{2}$, $5\sqrt{2}$, $2\sqrt{17}$ **11**. 3 **12**. $-4, 2$ **13**. 4

Exercise 9.2

1. (4, 3) **2**. (2, 4) **3**. (- 3, - 4) **4**. (- 2, 1) **5**. (- 2, 3) **6**. (1, - 1) **7**. $\left(\frac{5}{2}, 3\right)$

8. $\left(-\frac{1}{2}, -1\right)$ **9**. $\left(-\frac{3}{2}, 0\right)$ **10**. (- 1, 1) **11**. (4, 3) **12**. (- 3, - 2)

13. (a) (5, 3), $\left(-\frac{3}{2}, -\frac{3}{2}\right)$, $\left(\frac{1}{2}, -\frac{1}{2}\right)$ **14**. (a) (4, 2), (3, 2), (3, 1) **15**. (4, 8) 16. $x = 9, y = 5$

17. $x = 5, y = -3$

Exercise 9.3

1. $\frac{7}{4}$ **2**. -2 **3**. $\frac{1}{2}$ **4**. $\frac{3}{2}$ **5**. $-\frac{4}{5}$ **6**. $-\frac{5}{6}$ **7**. $\frac{1}{3}$ **8**. $\frac{3}{2}$ **12**. 14 **13**. 5 **14**. 2

15. 1 **16**. $a = 2, b = 1$

Exercise 9.4

1. $y = 3x + 4$ 2. $y = -2x + 5$ 3. $y = 4x$ 4. $3x + 4y + 16 = 0$ 5. $3y - 2x - 3 = 0$

6. $3x + 2y + 6 = 0$ 7. $\frac{3}{2}$, (0, - 2) 8. $\frac{3}{4}$, $\left(0, \frac{5}{2}\right)$ 9. -3, $\left(0, \frac{5}{2}\right)$ 10. $\frac{4}{5}$, (0, - 3)

11. $-\frac{4}{3}$, $\left(0, \frac{2}{3}\right)$ 12. $-\frac{1}{2}$, $\left(0, \frac{3}{2}\right)$ 13. (0, - 1) 14. (0, 4) 15. (0, 5) 16. (0, 4) 17. (0, - 5)

18. (0, - 4) 19. $y = 3x - 2$ 20. $y = -3x - 5$ 21. $y = -4x + 5$ 22. $y = 2x - 9$

23. $y = -7x - 12$ 24. $y = 4x - 5$ 25. $3x + 2y - 24 = 0$ 26. $5x - 3y + 9 = 0$

27. $2x - 5y - 18 = 0$ 28. $3x + 2y + 19 = 0$ 29. $4x - 3y + 1 = 0$

30. $3x + 5y - 13 = 0$ 31. $y = x + 1$ 32. $y = 2x - 9$ 33. $y = 3x - 17$

34. $y = -x + 4$ 35. $y = 3$ 36. $y = -2x + 3$ 37. $y = 3x + 10$ 38. $y = -3x + 17$

39. $y = 2x + 8$ 40. $3x + 2y + 2$ 41. $4x + 3y - 6 = 0$ 42. $3x + 4y + 22 = 0$

43. $5x + 4y - 32 = 0$ 44. $3x - 4y = 0$ 45. $2x - 3y - 16 = 0$

Exercise 9.5

1. parallel 2. Not 3. Parallel 4. Parallel 5. $4x - 3y - 7 = 0$

6. $5x - 2y + 13 = 0$ 7. $2x + 3y = 0$ 8. $3x - 2y + 4 = 0$ 9. $y = -2$

10. $x = 5$ 11. $y = 11$ 12. 3 13. $6x + 4y - 29 = 0$ 14. $-\frac{1}{3}$ 15. $\frac{1}{2}$ 16. – 2

17. 3 18. $-\frac{4}{3}$ 19. $\frac{2}{3}$ 20. perpendicular 21. perpendicular 22. Not 23. Not

24. $3x - 5y + 3 = 0$ 25. $y = 3x + 11$ 26. $2x - 3y + 12 = 0$ 27. $3x + 2y - 1 = 0$

28. $2x - 3y + 4 = 0$ 29. $3x + 2y - 12 = 0$ 30. $y = -x + 7$ 31. – 1 32. $\frac{9}{4}$

33. $3x + 2y - 12 = 0$ 34. 6.5

Review Exercises

1. 5.39 2. 5 3. 13 4. 3.61 5. 5.83 6. 7 7. 8 8. 5.39 9. 12.21

10. 7.21 11. 15 12. 9.22 13. (4, 1) 14. (0, 2) 15. $\left(\frac{9}{2}, \frac{7}{2}\right)$ 16. (1, 1)

17. (4, 3) 18. $\left(\frac{7}{2}, 6\right)$ 19. $\left(\frac{5}{2}, 0\right)$ 20. $\left(-\frac{5}{2}, -2\right)$ 21. (2, 3) 22. $\left(\frac{9}{2}, \frac{9}{2}\right)$

23. (- 2, - 1) 24. (2, 2) 25. – 2 26. $-\frac{4}{3}$ 27. $\frac{3}{2}$ 28. $\frac{5}{4}$ 29. $\frac{4}{3}$ 30. $\frac{3}{4}$

31. $-\frac{7}{2}$ 32. $\frac{2}{3}$ 36. $y = 3x - 2$ 37. $y = 4x - 11$ 38. $y = -2x + 6$

39. $3x - 4y + 18 = 0$ 40. $y = -3x + 6$ 41. $y = 2x + 3$ 42. $2x - 3y + 4 = 0$

43. $x - 4y - 10 = 0$ 44. $x - 2y + 7 = 0$ 45. $5x + 2y - 22 = 0$ 46. $y = x + 1$

47. $5x - 3y - 14 = 0$ 48. $y - 4 = 0$ 49. $y = -3x + 7$ 50. $4x + 3y + 17 = 0$

51. $2x - 3y - 1 = 0$ 52. $\frac{5}{3}$; (0, - 5) 53. $-\frac{3}{4}$; $\left(0, \frac{5}{2}\right)$ 54. 3, $\left(0, \frac{5}{2}\right)$ 55. $-\frac{4}{5}$; (0, 2)

56. $\frac{4}{3}$; $\left(0, \frac{8}{3}\right)$ 57. $\frac{1}{2}$; $\left(0, \frac{3}{2}\right)$ 58. $4x - 3y - 25 = 0$ 59. $5x - 2y + 9 = 0$

60. $2x + 3y + 3 = 0$ 61. $3x - 2y - 10 = 0$ 62. $3x + 5y - 27 = 0$

63. $y = 3x + 1$ 64. $2x + 3y = 0$ 65. $3x + 2y - 5 = 0$ 66. $2x - 3y - 4 = 0$

67. $3x - 2y = 0$

Chapter 10

Exercise 10.1

1. Real 2. Imaginary 3. Real 4. Imaginary 5. Real 6. Real 7. 64; 2

8. 25; 2 9. 0; 1 10. 0; 1 11. 37; 2 12. – 39; no 13. 25; 2 14. – 63; no

15. 0; 1 16. 1, 9 17. $-\frac{1}{3}$, 1 18. $k \le 2$ 19. $k > -\frac{1}{4}$ 20. 2

21. $x \le 1$ or $x \ge 5$

Exercise 10.2

1. $\frac{9}{4}$ 2. 6 3. $\frac{49}{4}$ 4. $\frac{49}{8}$ 5. $\frac{10}{3}$ 6. 4 7. $-\frac{17}{4}$ 8. – 10 9. 2

10. $-\frac{41}{8}$ 11. 1 12. $\frac{1}{2}$ 13. (3, 1) 14. $\left(-\frac{1}{2}, \frac{25}{4}\right)$ 15. (- 1, 3) 16. (- 2, 1)

17. (- 1, 4) 18. (2, 5)

Exercise 10.4

1. $-4 < x < 3$ 2. $x < -5$ or $x > 2$ 3. $x < -6$ or $x > 2$ 4. $-5 < x < 3$

5. $x \le 2$ or $x \ge 3$ 6. $-5 \le x \le -2$ 7. $-2 \le x \le 3$ 8. $-\frac{2}{3} \le x \le 4$

9. $-1 < x < \frac{3}{4}$ 10. $x \le -8$ or $x \ge 3$ 11. $x < -1$ or $x > 4$ 12. $-2 < x < 4$

13. $-3 \le x \le 4$ 14. $x \le -4$ or $x \ge -3$ 15. $-2 < x < 5$ 16. $x < -9$ or $x > 3$

17. $-3 \le x \le 4$ 18. $x < -3$ or $x > 6$ 19. $-\frac{3}{2} < x < -1$ 20. $x \le -\frac{5}{3}$ or $x \ge -1$

21. $-1 \le x \le \frac{1}{2}$ 22. $-\frac{2}{3} \le x \le 3$

Review Exercises

1. real roots 2. Imaginary roots 3. One real root 4. Real roots 5. Imaginary roots

6. real roots 7. 9; 2 8. 49; 2 9. 0; 1 10. 30; 2 11. 0; 1 12. 84; 2

13. 1, 5 14. 2 15. 0, 12 16. 2 17. $2 < k < 6$ 18. $k > \frac{1}{3}$ 19. $k > 3$

20. $\frac{21}{4}$ 21. $\frac{49}{4}$ 22. 10 23. $\frac{17}{4}$ 24. 8 25. 11 26. – 1 27. – 5 28. $-\frac{4}{3}$

29. – 5 30. $\frac{23}{8}$ 31. – 7 32. (- 3, - 4) 33. (4, - 1) 34. (- 1, 2) 35. (2, 3)

36. $\left(-\frac{3}{4}, -\frac{41}{8}\right)$ 37. $\left(\frac{2}{3}, -\frac{7}{3}\right)$ 44. $x \le -5$ or $x \ge 1$ 45. $-1 < x < 7$ 46. $-5 \le x \le 2$

47. $2 < x < 6$ 48. $-2 \le x \le \frac{4}{3}$ 49. $x \le -\frac{5}{2}$ or $x \ge 4$ 50. $4 < x < 7$

51. $x < -4$ or $x > -2$ 52. $-\frac{1}{3} \le x \le 1$ 53. $x \le -\frac{1}{2}$ or $x \ge \frac{2}{3}$ 54. $-7 < x < -2$

55. $-4 < x < \frac{3}{2}$

Chapter 11

Exercise 11.1

1. -4 2. 2 3. 0 4. -50 5. $-\frac{1}{2}$ 6. 7 7. 0 8. 1 9. -104 10. 70

11. $(x^2+3x+6);\ 7$ 12. $(x^2-x+2);\ 2$ 13. $(x^2+2x-3);\ 2$ 14. $(2x^2+x-4);\ 10$

15. $(3x^2+x+6);\ 4$ 16. $(4x^2+8x+19);\ 33$ 17. $(x^2-3x+4);\ 0$

18. $(x^2+2x+2);\ -10$ 19. $(2x^2-9x+20);\ -42$ 20. $(2x^2+2x-1);\ 2$

Exercise 11.1

1. 2 2. 0 3. 31 4. -10 5. 4 6. 4 7. $(x-3)(x-1)(x+1)$

8. $(x-5)(x-1)(x+1)$ 9. $(x-2)(x+1)(x+3)$ 10. $(x-2)(x+2)(2x+3)$

11. $(2x-1)(x-2)(x+1)$ 12. $(2x-1)(x+2)(x+3)$ 13. $(x-2)(x^2+3)$

14. $(x+1)(x+2)(x+3)$ 15. $(x-2)(x+3)(x+4)$ 16. $(x-2)(x+3)(2x+1)$

17. $(x-2)(x-1)(5x+7)$ 18. $(2x-1)(x+2)(x+3)$ 19. 4 20. -1 21. $-\frac{1}{2}$, 1

22. $p=1,\ q=-3$ 23. $a=2,\ b=-10;\ 2x+5$ 24. $a=-3,\ b=-5$

25. $a=5, b=-2$ 26. $p=1,\ q=2$ 27. $a=1,\ b=-3$

Review Exercises

1. x^2+x+2 2. x^2-1 3. $2x^2-x+3$ 4. x^2-1 5. 18 6. 0

7. 0 8. 2 9. $(x-4)(x-2)(x+1)$ 10. $(x-2)(3x-5)(x+1)$

11. $(x-2)(2x-1)(x+2)$ 12. $(x-2)(x-1)(x+3)$ 13. $(x-1);\ (x-2)$

14. $(2x-3);\ (x-1)$ 15. $(x-3)(x-1)(x+2)$ zeros: - 2, 1, 3

16. $(x-3)(x-2)(x+1)$ zeros: - 1, 2, 3 17. $(x-3)(x-2)(x-1)$ zeros: 1, 2, 3

18. $(x-2)(2x-1)(x+2)$ zeros: - 2, $\frac{1}{2}$, 2 19. $(x-3)(x+2)(2x+3)$ zeros: - 2, $-\frac{3}{2}$, 3

20. $(x-5)(x-3)(x-1)(x+1)$ zeros: - 1, 1, 3, 5 21. $a=-7,\ b=-27$

22. $p=-2$ zeros: - 1, $\frac{1}{2}$, 1 23.(a) $a=3,\ b=-4$ (b) $(x+2),\ (x+3)$ (c) 30

24. (a) $a=-2, b=-5$ (b) -2, 1, 3 25. (a) $\frac{2}{3}$, 2 (b) 6

26. $a=2,\ b=-1,\ c=-6$ zeros: $-\frac{3}{2}$, 2 27. $a=2,\ b=-3$

Chapter 12

Exercise 12.1

1. $-\frac{1}{2}$ 2. -4 3. 1 4. $-2, 2$ 5. $-1, \frac{1}{2}$ 6. $-3, 3$ 7. $\frac{3}{2}$

8. 2, 3 9. $-4, 2$ 10. -3 11. $-2, 2$ 12. $-\frac{3}{2}, 2$

Exercise 12.2

1. $\frac{2}{x+1}+\frac{3}{x+2}$ 2. $\frac{-2}{x-1}+\frac{3}{x-2}$ 3. $\frac{4}{3(x-4)}-\frac{1}{3(x-1)}$ 4. $\frac{-3}{2(x+1)}+\frac{3}{2(x-1)}$

5. $\frac{1}{2(x-2)}+\frac{1}{2(x+2)}$ 6. $\frac{5}{3(x-1)}+\frac{4}{3(x+2)}$ 7. $\frac{-7}{x-2}+\frac{9}{x-3}$ 8. $\frac{-1}{5(x-2)}-\frac{9}{5(x+3)}$

9. $\frac{1}{x-1}+\frac{x-1}{x^2+1}$ 10. $\frac{-x}{x^2+1}+\frac{1}{x-2}$ 11. $1+\frac{1}{2(x-1)}-\frac{1}{2(x+1)}$ 12. $1-\frac{7}{4(x+3)}-\frac{1}{4(x-1)}$

13. $x+\frac{1}{2(x-1)}+\frac{1}{2(x+1)}$ 14. $\frac{1}{x+1}+\frac{2}{(x+1)^2}$ 15. $\frac{1}{x+2}-\frac{3}{(x+2)^2}$ 16. $\frac{3}{x-2}-\frac{2}{(x-2)^2}$

Review Exercises

1. $\frac{2}{3}$ 2. $-\frac{3}{2}$ 3. 1, 5 4. $-2, 2$ 5. $-1, 1$ 6. $-2, -1, 2$ 7. $-2, 3$

8. 2, 3 9. $-3, 3$ 10. $-2, \frac{3}{2}$ 17. $\frac{-3}{x-2}+\frac{4}{x-5}$ 18. $\frac{1}{x-1}+\frac{2}{3x+4}$ 19. $\frac{3}{x-7}+\frac{2}{x+4}$

20. $2+\frac{3}{x-4}+\frac{5}{x+3}$ 21. $\frac{-1}{x-2}+\frac{8}{x+5}$ 22. $\frac{5}{x-3}-\frac{2}{x-1}$ 23. $1+\frac{1}{x-2}-\frac{1}{x+2}$

24. $2-\frac{3}{x-2}+\frac{1}{x+1}$ 25. $\frac{2}{1+x}+\frac{3}{(1+x)^2}$ 26. $\frac{1}{2(x-1)}+\frac{x+1}{2(x^2+1)}$

Chapter 13

Exercise 13.1

1. 2 2. $\frac{3}{2}$ 3. $\frac{5}{7}$ 4. 2 5. -2 6. $\frac{7}{3}$ 7. 5 8. 27 9. 1 10. -3

11. -1 12. $\frac{1}{9}$

Exercise 13.2

1. $log_5 125 = 3$ 2. $log_2 4 = 2$ 3. $log_{32} 2 = \frac{1}{5}$ 4. $log_e 1 = 0$ 5. $log_{27} \frac{1}{3} = -\frac{1}{3}$

6. $log_{10} 1000 = 3$ 7. $8 = 2^3$ 8. $64 = 4^3$ 9. $9 = 9^1$ 10. $\frac{1}{16} = 4^{-2}$ 11. $\frac{1}{27} = 9^{-\frac{3}{2}}$

12. 5 13. 5 14. 3 15. -7 16. 3 17. -2 18. $\frac{1}{2}$ 19. 6 20. -2

21. -18 22. -3 23. -2

Exercise 13.3

1. $3log_a x + 2log_a y$ 2. $2log_a x - log_a y$ 3. $2logx + \frac{1}{3}logy$ 4. $-5log_2 x$

5. $3Inx + \frac{1}{2}Iny$ 6. $2log_3 x + 3log_3 y - 2log_3 z$ 7. $2log_5 x + log_5 y - \frac{1}{3}log_5 z$

8. $4log_{10}x + \frac{1}{3}log_{10}y - 3log_{10}z$ 9. $\frac{1}{4}logx + \frac{3}{4}logy - \frac{5}{4}logz$

10. $\frac{1}{3}logx + \frac{2}{3}logy - \frac{4}{3}logz$ 11. $2logx + 3logy - \frac{5}{2}logz$

12. $\frac{2}{3}log_5 10 + \frac{1}{3}log_5 x - \frac{2}{3}log_5 y - \frac{4}{3}log_5 z$ 13. $log_2 x^2 y^3$ 14. $log_3 \frac{x}{y^2}$

15. $log\frac{x^3}{y}$ 16. $log_5 x\sqrt[3]{y^2}$ 17. $log\frac{\sqrt[3]{x^2}}{y^2}$ 18. $In\frac{x^3y^2}{z}$ 19. $log_3\frac{x^2}{y^3z}$ 20. $log_5\frac{\sqrt{x}}{z^3\sqrt[3]{y^2}}$

21. $log\sqrt{\left(\frac{x^2y^4}{y^6}\right)^3}$ 22. $log_2\sqrt[4]{\frac{x^2y^3}{z^4}}$ 23. $log\frac{1000x^2}{\sqrt[4]{y}}$ 24. $log\frac{10\sqrt[3]{x^2}}{y^3}$ 25. 5 26. 7

27. 5 28. 3 29. – 2 30. 3 31. 4 32. – 3 33. $-\frac{1}{3}$ 34. $\frac{1}{3}$ 35. $-\frac{5}{2}$

36. 3 37. $\frac{3}{2}$ 38. $\frac{3}{2}$ 39. 2 40. $\frac{3}{4}$ 41. $\frac{5}{6}$ 42. $\frac{1}{2}$ 43. $\frac{2}{5}$ 44. – 5 45. $-\frac{5}{2}$

46. 3 47. 0, $\frac{1}{3}$ 48. – 6 49. 1.29 50. 2.16 51. 0.63 52. 0.67 53. 1.47

54. 2.34 55. 3.82 56. 2.63

Exercise 13.4

1. 4.70 2. 1.585 3. 1.431 4. 1.893 5. 1.277 6. 2.708 7. 1.667

8. 0.9464 9. 1.285 10. 1.723

Exercise 13.5

1. 81 2. 5 3. 2 4. $\frac{5}{2}$ 5. 2 6. 8 7. $\frac{2}{3}$ 8. 125 9. 7 10. 13 11. 8

12. 20 13. 9 14. 4 15. 2 16. 3 17. $\frac{1}{3}$, 9 18. 2 19. 1 20. 2 21. $\frac{3}{2}$

22. 6 23. 3 24. $\frac{12}{19}$ 25. $\frac{15}{4}$ 26. 6 27. 6 28. 3, 5 29. 4, 6 30. 1

Review Exercises

1. 10 2. $\frac{2}{3}$ 3. $\frac{4}{3}$ 4. 0 5. 2 6. – 3 7. $2logx + logy - 3logz$

8. $5logx + 3logy - 2logz$ 9. $2log_2 3 + 2log_2 x + 3log_2 y - 4log_2 z$

10. $\frac{3}{2}Inx + \frac{5}{2}Iny - \frac{3}{2}Inz$ 11. $\frac{5}{2}log_5 x - \frac{3}{2}log_5 y - \frac{7}{2}log_5 z$ 12. $\frac{1}{3}logx + \frac{2}{3}logy - \frac{4}{3}logz$

13. $logx^3y^5$ 14. $log\frac{x^2}{y^3}$ 15. $log\frac{x^5y^3}{z^2}$ 16. $log\sqrt[5]{\frac{x^3}{y^4}}$ 17. $log\frac{\sqrt[3]{x^2y^4}}{z^3}$ 18. $Inx^2\sqrt[4]{y}$

19. $In\frac{z^2\sqrt{x}}{y^{\ 3}}$ 20. $log_2\frac{x^3}{32y^2}$ 21. 3 22. 6 23. – 3 24. 1 25. – 3 26. $\frac{1}{3}$

27. $\frac{4}{3}$ 28. 2 29. $\frac{2}{3}$ 30. $\frac{7}{5}$ 31. $\frac{4}{3}$ 32. $\frac{3}{4}$ 33. $\frac{3}{2}$ 34. 4.106 35. 1.292

36. 1.585 37. 1.608 38. 0.7337 39. 0.8549 40. 9 41. 3 42. 5 43. 36

44. $\frac{1}{4}$ 45. 2 46. $\frac{7}{4}$ 47. 6 48. 2 49. $\frac{1}{2}$ 50. 2 51. 4.1625 52. $\frac{1}{5}$, 25

53. 1.1761 54. 0.2552 55. $x = 4,\ y = \frac{1}{2}$ 56. $x = 9,\ y = 3$ 57. $a = b^2$

58. (a) 6 (b) $-\frac{3}{2}$

Chapter 14

Exercise 14.1

1. $y = 5x$ 2. $p = 4r^2$ 3. $y = 45\sqrt[3]{r}$ 4. $y = 3x$ 5. 18 6. 24 7. 9 8. 125

9. 3 10. 312 grams 11. $1512 12. $2.88 13. 450 cm^3 14. 627.2 m 15. 12.5 inches

Exercise 14.2

1.5 2. 4 3. 4 4. ±3/5 5. ± 2/3 6. 25

7. 4 8. 25/9 9. 3/2 10. 6 11. ± 1/5 12. 4/9

Exercise 14.3

1. 30 2. 90 3. 10 4. ± 3 5. 25

Exercise 14.4

1. 8 2. 8 3. 15 4. ± 4 5. 16 6. 49 7. 4 8. ± 6 9. ± 4 10. 60

Review Exercises

1. $p = 48/\sqrt{r}$ 2. $w = 4x^2y$ 3. $p = 10q/\sqrt{r}$ 4. $w = 15xy/z$ 5. 33

6. 50 7. 1/3 8. 16 9. 100 10. ± 4 11. 4 12. ± 9

Chapter 15

Exercise 15.1

1. 3 2. 4 3. – 6 4. – 12 5. ½ 6. ¾ 7. arithmetic; 2 8. Not

9. arithmetic; - 2 10. Not 11. arithmetic; 4 12. arithmetic; 3/2 13. 16, 19

14. 16, 21 15. 0, - 2 16. 9, 23/2 17. 1, 7/4 18. 13/4, 3 19. 114 20. – 60

21. 26 ½ 22. 2 23. – 2 24. 46 25. 23 26. 13 27. 13 28. 41

29. 31 30. 20 31. 3, 4 32. – 5 33. 5, 50 34. 2, 5, 8 35. 5, 7, 9

36. $1.65 37. $33400 38. $300 39. 12 40. 42 41. 2550 42. 648

43. 2091 44. 441 45. 448 46. 147 47. 456 48. – 40 49. 115

50. – 536 51. 111 52. 405 53. 1683 54. 3750 55. 4/5 56. 2

57. 8, 4, 540 58. 24 59. 52, 12 60. $116.25 61. $134700 62. $10950, $174750 63. 12 64. 1560 65. 126 66. 210 67. 480 68. 79.5 m

69. $39300

Exercise 15.2

1. 2 2. 3 3. 1/10 4. – 1/3 5. 1/5 6. – 1/3 7. geometric 8. Not

9. geometric 10. geometric 11. Not 12. geometric 13. 32, 64 14. ¾ , 3/8

15. – 40, 80 16. – 1/8, 1/16 17. 0.008, 0.0016 18. 486 19. 1/8 20. 2/27

21. 1024 22. 1/8 23. – 3/8 24. 8 25. 8 26. 9 27. 8 28. 10 29. 6

30. 6, 2 31. ± 3, 1/9 32. 2 33. 12, 27 34. $18225 35. 7430 36. $1934.84

37. $13500 38. 510 39. 13.48 40. 6.248 41. 5.25 42. 3.996 43. – 255

44. $\frac{1}{5}, -1, -\frac{1}{5}, -\frac{1}{25}, -\frac{1}{125}$; $-\frac{5}{4}\left[1-\left(\frac{1}{5}\right)^n\right]$ 45. 3, $\frac{1}{9}, \frac{1}{3}, 1, 3$; $\frac{1}{18}[3^n - 1]$

46. 2, $\frac{1}{2}$, 1, 2, 4; $\frac{1}{2}[2^n - 1]$ 47. $\frac{3}{2}$, $\frac{1}{2}, \frac{3}{4}, \frac{9}{8}, \frac{27}{16}$; $\left[\left(\frac{3}{2}\right)^n - 1\right]$ 48. 96, 765

49. – 3, 2, 183, 93 50. 16, 211 51. 7 52. 9 53. $1257.79 54. $1628.89

55. $99197.86 56. 72 57. 3493

Exercise 15.3

1. 42 2. 60 3. 75 4. 136 5. – 43 6. 13.44 7. 382.5 8. 1524

9. 27 10. 24 11. 40/3 12. 3/2 13. 3/2 14. 1/3 15. 10 16. 5/9 17. 18

18. 5/2 19. 1/5 20. 10 21. 2/3 22. 19/30 23. 3/11 24. 2/11 25. 2/9

26. 3(8/11) 27. ½ 28. 1/3, 2/3, 18, 9 29. 100

Exercise 15.4

1. 1, 2, 4, 7, 11 2. 1, 3, 5, 7, 9 3. 1, 0, - 1, - 4, - 17 4. – 2, - 1, 1, 4, 8

5. – 1, 1, 5, 13, 29 6. 1, 0, 2, 1, 3 7. 1, 2, 6, 15, 31 8. – 3, - 5, - 8, - 13, - 22

9. 1, 2, 4, 5, 7 10. 1, 1, 2, 3, 5 11. 1, 4, 14, 45, 139 12. – 1, 1, 0, 2, 2

13. 2, 5, 8, 11, 14; 3n – 1 14. 8, 14, 20, 26, 32; 6n + 2 15. 3, 1, - 1, - 3, - 5; 5 – 2n

16. 21, 16, 11, 6, 1; 26 – 5n 17. 3, 8, 13, 18, 23; 5n – 2 18. – 12, - 7, - 2, 3, 8; 5n – 17

19. 1, 2, 4, 8, 16; 2^{n-1} 20. 4, - 8, 16, - 32, 64; $(-2)^{n+1}$ 21. 5, 9, 13; 860

22 k = 1, m = 2; $a_0 = 1$, $a_4 = 9$

Review Exercises

1. arithmetic; 4 2. Not 3. Arithmetic; - 3 4. Arithmetic; 0.7 5. Not

6. arithmetic; 10 7. Arithmetic; 4 8. Not 9. Geometric; - 2 10. Not

11. Geometric; ½ 12. geometric; – 2/3 13. Not 14. Geometric; 3 15. Not

16. Geometric; 3/5 17.45 18. 384 19. – 1/3 20. 6 21. 16/9 22. 4/81

23. 72 24. 3/125 25. 10 26. 10 27. 16 28. 15 29. 6 30. 12 31. 5

32. 10 33. – 14762 34. 522 35. 26.79 36. 1850 37. – 4095

38. 13120 39. 12200 40. 78 41. 63 42. 4080 43. 230 44. 2016 45. 72

46. 123 47. 13116 48. – 28 49. 27/2 50. 48 51. 5/9 52. 3 53. 5/9 54. 7/9

55. 14/15 56. 31/60 57. 9/11 58. 18/37 59. – 2, - 1, 1, 4, 8 60. – 1, 1, 5, 13, 29

61. – 2, - 5, - 13, - 36, - 104 62. 1, 2, 6, 15, 31 63. 10, 3 64. 7 65. $4n + 2$

66. (a) $6n + 3$ (b) 6 67. 15 68. (a) 3^{4-n} (b) $\frac{81\left[1-\left(\frac{1}{3}\right)^n\right]}{2}$ (c) $\frac{81}{2}$

69. (a) $\frac{1}{2}$ (b) 8 (c) $16\left[1-\left(\frac{1}{2}\right)^n\right]$ 70. 7 71. (a) 2, 6, 18, 54 (b) 7 72. $\frac{7}{4}$

73. (a) $a_2 = \frac{7}{4}$, $a_3 = \frac{37}{16}$, $a_4 = \frac{175}{64}$ (b) $4\left[1-\left(\frac{3}{4}\right)^n\right]$ (c) 4

74. (a) (i) $a_1 = 9$, $a_2 = 13$, $a_3 = 17$ (ii) $4n + 1$ (b) $2n^2 + 3n$; 324

75. $\frac{1}{2}n(2n + 8)$; 8 76. \$25.42 77. \$20850; \$168000 78. \$39150.47; \$150934.71

79. 182 80. (a) 5.25 m (b) 79.5 m 81. 7 hours 82. \$39150.47 83. \$15938.48

84. \$7596.46 85. 18.4°

Chapter 16

Exercise 16.1

1. $1 + 12x + 60x^2 + 160x^3 + 240x^4 + 192x^5 + 64x^6$

2. $x^4 - 8x^3y + 24x^2y^2 - 32xy^3 + 16y^4$ 3. $1 - 5x + 10x^2 - 10x^3 + 5x^4 - x^5$

4. $16x^4 + 96x^3y + 216x^2y^2 + 216xy^3 + 81y^4$ 5. $64x^3 - 48x^2 + 12x - 1$

6. $81x^4 + 108x^3y + 54x^2y^3 + 12xy^3 + y^4$ 7. $1 - 12x + 54x^2 - 108x^3 + 81x^4$

8. $64a^6 + 192a^5x + 240a^4x^2 + 160a^3x^3 + 60a^2x^4 + 12ax^5 + x^6$

9. $16a^4 - 96a^3x + 216a^2x^2 - 216xa^3 + 81x^4$

10. $1 - 5x + 10x^2 - 10x^3 + 5x^4 - x^5$; 0.904

11. $1 + 4x + 6x^2 + 4x^3 + x^4$; 1.0406

12. $1 - 10x + 40x^2 - 80x^3 + 80x^4 - 32x^5$; 0.904

Exercise 16.2

1. 84 2. 1 3. 10 4. 15 5. 35 6. 252 7. 220 8. 792 9. $240x^2$

10. $-22680x^4$ 11. $-\frac{63}{8}x^5$ 12. $4032x^4$ 13. $90720x^4y^4$ 14. $-4320x^3$ 15. $7920x^4y^8$

16. $-5376x^3$ 17. $1180980x^2y^8$ 18. $\frac{945}{16}x^2$ 19. $1 - 10x + 45x^2 - 120x^3$

20. $1 + 18x + 144x^2 + 672x^3$ 21. $1024 - 5120x + 11520x^2 - 15360x^3$

22. $1 + 3x + \frac{15}{4}x^2 + \frac{5}{2}x^3$ 23. $2187 - 3402x + 2268x^2 - 840x^3$

24. $128 + 224x + 168x^2 + 70x^3$ 25. 1.127 26. 0.817 27. 535.506 28. 3856.886

29. 240 30. $40x$ 31. $70\left(\frac{x}{y}\right)^4$ 32. $1 - 20x + 180x^2 - 960x^3$; 0.8170

33. $1 + \frac{6}{5}x + \frac{3}{5}x^2 + \frac{4}{25}x^3$; 1.062 34. $1 + 7x + 14x^2$ 35. $1 - 12x + 78x^2$

36. $1 - 6x + 14x^2 - 14x^3$ 37. $1 + 9x + 30x^2 + 40x^3$

Exercise 16.3

1. $1 + 2x + 3x^2 + 4x^3$ 2. $1 - x + 2x^2 - \frac{14}{3}x^3$ 3. $1 - \frac{1}{3}x - \frac{1}{9}x^2 - \frac{5}{81}x^3$

4. $1 + x - \frac{1}{2}x^2 + \frac{1}{2}x^3$ 5. $1 + x + \frac{3}{4}x^2 + \frac{1}{2}x^3$ 6. $1 - \frac{2}{3}x + \frac{5}{9}x^2 - \frac{40}{81}x^3$

7. $1 - 3x + 9x^2 - 27x^3$ 8. $1 - \frac{1}{2}x + \frac{3}{8}x^2 - \frac{5}{16}x^3$ 9. $1 - x + \frac{3}{4}x^2 - \frac{1}{2}x^3$

10. $1 - x + \frac{3}{2}x^2 - \frac{5}{2}x^3$ 11. $\sqrt{2}\left(1 - \frac{1}{4}x - \frac{1}{32}x^2 - \frac{1}{128}x^3\right)$ 12. $\frac{1}{3} - \frac{1}{9}x + \frac{1}{27}x^2 - \frac{1}{81}x^3$

13. 1.0050 14. 0.9426 15. 0.999 16. 0.9899 17. $1 + \frac{1}{3}x - \frac{1}{9}x^2 + \frac{5}{81}x^3$; 4.017

18. $1 + \frac{1}{2}x - \frac{1}{8}x^2 + \frac{1}{16}x^3$; 4.123

Review Exercises

1. 1 2. 15 3. 70 4. 84 5. 792 6. 210 7. 35 8. 455

9. $x^3 + 6x^2 + 12x + 8$ 10. $x^5 + 15x^4 + 90x^3 + 270x^2 + 405x + 243$

11. $x^4 - 4x^3y + 6x^2y^2 - 4xy^3 + y^4$ 12. $32x^5 - 80x^4 + 80x^3 - 40x^2 + 10x - 1$

13. $64x^6 + 192x^5y + 240x^4y^2 + 160x^3y^3 + 60x^2y^4 + 12xy^5 + y^6$

14. $1 - 12x + 60x^2 - 160x^3 + 160x^4 - 192x^5 + 64x^6$ 15. $x^3 + 12x^2 + 48x + 64$

16. $32x^5 - 80x^4y + 80x^3y^2 - 40x^2y^3 + 10xy^4 - y^5$

17. $81x^4 + 216x^3y + 216x^2y^2 + 96xy^3 + 16y^4$ 18. $512 + 230x + 4608x^2 + 5376x^3$

19. $2187 - 5103x + 5103x^2 - 2835x^3$ 20. $1 + 21x + 264x^2 + 1760x^3$

21. $1 - 24x + 252x^2 - 1512x^3$ 22. $-120x^7y^3$ 23. $5940x^3y^9$ 24. $5103x^5y^2$

25. $-129024a^4b^5$ 26. $1792x^3y^5$ 27. $-5940x^9y^3$ 28. – 120 29. 1365 30. – 1760

31. 120 32. 54 33. – 5940 34. 240 35. $\frac{7}{16}$ 36. 11520

37. $x^4 + 4x^3y + 6x^2y^2 + 4xy^3 + y^4$; 15.682

38. (a) $1 + 7x + 21x^2 + 35x^3 + 35x^4 + 21x^5 + 7x^6 + x^7$ (b) 0.97919

39. (a) $1 - \frac{8}{3}x + \frac{28}{9}x^2 - \frac{56}{27}x^3 + \frac{70}{81}x^4$ (b) 0.99734

40.$1 - 12x + 60x^2 - 160x^3 + 240x^4 - 192x^5 + 64x^6$; 0.690 41. $\pm\frac{1}{4}$ 42. ± 3

43. $192 + 80x - 3x^2 - \frac{15}{2}x^3$ 44. $256 - 512x - 256x^2 + 1792x^3$

45. $64 - 64x - 80x^2 - 128x^3$ 46. $1 + \frac{1}{3}x - \frac{1}{9}x^2 + \frac{5}{81}x^3$; 10.0332

47. $1 + x - \frac{1}{2}x^2 + \frac{1}{2}x^3$; 1.732 48. $1 + \frac{1}{4}x + \frac{3}{32}x^2 + \frac{5}{128}x^3$; 3.1623

Chapter 17

Exercise 17.1

1. not 2. Circle 3. Not 4. Circle 5. Not 6. Circle 7. Not 8. Circle

9. $x^2 + y^2 - 6x - 4y - 3 = 0$ 10. $x^2 + y^2 - 8x - 6y - 11 = 0$ 11. $x^2 + y^2 + 4x - 6y - 12 = 0$

12. $x^2 + y^2 - 4x + 10y + 20 = 0$ 13. $x^2 + y^2 + 10x + 21 = 0$

14. $x^2 + y^2 + 10x + 6y + 9 = 0$ 15. $x^2 + y^2 + 6x + 8 = 0$

16. $4x^2 + 4y^2 - 40x + 16y + 107 = 0$ 17. $x^2 + y^2 + 6x - 8y = 0$

18. $x^2 + y^2 + 6x + 4y + 36 = 0$ 19. $x^2 + y^2 - 10x - 8y - 8 = 0$

20. $9x^2 + 9y^2 + 36x + 18y + 44 = 0$ 21. $(3, -1); 4$ 22. $(2, 1); 2$

23. $(3, -4); 6$ 24. $(-3, 2); 4$ 25. $(1, 3); 4$ 26. $(1, -3); 5$ 27. $(2, -1); 2$

28. $(-4, -2); 5$ 29. $(-1, 0); 4$ 30. $(-1, -3); 2$ 31. $(0, 4); 5$ 32. $(7, -4); 3$

33. $\left(\frac{3}{4}, -\frac{1}{2}\right);\ \frac{1}{4}\sqrt{5}$ 34. $\left(-1, \frac{1}{2}\right);\ \frac{1}{2}\sqrt{5}$ 35. $\left(-1, \frac{1}{2}\right); 2$ 36. $\left(\frac{2}{3},\ 1\right);\ 2$

37. $(2, -1);\ \frac{3}{2}$ 38. $(-3, 2);\ \frac{3}{2}\sqrt{2}$ 39. $\left(\frac{2}{3},\ \frac{1}{3}\right);\ 2$ 40. $(-2, 3);\ \frac{2}{3}\sqrt{3}$

41. $x^2 + y^2 - 2x - 4y - 3 = 0$ 42. $x^2 + y^2 + 4x - 6y + 3 = 0$

43. $x^2 + y^2 - 4x - 6y - 13 = 0$ 44. $x^2 + y^2 - 2x + 8y - 8 = 0$

45. $x^2 + y^2 + 8x - 8y + 16 = 0$ 46. $x^2 + y^2 - 2x - 6y - 3 = 0$

47. $x^2 + y^2 + 6x - 4y + 4 = 0$ 48. $x^2 + y^2 - 10x - 4y + 4 = 0$

Exercise 17.2

1. $x^2 + y^2 - 4x - 2y - 4 = 0$ 2. $x^2 + y^2 + 6x + 4y - 12 = 0$ 3. $8x + 2y + 5 = 0$

4. $16x - 2y - 41 = 0$ 5. $x + 1 = 0$ 6. $y^2 = 6x + 9$ 7. $x^2 = -4y$

8. $3x^2 + 3y^2 - 6x - 9 = 0$ 9. $3x^2 + 3y^2 - 38x + 36y + 159 = 0$

10. $x^2 - 3y^2 + 48 = 0$ 11. $x^2 + y^2 - 2x - 3y = 0$ 12. $x^2 + y^2 - 8x - 2y - 13 = 0$

13. $x^2 + y^2 - 5x - 3y - 4 = 0$; circle, center $\left(\frac{5}{2}, \frac{3}{2}\right)$ and radius $\frac{5\sqrt{2}}{2}$

14. $3x^2 + 4y^2 - 14x + 11 = 0$ 15. $x^2 + y^2 = 4$ 16. $x^2 + y^2 = 25$

Exercise 17.3

1. $(1, 0)$ 2. $\left(-\frac{3}{2}, 0\right)$ 3. $\left(\frac{5}{2}, 0\right)$ 4. $\left(-\frac{5}{4}, 0\right)$ 5. $(0, -1)$ 6. $\left(0,\ \frac{3}{2}\right)$

7. $(0, -3)$ 8. $\left(0,\ \frac{5}{4}\right)$ 9. $x = -2$ 10. $x = 1$ 11. $x = -3$ 12. $x = \frac{3}{4}$

13. $y = 2$ 14. $y = -1$ 15. $y = 4$ 16. $y = -\frac{3}{4}$ 17. $y^2 = 12x$ 18. $y^2 = -8x$

19. $y^2 = 28$ 20. $y^2 = -16x$ 21. $x^2 = 4y$ 22. $x^2 = 8y$ 23. $x^2 = -12y$

24. $x^2 = -8y$ 25. Vertex $(1, -2)$; focus $(-1, -2)$; directrix $x = 3$

26. Vertex $(-3, 2)$; focus $\left(-3,\ \frac{7}{4}\right)$; directrix $y = \frac{9}{4}$

27. Vertex $(1,\ 1)$; focus $(1,2)$; directrix $y = 0$

28. Vertex $(8,-1)$; focus $(9,-1)$; directrix $x = 7$

29. Vertex $(-3,-2)$; focus $(-3,-4)$; directrix $y = 0$

30. Vertex $\left(-\frac{3}{2},\ 3\right)$; focus $(0,3)$; directrix $x = -3$

31. $(x+2)^2 = -4(y+1)$ 32. $(x-3)^2 = 8(y-3)$ 33. $(y-4)^2 = -4(x+2)$

34. $(y-3)^2 = 8(x-4)$ 35. $(y-2)^2 = -8(x-4)$ 36. $(x+1)^2 = -8(y-3)$

37. $x^2 = 8(y-3)$ 38. $(y+1)^2 = 4(x-3)$

Review Exercises

1. $x^2 + y^2 - 8x - 6y - 75 = 0$ 2. $x^2 + y^2 + 4x - 10y - 7 = 0$

3. $x^2 + y^2 - 10x + 6y - 47 = 0$ 4. $4x^2 + 4y^2 + 40x + 16y - 91 = 0$

5. $x^2 + y^2 + 4x - 2y + 1 = 0$ 6. $x^2 + y^2 - 6x - 4y - 4 = 0$

7. $x^2 + y^2 - 10x - 6y + 9 = 0$ 8. $x^2 + y^2 + 6x + 10y + 26 = 0$

9. Center: $(2,1)$; radius: 2 10. Center: $(-3,2)$; radius: 4 11. Center: $(1,-3)$; radius: 5

12. Center: $(7,-4)$; radius: 3 13. Center: $\left(-\frac{2}{3},\frac{1}{3}\right)$; radius: 1

14. $x^2 + y^2 - 8x - 10y + 21 = 0$ 15. 4

16. $3x^2 + 3y^2 - 16x + 24y + 52 = 0$; Center: $\left(\frac{8}{3},-4\right)$; radius: $\frac{2}{3}\sqrt{13}$

17. $8x^2 + 8y^2 - 2x + 40y + 31 = 0$

18. $x^2 + y^2 - x - 4y + 1 = 0$; Center: $\left(\frac{1}{2},2\right)$; radius: $\frac{2}{3}\sqrt{13}$

19. $x^2 + y^2 - 2x - 6 = 0$ 20. $x^2 = -3y$ 21. $y^2 = 12x$ 22. $x^2 = 4$ 23. $y^2 = -16x$

24. Focus: $\left(0,\frac{1}{2}\right)$; directrix: $y = -\frac{1}{2}$ 25. Focus: $\left(\frac{3}{4},0,\right)$; directrix: $x = -\frac{3}{4}$

26. Focus: $\left(0,-\frac{5}{2}\right)$; directrix: $y = \frac{5}{2}$ 27. Focus: $\left(\frac{1}{4},0\right)$; directrix: $x = -\frac{1}{4}$

28. Vertex: $(1,-2)$; Focus: $(1,-4)$; directrix: $y = 0$

29. Vertex: $(-3,\ 2)$; Focus: $\left(-\frac{13}{4},2\right)$; directrix: $x = -\frac{11}{4}$

30. Vertex: $(8,-1)$; Focus: $(9,-1)$; directrix: $x = 7$

31. $(x+1)^2 = -8(y-2)$ 32. $(y-1)^2 = -12(x+2)$

Chapter 18

Exercise 18.1

1. $x=3, y=1, z=2$ 2. $x=2, y=3$

3. $\begin{bmatrix} 11 & 7 \\ 2 & 5 \end{bmatrix}$ 4. $\begin{bmatrix} 6 & 1 \\ 9 & 1 \end{bmatrix}$ 5. $\begin{bmatrix} 5 & 5 \\ 1 & 1 \end{bmatrix}$ 6. $\begin{bmatrix} -2 & 1 & -1 \\ 2 & 3 & -3 \end{bmatrix}$ 7. $\begin{bmatrix} 1 & -1 & 1 \\ 2 & 4 & -2 \end{bmatrix}$

8. $\begin{bmatrix} -1 & -1 \\ 5 & 2 \\ 1 & -1 \end{bmatrix}$ 9. $\begin{bmatrix} 2 & 2 & -4 \\ 2 & 4 & 2 \\ -1 & 1 & -1 \end{bmatrix}$ 10. $a=1, b=-1, c=-5, d=3$

11. $w=-1, x=-2, y=5, z=3$ 12. $x=2, y=1$ 13. $x=1, y=1$

14. $\begin{bmatrix} 2 & -3 \\ -1 & 4 \end{bmatrix}$ 15. $\begin{bmatrix} 3 & 2 \\ -5 & 1 \end{bmatrix}$ 16. $\begin{bmatrix} -6 & 0 \\ 0 & 6 \end{bmatrix}$ 17. $\begin{bmatrix} -3 & 2 & 0 \\ 1 & -5 & 2 \end{bmatrix}$ 18. $\begin{bmatrix} 4 & 0 \\ -3 & -1 \\ 2 & 1 \end{bmatrix}$

19. $\begin{bmatrix} -2 & 5 & -3 \\ 3 & 0 & 2 \\ -4 & 3 & -2 \end{bmatrix}$ 20. $\begin{bmatrix} 1 & 1 \\ 0 & 1 \end{bmatrix}$ 21. $\begin{bmatrix} 3 & -6 \\ -2 & 3 \end{bmatrix}$ 22. $\begin{bmatrix} 7 & -4 \\ -3 & 4 \end{bmatrix}$ 23. $\begin{bmatrix} 0 & 4 \\ -3 & 2 \end{bmatrix}$

24. $\begin{bmatrix} 1 & 8 & -3 \\ 5 & -6 & -3 \end{bmatrix}$ 25. $\begin{bmatrix} 8 & 4 & 4 \\ 1 & -1 & 1 \end{bmatrix}$ 26. $\begin{bmatrix} 11 & 1 \\ -10 & 1 \\ 3 & -3 \end{bmatrix}$ 27. $\begin{bmatrix} 5 & 1 & -1 \\ -4 & 2 & 2 \\ 7 & -12 & 3 \end{bmatrix}$

28. $\begin{bmatrix} 3 & 6 \\ -6 & 3 \end{bmatrix}$ 29. $\begin{bmatrix} 6 & -2 \\ -2 & 4 \end{bmatrix}$ 30. $\begin{bmatrix} -4 & 0 \\ 0 & 4 \end{bmatrix}$ 31. $\begin{bmatrix} 1 & -3 \\ -2 & 0 \end{bmatrix}$ 32. $\begin{bmatrix} -4 & 0 \\ 8 & -6 \end{bmatrix}$

33. $\begin{bmatrix} -4 & 2 & 6 \\ 0 & -6 & 4 \end{bmatrix}$ 34. $\begin{bmatrix} 0 & 3 \\ -3 & 6 \\ -6 & 9 \end{bmatrix}$ 35. $\begin{bmatrix} -1 & 0 & 2 \\ 1 & -3 & 1 \\ 0 & 1 & -2 \end{bmatrix}$ 36. $\begin{bmatrix} -8 & 5 \\ 5 & 5 \end{bmatrix}$

37. $\begin{bmatrix} 4 & -11 \\ 1 & -4 \end{bmatrix}$ 38. $\begin{bmatrix} 9 & -12 \\ -4 & 4 \end{bmatrix}$ 39. $\begin{bmatrix} 2 & -5 \\ -1 & 12 \end{bmatrix}$ 40. $\begin{bmatrix} -5 & 12 \\ 0 & -1 \end{bmatrix}$

Exercise 18.2

1. $\begin{bmatrix} 11 & 0 \\ 12 & -3 \end{bmatrix}$ 2. $\begin{bmatrix} -7 & -2 \\ -2 & -8 \end{bmatrix}$ 3. $\begin{bmatrix} 6 & -14 \\ 10 & -16 \end{bmatrix}$ 4. $\begin{bmatrix} 28 & 6 \\ 5 & 0 \end{bmatrix}$ 5. $\begin{bmatrix} 8 \\ 14 \end{bmatrix}$ 6. $\begin{bmatrix} -1 \\ -1 \end{bmatrix}$

7. $\begin{bmatrix} -14 \\ 3 \end{bmatrix}$ 8. $\begin{bmatrix} -1 \\ 3 \end{bmatrix}$ 9. $\begin{bmatrix} -10 & -4 \\ -7 & -4 \end{bmatrix}$ 10. $\begin{bmatrix} 8 & -4 \\ 8 & -2 \\ -3 & 1 \end{bmatrix}$ 11. $\begin{bmatrix} 1 & -7 \\ 4 & -2 \\ 2 & -8 \end{bmatrix}$

12. $\begin{bmatrix} 11 & -7 & 5 \\ 11 & 0 & 1 \\ 6 & 3 & 0 \end{bmatrix}$ 13. $\begin{bmatrix} 0 & 0 \\ 0 & 0 \end{bmatrix}$ 14. $\begin{bmatrix} 0 & 0 \\ 0 & 0 \end{bmatrix}$ 15. $\begin{bmatrix} -2 & -3 \\ 9 & 1 \end{bmatrix}$; $\begin{bmatrix} -11 & -4 \\ 12 & -7 \end{bmatrix}$

16. $\begin{bmatrix} 7 & 2 \\ -4 & -1 \end{bmatrix}$; $\begin{bmatrix} -19 & -5 \\ 10 & 3 \end{bmatrix}$ 17. $\begin{bmatrix} 1 & 0 \\ 0 & 1 \end{bmatrix}$; $\begin{bmatrix} 1 & 0 \\ 2 & -1 \end{bmatrix}$ 18. $x=2, y=-5, z=-11$

19. $x=3, y=-2, z=7$ 20. $a=-2, b=1, c=-1, d=2$

21. $a=2, b=-3, c=3, d=-2$

Exercise 18.3

5. – 1 6. 2 7. 2 8. – 1 9. 0 10. – 2 11. – 7 12. – 14 13. 98

14. 98 15. 98 16. – 5 17. ± 1 18. 1, 6 19. Has inverse 20. Has inverse

21. Has no inverse 22. Has no inverse 23. Has inverse 24. Has inverse

25. $\begin{bmatrix} 0 & \frac{1}{2} \\ \frac{1}{3} & \frac{1}{3} \end{bmatrix}$ 26. $\begin{bmatrix} \frac{3}{2} & -\frac{5}{2} \\ -1 & 2 \end{bmatrix}$ 27. $\begin{bmatrix} -\frac{3}{2} & -\frac{1}{2} \\ 2 & 1 \end{bmatrix}$ 28. $\begin{bmatrix} -\frac{5}{2} & 2 \\ \frac{3}{2} & -1 \end{bmatrix}$ 29. $\begin{bmatrix} \frac{1}{5} & \frac{1}{5} \\ 0 & \frac{1}{2} \end{bmatrix}$

30. $\begin{bmatrix} \frac{1}{4} & -\frac{1}{4} \\ \frac{1}{8} & \frac{3}{8} \end{bmatrix}$ 31. $x = -1, y = 2$ 32. $x = 1, y = 0$ 33. $x = 2, y = 3$

34. $x = -2,\ y = 2$ 35. $x = 3, y = -2$ 36. $x = 2, y = 3$

Exercise 18.4

1. 7 2. – 2 3. 14 4. – 4 5. 25 6. 14 7. – 51 8. – 2 9. – 2 10. 6

11. $\begin{bmatrix} \frac{1}{6} & -\frac{1}{3} & 0 \\ \frac{1}{6} & \frac{1}{6} & 0 \\ -\frac{7}{6} & -\frac{1}{6} & \frac{1}{2} \end{bmatrix}$ 12. $\begin{bmatrix} -\frac{3}{2} & \frac{3}{2} & 1 \\ \frac{9}{2} & \frac{-7}{2} & -3 \\ -1 & 1 & 1 \end{bmatrix}$ 13. $\begin{bmatrix} \frac{4}{3} & \frac{1}{3} & \frac{-4}{3} \\ \frac{1}{3} & \frac{1}{6} & -\frac{5}{6} \\ \frac{-2}{3} & \frac{1}{6} & \frac{7}{6} \end{bmatrix}$ 15. $\begin{bmatrix} -\frac{4}{3} & -\frac{5}{3} & 1 \\ -\frac{4}{3} & -\frac{8}{3} & 1 \\ \frac{1}{3} & \frac{2}{3} & 0 \end{bmatrix}$

17. $\begin{bmatrix} \frac{3}{5} & -\frac{1}{5} & -\frac{4}{5} \\ -\frac{1}{5} & \frac{2}{5} & \frac{3}{5} \\ \frac{2}{5} & \frac{4}{5} & \frac{1}{5} \end{bmatrix}$ 19. $\begin{bmatrix} \frac{-4}{9} & \frac{1}{9} & \frac{1}{9} \\ \frac{4}{3} & -\frac{2}{3} & \frac{1}{3} \\ \frac{7}{9} & -\frac{5}{9} & \frac{4}{9} \end{bmatrix}$

Exercise 18.5

1. $3x - 4y + 21 = 0$ 2. $3x + 11y + 35 = 0$ 3. $5x + 4y - 23 = 0$

4. $x - 5y - 19 = 0$ 5. $7x - 3y + 8 = 0$ 6. $3x + 2y + 5 = 0$ 7. $\frac{23}{2}$

8. 42 9. 3 10. 38 11. $\frac{31}{2}$ 12. 27

Exercise 18.6

1. (1, 2, 3) 2. (2, 1, - 1) 3. (0, 5, 3) 4. (1, 1, 1) 5. (-1, 3, 2) 6. (5, 8, - 2)

Review Exercises

1. $\begin{bmatrix} 4 & 1 \\ 4 & 6 \end{bmatrix}$ 2. $\begin{bmatrix} 5 & 2 \\ 8 & 0 \end{bmatrix}$ 3. $\begin{bmatrix} -4 & -5 \\ -12 & 14 \end{bmatrix}$ 4. $\begin{bmatrix} 3 & 5 \\ 4 & 2 \end{bmatrix}$ 5. $\begin{bmatrix} -3 & -4 \\ -8 & 8 \end{bmatrix}$

6. $\begin{bmatrix} -6 & 13 \\ -4 & -14 \end{bmatrix}$ 7. $\begin{bmatrix} 0 & -5 \\ -4 & 10 \end{bmatrix}$ 8. $\begin{bmatrix} 6 & 15 \\ 12 & -6 \end{bmatrix}$ 9. $\begin{bmatrix} -23 & -6 \\ -40 & 8 \end{bmatrix}$ 10. $\begin{bmatrix} -6 & 3 \\ -12 & 6 \end{bmatrix}$

11. $\begin{bmatrix} 2 & 10 \\ 0 & 8 \end{bmatrix}$ 12. $\begin{bmatrix} 2 & -9 \\ 4 & -2 \end{bmatrix}$ 13. $\begin{bmatrix} 12 & 10 \\ 10 & -8 \end{bmatrix}$ 14. $\begin{bmatrix} 4 \\ 2 \end{bmatrix}$ 15. $\begin{bmatrix} 20 & -8 \\ 2 & 2 \end{bmatrix}$ 16. $\begin{bmatrix} -2 \\ 7 \end{bmatrix}$

17. $\begin{bmatrix}16 & -4\\18 & -10\end{bmatrix}$ 18. $\begin{bmatrix}12\\1\end{bmatrix}$ 19. $\begin{bmatrix}22 & 12\\12 & 4\end{bmatrix}$ 20. $\begin{bmatrix}-2 & 14\\-4 & 0\end{bmatrix}$ 21. $\begin{bmatrix}16\\-4\end{bmatrix}$ 22. $\begin{bmatrix}9 & -1\\0 & 4\end{bmatrix}$

23. $\begin{bmatrix}-27 & 7\\0 & 8\end{bmatrix}$ 24. $\begin{bmatrix}10 & -6\\9 & -5\end{bmatrix}$ 25. $\begin{bmatrix}-3 & 4\\-4 & -3\end{bmatrix}$ 26. $\begin{bmatrix}22 & 14\\19 & 13\end{bmatrix}$ 27. $\begin{bmatrix}0 & 2\\-25 & 11\end{bmatrix}$

28. $a = 3, b = 1, c = 1, d = -2$ 29. $x = -5, y = 9$ 30. $x = 4, y = 5$ 31. 2

32. 7 33. 28 34. 14 35. 2 36. – 6 37. 2 38. – 26 39. 0 40. – 5 41. ± 1

42. 1, 6 43. $-\frac{3}{2}$, 4 44. – 1, - 2 45. 2, 3

46. $\begin{bmatrix}1 & -1\\-1 & 2\end{bmatrix}$ 47. $\begin{bmatrix}1 & 1\\2 & 3\end{bmatrix}$ 48. $\begin{bmatrix}1 & -\frac{5}{2}\\-1 & 3\end{bmatrix}$ 49. $\begin{bmatrix}-1 & -\frac{1}{2}\\-3 & -2\end{bmatrix}$ 50. $\begin{bmatrix}2 & 1\\-\frac{3}{2} & -\frac{1}{2}\end{bmatrix}$

51. $\begin{bmatrix}-3 & 2\\2 & -1\end{bmatrix}$ 52. $\begin{bmatrix}\frac{1}{2} & \frac{1}{2} & -\frac{1}{2}\\-1 & 0 & 1\\\frac{3}{4} & -\frac{1}{4} & -\frac{1}{4}\end{bmatrix}$ 53. $\begin{bmatrix}-\frac{1}{4} & 0 & \frac{1}{4}\\\frac{3}{4} & 0 & \frac{1}{4}\\\frac{1}{8} & \frac{1}{2} & -\frac{1}{8}\end{bmatrix}$ 54. $\begin{bmatrix}-\frac{1}{2} & \frac{1}{2} & \frac{1}{2}\\-\frac{7}{2} & \frac{3}{2} & \frac{5}{2}\\-5 & 2 & 4\end{bmatrix}$

55. (3, - 4) 56. (3, 2) 57. (- 1, 2) 58. $\left(-3, \frac{1}{2}, 1\right)$ 59. (5, 2, - 4) 60. (- 2, 4, 1)

Chapter 19

Exercise 19.1

1. $\frac{\pi}{6}$ 2. $\frac{\pi}{3}$ 3. $\frac{\pi}{2}$ 4. $\frac{2\pi}{3}$ 5. $\frac{5\pi}{6}$ 6. $-\frac{7\pi}{6}$ 7. $\frac{5\pi}{4}$ 8. $\frac{3\pi}{2}$ 9. $\frac{9\pi}{4}$ 10. $-\frac{4\pi}{3}$

11. $-\frac{11\pi}{6}$ 12. $\frac{5\pi}{3}$ 13. 36° 14. 135° 15. 120° 16. 540° 17. - 225°

18. 144° 19. - 252° 20. - 480° 21. 80° 22. 18° 23. - 720° 24. 450°

Exercise 19.2

1. 70° 2. 20° 3. 80° 4. 40° 5. 36° 6. 60° 7. 10° 8. 23° 9. 87°

10. 41° 11. 40° 12. 10° 13. Quadrant III 14. Quadrant III 15. Quadrant IV

16. Quadrant II 17. Quadrant III 18. Quadrant I 19. Quadrant I 20. Quadrant IV

21. Quadrant II 22. Quadrant III 23. Quadrant I 24. Quadrant II 25. $-\frac{\sqrt{3}}{2}$

26. $-\sqrt{3}$ 27. $-\frac{1}{2}$ 28. 1 29. $\frac{\sqrt{2}}{2}$ 30. 0 31. $\frac{2\sqrt{3}}{3}$ 32. $-\frac{2\sqrt{3}}{3}$ 33. 1

34. $-\sqrt{3}$ 35. $\frac{\sqrt{2}}{2}$ 36. 1 37. – 2 38. 1 39. 0 40. 0 41. $\frac{\sqrt{2}}{2}$ 42. $\frac{\sqrt{3}}{3}$

43. $\frac{2\sqrt{3}}{3}$ 44. – 1 45. – 1 46. $-\sqrt{3}$ 47. – 1 48. $-\frac{1}{2}$ 49. 1 50. $-\frac{\sqrt{3}}{3}$

51. $\frac{2\sqrt{3}}{3}$ 52. $-\frac{1}{2}$ 53. $-\frac{\sqrt{2}}{2}$ 54. $\sqrt{3}$

Exercise 19.3

1. $\cos\theta$ 2. $\csc\theta$ 3. $tan^2\theta$ 4. $sec^2\theta$ 5. $1+\cos\theta$ 6. $\csc\theta+1$ 7. $tan^2\theta$

8. $\tan\theta$ 9. 1 10. 2 11. $\csc\theta$ 12. $2\sin\theta\cos\theta$ 13. $sin^2\theta$ 14. $2sec^2\theta$ 15. $2\csc\theta$

Exercise 19.4

1. $\frac{\sqrt{2}}{4}(\sqrt{3}+1)$ 2. $\frac{\sqrt{2}}{4}(\sqrt{3}+1)$ 3. $2+\sqrt{3}$ 4. $-2-\sqrt{3}$ 5. $\frac{\sqrt{2}}{4}(\sqrt{3}-1)$ 6. $2-\sqrt{3}$

7. $\frac{\sqrt{2}}{4}(1-\sqrt{3})$ 8. $\frac{\sqrt{2}}{4}(\sqrt{3}-1)$ 9. $2-\sqrt{3}$ 10. $\frac{\sqrt{2}}{4}(1-\sqrt{3})$ 11. $\cos x$ 12. $\cos x$

13. $-\tan x$ 14. $-\sin x$ 15. $\frac{1}{2}$ 16. $\frac{1}{2}$ 17. $\frac{\sqrt{2}}{2}$ 18. $-\frac{\sqrt{2}}{2}$ 19. $\frac{\sqrt{3}}{3}$ 20. $\frac{\sqrt{3}}{2}$

21. $\cos(60° - x)$; $\sin(x+30°)$ 22. $\sin(x-45°)$ 23. $\tan(60°+x)$ 24. $-\frac{\sqrt{2}}{2}$

25. $-\frac{16}{65}$ 26. $-\frac{63}{65}$ 27. $\frac{16}{65}$ 28. $\frac{63}{65}$ 29. $\frac{16}{65}$ 30. $\frac{33}{56}$ 31. $-\frac{16}{65}$ 32. $\frac{56}{33}$

33. $-\frac{63}{65}$ 39. $\frac{1}{2}$ 40. $\frac{\sqrt{3}}{2}$ 41. $\frac{\sqrt{2}}{2}$ 42. $-\frac{\sqrt{2}}{2}$ 43. $-\frac{\sqrt{3}}{2}$ 44. $\frac{\sqrt{3}}{3}$ 45. $\frac{120}{169}, \frac{119}{169}$

46. $\frac{240}{289}, \frac{161}{289}$ 47. $\frac{120}{169}, -\frac{119}{169}$

Exercise 19.5

1. 60°, 300° 2. 45°, 225° 3. 45°, 135°, 225°, 315° 4. 60°, 120°, 240°, 300°

5. 60°, 120°, 240°, 300° 6. 60°, 120°, 240°, 300° 7. 270° 8. 30°, 150°, 210°, 330°

9. 90°, 120°, 240°, 270° 10. $0, \frac{2\pi}{3}, \frac{4\pi}{3}, 2\pi$ 11. $\frac{\pi}{3}, \frac{5\pi}{6}$ 12. $\frac{\pi}{3}, \frac{5\pi}{6}$ 13. $\frac{\pi}{4}, \frac{3\pi}{4}, \frac{5\pi}{4}, \frac{7\pi}{4}$

14. $0, \frac{3\pi}{4}, \pi, \frac{7\pi}{4}, 2\pi$ 15. $\frac{\pi}{6}, \frac{5\pi}{6}$ 16. $\pm\frac{2\pi}{3}$ 17. $-\frac{3\pi}{4}, \frac{\pi}{4}$ 18. $-\frac{\pi}{6}, \frac{2\pi}{3}$ 19. $0, \frac{2\pi}{3}$

20. $\frac{\pi}{6}, \frac{2\pi}{3}$ 21. $0, \pm\frac{\pi}{3}, \pm\pi$ 22. $\pm\frac{\pi}{4}, \pm\frac{3\pi}{4}$ 23. $\pm\frac{\pi}{3}, \pm\frac{2\pi}{3}$ 24. $\pm\frac{\pi}{4}, \pm\frac{3\pi}{4}$

25. $0, \pm\frac{\pi}{2}, \pm\pi$ 26. $-\frac{3\pi}{4}, 0, \frac{\pi}{4}, \pm\pi$ 27. $0, \pm\frac{2\pi}{3}$ 28. $-\frac{5\pi}{6}, -\frac{\pi}{2}, -\frac{\pi}{6}$ 29. $-\frac{\pi}{2}$

30. $-\frac{\pi}{4}, \frac{3\pi}{4}$ 31. $-\frac{\pi}{2}, \frac{\pi}{6}, \frac{5\pi}{6}$ 32. $\frac{\pi}{2}, \frac{7\pi}{6}, \frac{11\pi}{6}$ 33. $0, \pi, 2\pi$ 34. $\frac{\pi}{6}, \frac{5\pi}{6}, \frac{3\pi}{2}$

35. $0, \frac{\pi}{3}, \frac{5\pi}{3}, 2\pi$ 36. $\frac{\pi}{6}, \frac{5\pi}{6}, \frac{7\pi}{6}, \frac{11\pi}{6}$ 37. $\frac{\pi}{3}, \frac{2\pi}{3}, \frac{4\pi}{3}, \frac{5\pi}{3}$ 38. $0, \frac{2\pi}{3}, \frac{4\pi}{3}, 2\pi$

39. $0, \frac{\pi}{3}, \pi, \frac{5\pi}{3}, 2\pi$

Exercise 19.6

1. max: 3, 180°; min: - 3, 540° 2. max: 2, 180°; min: - 2, 0°

3. max: 5, 0°; min: 1, 180° 4. max: 3, 90°; min: - 1, 30°

5. max: 2, 90°; min: - 2, 270° 6. max: $-\frac{1}{2}$, 0°; min: $-\frac{3}{2}$, 180°

7. max: 4, 180°; min: - 2, 0° 8. max: $\frac{7}{3}$, 270°; min: $\frac{5}{3}$, 90°

9. max: 1, 90°; min: $\frac{1}{5}$, 270° 10. max: 1, 0°; min: - 3, 180°

11. max: 2, 90°; min: $\frac{2}{7}$, 2700° 12. max: 4, 0°; min: $\frac{4}{5}$, 180°

13. amplitude: 4 period: $\frac{\pi}{3}$ 14. amplitude: 6 period: $\frac{\pi}{2}$

15. amplitude: 3 period: $\frac{2\pi}{5}$ 16. amplitude: 5 period: 4π

17. amplitude: 1 period: $\frac{2\pi}{3}$ 18. amplitude: 2 period: $\frac{8\pi}{3}$

25. amplitude: 2 period: 2π phase shift: 0 vertical shift: 0

26. amplitude: 1 period: $\frac{4\pi}{3}$ phase shift: 0 vertical shift: 0

27. amplitude: $\frac{1}{2}$ period: 2π phase shift: π vertical shift: 0

28. amplitude: 4 period: 4π phase shift: -2π vertical shift: 2

29. amplitude: 4 period: 2π phase shift: $-\frac{\pi}{4}$ vertical shift: 0

30. amplitude: 3 period: π phase shift: $\frac{\pi}{6}$ vertical shift: - 2

Exercise 19.7

1. $\frac{\pi}{6}$ 2. $\frac{\pi}{3}$ 3. $\frac{\pi}{4}$ 4. 0 5. $\frac{\pi}{2}$ 6. $-\frac{\pi}{2}$ 7. $-\frac{\pi}{3}$ 8. $\frac{5\pi}{6}$ 9. $-\frac{\pi}{6}$

10. $\frac{5}{13}$ 11. $\frac{15}{17}$ 12. $\frac{5}{4}$ 13. $-\frac{140}{221}$ 14. $\frac{31}{8}$

Review Exercises

1. $\frac{5\pi}{12}$ 2. $-\frac{\pi}{4}$ 3. $-\frac{7\pi}{12}$ 4. $\frac{7\pi}{4}$ 5. 60° 6. 495° 7. – 420°

8. 900° 9. 15° 10. 50° 11. 38° 12. 80° 13. Quadrant III 14. Quadrant II

15. Quadrant IV 16.Quadrant I 17. – 1 18. $-\frac{\sqrt{3}}{2}$ 19. $-\frac{\sqrt{3}}{2}$ 20. $\frac{1}{2}$ 21. $\sin\theta$

22. $cot^2\theta$ 23. 2 24. 1 29. $0, \frac{3\pi}{4}, \pi, \frac{7\pi}{4}, 2\pi$ 30. $\frac{3\pi}{3}, \frac{2\pi}{3}, \frac{4\pi}{3}, \frac{5\pi}{3}$ 31. $\frac{7\pi}{6}, \frac{11\pi}{6}$

32. $\frac{\pi}{3}, \frac{5\pi}{3}$ 33. $\frac{\pi}{3}, \frac{3\pi}{4}, \frac{4\pi}{3}, \frac{7\pi}{4}$ 34. $\frac{\pi}{6}, \frac{5\pi}{6}$ 35. $\pm\frac{2\pi}{3}, \pm\pi$ 36. $\pm\pi$ 37. $\pm\frac{\pi}{3}, \pm\pi$

38. $\pm\frac{\pi}{3}$ 39. $\pm\frac{\pi}{4}, \pm\frac{3\pi}{4}$ 40. $\pm\frac{\pi}{4}, \pm\frac{3\pi}{4}$ 41. $\frac{\pi}{2}, \frac{7\pi}{6}, \frac{3\pi}{2}, \frac{11\pi}{6}$ 42. $\frac{\pi}{3}, \frac{2\pi}{3}, \frac{4\pi}{3}, \frac{5\pi}{3}$

43. $\frac{2\pi}{3}, \frac{4\pi}{3}$ 44. $\frac{\pi}{2}, \frac{3\pi}{2}$ 45. max: 2, 0°; min: - 2, 360° 46. max: 2, 270°; min: - 2, 90°

47. max: 5, 270°; min: 1, 90° 48. max: 3, 60°; min: - 1, 0°

49. amplitude: 5 period: π 50. amplitude: 3 period: π

51. amplitude: 2 period: $\frac{2\pi}{3}$ 52. amplitude: 2 period: $\frac{\pi}{2}$

53. amplitude: 5 period: π phase shift: 0 vertical shift: 0

54. amplitude: 3 period: $\frac{\pi}{2}$ phase shift: 0 vertical shift: 0

55. amplitude: 2 period: $\frac{2\pi}{3}$ phase shift: $\frac{\pi}{3}$ vertical shift: 0

56. amplitude: 1 period: 2π phase shift: $-\frac{\pi}{4}$ vertical shift: 0

57. amplitude: 3 period: π phase shift: 0 vertical shift: 2

58. amplitude: 2 period: $\frac{2\pi}{3}$ phase shift: $\frac{2\pi}{3}$ vertical shift: -1 61. $-\frac{\pi}{6}$ 62. $\frac{2\pi}{3}$

63. $\frac{\pi}{6}$ 64. $-\frac{\pi}{3}$ 65. $\frac{\pi}{4}$ 66. $\frac{\pi}{3}$ 67. $\frac{12}{13}$ 68. $-\frac{5}{13}$ 69. $\frac{15}{8}$ 70. $\frac{1}{2}$

Chapter 20

Exercise 20.1

1. $\langle 3, -1\rangle$ 2. $\langle -1, 2\rangle$ 3. $\langle 0, 1\rangle$ 4. $\langle -1, -1\rangle$ 5. $\langle -6, 7\rangle$ 6. $\langle 3, -2\rangle$

7. $x = 5, y = 5$ 8. $x = -3, y = 2$ 9. $x = 1, y = 2$ 10. $x = 2, y = -1$

11. $x = -3, y = 4$ 12. $\langle -1, -3\rangle$ 13. $\langle -3, 4\rangle$ 14. $\langle 2, -5\rangle$ 15. $\langle 2, 0\rangle$

16. $\langle -2, -3\rangle$ 17. $\langle 2, -4\rangle$ 18. $\langle -6, 3\rangle$ 19. $\langle 4, -1\rangle$ 20. $\langle -4, 1\rangle$ 21. $\langle 2, 3\rangle$

22. $\langle 7, -1\rangle$ 23. $\langle 0, 4\rangle$ 24. $\langle -2, -9\rangle$ 25. $\langle -3, -1\rangle$ 26. $\langle 6, 5\rangle$ 27. $\langle 1, -1\rangle$

28. $\langle 2, 4\rangle$ 29. $\langle -1, -5\rangle$ 30. $\langle -1, 1\rangle$ 31. $\langle 6, -2\rangle$ 32. $\langle 3, -6\rangle$ 33. $\langle -1, 3\rangle$

34. $\langle -2, 4\rangle$ 35. $\langle 6, 0\rangle$ 36. $\langle -2, -4\rangle$ 37. $\langle 6, -9\rangle$ 38. $\langle 7, 0\rangle$ 39. $\langle 1, -10\rangle$

40. $\langle 7, 8\rangle$ 41. $x = 3, y = 2$ 42. $x = 1, y = 1$ 43. $k = 2, m = 3$ 44. Parallel

45. Not parallel 46. Not parallel 47. Parallel 48. 5 49. $2\sqrt{5}$ 50. $\sqrt{34}$

51. 13 52. $\sqrt{85}$ 53. $\sqrt{73}$ 54. $5\sqrt{10}$ 55. $\sqrt{145}$ 56. $\langle -\frac{4}{5}, \frac{3}{5}\rangle$ 57. $\langle \frac{15}{17}, \frac{8}{17}\rangle$

58. $\langle -\frac{12}{13}, -\frac{5}{13}\rangle$ 59. $\langle \frac{7}{25}, -\frac{24}{25}\rangle$ 60. $\langle \frac{3\sqrt{10}}{10}, -\frac{\sqrt{10}}{10}\rangle$ 61. $\langle -\frac{\sqrt{2}}{2}, \frac{\sqrt{2}}{2}\rangle$ 62. $\langle 2, 1\rangle$

63. $\langle\frac{2\sqrt{5}}{5}, \frac{\sqrt{5}}{5}\rangle$ 63. $\langle 1, 0\rangle$ 64. $4i + 3j$ 65. $3i - 2j$ 66. $-5i - 2j$ 67. $5j$

68. $-7i$ 69. $-i - j$ 70. $-3i - 7j$ 71. $12i + 5j$ 72. $-13i + 12j$ 73. $\sqrt{10}$

74. 5 75. 12 76. 17 77. $\frac{\sqrt{2}}{2}i + \frac{\sqrt{2}}{2}j$ 78. $-\frac{3\sqrt{13}}{13}i + \frac{2\sqrt{13}}{13}j$ 79. $\frac{2\sqrt{5}}{5}i - \frac{\sqrt{5}}{5}j$

80. $-\frac{\sqrt{2}}{10}i - \frac{7\sqrt{2}}{10}j$ 81. $\langle -9, 1\rangle$ 82. $\langle -3, -6\rangle$ 83. $\langle 8, 2\rangle$ 84. $\langle -3, -3\rangle$ 85. $2i + j$

86. $2i - 3j$ 87. $-i - 5j$ 88. $-4i + j$ 89. $x = 5, y = 3$ 90. (6, - 2) 91. (1, - 7)

92. $m = 1, n = 2$ 93. $3i + 4j$ 94. $17i - 2j$ 95. $\langle 6, 1\rangle$

97. (a) $c - b$ (b) $\frac{1}{2}b$ (c) $\frac{1}{2}c$ (d) $\frac{1}{2}(c - b)$ 99. (a) $\overrightarrow{EF} = \frac{1}{2}\overrightarrow{AC}$ (b) $\overrightarrow{HG} = \frac{1}{2}\overrightarrow{AC}$

105. $\frac{1}{3}(a + c)$ 106. 085°, 602.1 km 107. 237.5 km h^{-1}, 127° 108. 152°, 3.47 hours

109. 107°, 3.97 hours 110. 049°, 227 km h^{-1} 111. 139°, 97 km

112. (a) 207° (b) 2.44 hours

Exercise 20.2

1. 11 2. 9 3. – 8 4. – 3 5. 5 6. 0 7. – 1 8. – 2 9. $\langle -4, 3\rangle$ 10. $\langle 6, -3\rangle$

11. 0 12. – 2 19. 47.7° 20. 75.7° 21. 82.9° 22. 78.7° 23. 75.1° 24. 70.3°

25. $\frac{1}{17}\langle 56, 14\rangle, \frac{1}{17}\langle -5, 20\rangle$ 26. $\langle -1, 2\rangle, \langle 2, 1\rangle$ 27. $\frac{1}{17}\langle -14, -56\rangle, \frac{1}{17}\langle -20, 5\rangle$

28. $\frac{1}{5}\langle -8, 16\rangle, \frac{1}{5}\langle -2, -1\rangle$ 29. 60 N 30. 315 joules

Review Exercises

1. $i + 8j$ 2. $7i - 15j$ 3. $13i$ 4. $x = -1, y = 3$ 5. $m = 2, n = 3$

6. $p = -1, q = -15$ 7. $m = 3, n = -2$ 8. $\sqrt{41}$ 9. $\sqrt{221}$ 10. $\sqrt{65}$

11. 5, 53.1° 12. 13, 67.4° 13. $\sqrt{13}$, 56.3° 14. $\sqrt{41}$, 51.3° 15. $-\frac{3}{5}i + \frac{4}{5}j$

16. $\frac{3\sqrt{10}}{10}i - \frac{\sqrt{10}}{10}j$ 17. $\frac{2\sqrt{5}}{5}i + \frac{\sqrt{5}}{5}j$ 18. $-\frac{5}{13}i + \frac{12}{13}j$ 19. $\overrightarrow{BA} = \langle -4, 4\rangle$, $\overrightarrow{BA} = \langle 1, 5\rangle$

20. $\overrightarrow{AB} = -i + 3j$, $\overrightarrow{BC} = -3i + 3j$ $\overrightarrow{CA} = 4i - 6j$ 21. $-2i - 5j$ 22. $6i + 3j$

23. $-2i + 17j$ 24. $i + 6j$ 25. $m = 1, n = 2$ 26. 5, 36.9° 29. $x = 1, y = -9$ 30. (a) $\frac{1}{3}a$

31. (a) $b = a + c$ (b) $\overrightarrow{AC} = c - a$ $\overrightarrow{AY} = \frac{1}{2}(c - a)$ $\overrightarrow{OX} = \frac{1}{2}b$ $\overrightarrow{Oy} = \frac{1}{2}b$

(c) X and Y are the same point.

32. 0 33. 4 34. – 7 35. 90° 36. 53.1° 37. 24, 26.6° 39. 109.4°

40. 154.9 km h $^{-1}$, 138.9° 41. 164.9 km h $^{-1}$, 303.9° 42. 037°, 40 s 43. 40 N

44. 103.9 joules

www.ingramcontent.com/pod-product-compliance
Lightning Source LLC
Chambersburg PA
CBHW082201220526
45470CB00010B/3003

* 9 7 9 8 3 2 7 0 3 5 6 6 9 *